AF700082

PUBLIC ECONOMICS AND THE ENVIRONMENT IN AN IMPERFECT WORLD

NATURAL RESOURCE MANAGEMENT AND POLICY

Editors:

Ariel Dinar
Agricultural and Natural Resources Dept.
The World Bank
1818 H Street, NW
Washington, DC 20433

David Zilberman
Dept. of Agricultural and
Resource Economics
Univ. of California, Berkeley
Berkeley, CA 94720

EDITORIAL STATEMENT

There is a growing awareness to the role that natural resources such as water, land, forests and environmental amenities play in our lives. There are many competing uses for natural resources, and society is challenged to manage them for improving social well being . Furthermore, there may be dire consequences to natural resources mismanagement. Renewable resources such as water, land and the environment are linked, and decisions made with regard to one may affect the others. Policy and management of natural resources now require interdisciplinary approach including natural and social sciences to correctly address our society preferences.

This series provides a collection of works containing most recent findings on economics, management and policy of renewable biological resources such as water, land, crop protection, sustainable agriculture, technology, and environmental health. It incorporates modern thinking and techniques of economics and management. Books in this series will incorporate knowledge and models of natural phenomena with economics and managerial decision frameworks to assess alternative options for managing natural resources and environment.

Public intervention is essential in preventing and restoring environmental degradation. The Bovenberg-Cnossen book contributes to our understanding of the interface between public economics and environmental economics. The book emphasizes the role of incentives such as taxes and subsidies, and institutions, regulations and agency coordination in reducing environmental pollution, and their effects on welfare.

The Series Editors

Recently Published Books in the Series

Spubler, Nicolas and Sabbaghi, Asghar:
Economics of Water Resources: From Regulation to Privatization
Hueth, Darrell L. and Furtan, William H.:
Economics of Agricultural Crop Insurance: Theory and Evidence
Gowdy, John:
Coevolutionary Economics: The Economy, Society and the Environment
Schmitz, A., Moulton, K., Buckwell, A. and Davidova, S.:
Privatization of Agriculture in New Market Economies: Lessons from Bulgaria

PUBLIC ECONOMICS AND THE ENVIRONMENT IN AN IMPERFECT WORLD

edited by

Lans Bovenberg
Tilburg University

and

Sijbren Cnossen
Erasmus University Rotterdam

Kluwer Academic Publishers
Boston/London/Dordrecht

Distributors for North America:
Kluwer Academic Publishers
101 Philip Drive
Assinippi Park
Norwell, Massachusetts 02061 USA

Distributors for all other countries:
Kluwer Academic Publishers Group
Distribution Centre
Post Office Box 322
3300 AH Dordrecht, THE NETHERLANDS

Library of Congress Cataloging-in-Publication Data

A C.I.P. Catalogue record for this book is available from
the Library of Congress.

ISBN 0-7923-9618-9

Copyright © 1995 by Kluwer Academic Publishers

All rights reserved. No part of this publication may be reproduced, stored in a retrieval system or transmitted in any form or by any means, mechanical, photo-copying, recording, or otherwise, without the prior written permission of the publisher, Kluwer Academic Publishers, 101 Philip Drive, Assinippi Park, Norwell, Massachusetts 02061.

Printed on acid-free paper.

Printed in the United States of America

ACKNOWLEDGEMENTS

The papers contained in this volume were selected from 64 papers presented at the 50th Congress of the *International Institute of Public Finance*, held at Harvard University, Cambridge, Massachusetts, in August 1995. Subsequent to the Congress, the papers were refereed and edited for publication. In addition, 13 papers were published in a Special Issue of *International Tax and Public Finance*. The companion issue emphasizes theoretical aspects of the interface of public economics and the environment. This volume focuses on empirical work and policy issues.

The production of this volume was greatly facilitated by the extensive and able cooperation we received from many friends and colleagues -- too numerous to list individually. We are grateful to the discussants and session chairmen of the Congress who helped us with the initial review. We owe a very special debt of gratitude to the referees for their thorough and timely evaluations of the (revised) papers. Their insightful and stimulating comments helped authors make a good paper even better. Many thanks are due the authors who cheerfully kept revising their papers, responding in detail to the points raised by the referees and us.

Tribute should be paid to Glenn Jenkins and Rita Gilligan for their superb organization of the IIPF Congress on Harvard's campus. The friendly environment that they created was readily internalized. For indispensable secretarial assistance, we thank Dennis Blansjaar, Angelique Lieverst and Marja Sjerp at Erasmus University Rotterdam, and Georgina Broeken at CentER, Tilburg University.

Lans Bovenberg
Sijbren Cnossen

CONTENTS

Acknowledgements v

I. INTRODUCTION

1. **Public Economics and the Environment in an Imperfect World: An Introductory Summary** 3
Lans Bovenberg and *Sijbren Cnossen*
Introduction 3
An Imperfect World 5
Assessment Problems 7
Institutional Aspects 8
The Need for Coordination 10
Efficiency and Distribution Issues 12
Innovation and Sustained Economic Growth 15
Whither Environmental Economics? 16

2. **Public Finance and the Environment** 19
Agnar Sandmo
Introduction 19
Goods, Prices and Markets 20
The Environment as a Public Good 21
Pigovian Taxes 23
Redistribution and Tax Incidence 27
The Transnational Dimension 29
Short-run Stabilization Issues 30
The Political Economy of Environmental Policy 32

3. **The Political Economy of Implementing Environmental Taxes** 37
Mark Pearson
Introduction 37
Technical Flaws in Environmental Taxes 39
"Gainers and Losers" 44
What are Environmental Taxes for? 52
Conclusions 54

II. AN IMPERFECT WORLD

4. **Taxing Bads by Taxing Goods: Towards Efficient Pollution Control with Presumptive Charges** 61
Gunnar S. Eskeland and *Shantayanan Devarajan*
Introduction 61
The Economics of Pollution Control: a Basic Analytical Framework 62
Pollution from Mobile Sources 71
Control of Pollution from Industrial Sources 87
Conclusion 107

5. **Global Climate Change, Energy Subsidies and National Carbon Taxes** 113
Bjørn Larsen and *Anwar Shah*
Introduction 113
Subsidies to Energy and Energy Complements 114
National Carbon Taxes 121
Summary and Policy Implications 130

III. ASSESSMENT PROBLEMS

6. **How a Fee Per-unit Garbage Affects Aggregate Recycling in a Model with Heterogeneous Households** 135
Thomas C. Kinnaman and *Don Fullerton*
Introduction 135
A Model of Household Disposal Choice 137
The Model with a Littering Option 147
Implications and Conclusion 157

7. **Environmental Taxes on Intermediate and Final Goods When Both Can be Imported** 161
James M. Poterba and *Julio J. Rotemberg*
Introduction 161
Current Treatment of Imports under US Environmental Excise Taxes 163
Import Neutrality without Joint Production 165
Import Neutrality with Joint Production 166
Applications 169

IV. INSTITUTIONAL ASPECTS

8. **Are Incentive Instruments as Good as Economists Believe? Some New Considerations** 173
Hannelore Weck-Hannemann and *Bruno S. Frey*
The Issue 173
The Crowding-Out Effect 174
Relative Price and Crowding-Out Effects in the Environment 176
Consequences for Environmental Policy 181

9. **The Political Economy of the Environment in Developing Countries: Market Failure and Institutional Response** 187
Eduardo Wiesner
Introduction 187
The Choice of Policy Instruments 188
Political Economy and the Environment in Developing Countries 191
A Strategy of Managing the Environment in Developing Countries 192
Summary and Policy Conclusions 199

10. **Pollution Taxes as a Source of Budgetary Revenues in Economies in Transition** 203
Tomasz Żylicz
Pollution Taxes in Poland 203
The Administration of the Present Pollution Tax System 205
Other Environment-Related Fiscal Instruments 207
The Current Role of Pollution Taxes 209
To Earmark or Not to Earmark? 211
Pollution Taxes and the State Budget 212
The Future of Pollution Taxes in Economies in Transition 214
The Future of Environmental Funds in Poland 215
Summary and Conclusions 216

V. THE NEED FOR COORDINATION

11. **The Role of the European Union in Environmental Taxation** 221
Stephen Smith
Introduction 221
"Subsidiarity" in the European Union 223
Application of the Principle of Subsidiarity to Environmental Issues 226
Integration with Other EU Policies 233
Conclusions 235

12. **Pure Global Externalities: International Efficiency and Equity** 237
Peggy B. Musgrave
Pollution as a Social Bad 237
A Simple Model 238
Data Base and Assumptions 239
Alternative International Arrangements 243
Simulation of Alternative Agreements 246
Conclusion 257

13. **International Coordination of Environmental Policies and Stability of Global Environmental Agreements** 261
Carlo Carraro and *Domenico Siniscalco*
Introduction 261
Some Recent Experience in Environmental Policy Coordination 262
Profitability and Stability of International Environmental Agreements 263
Stability and Transfers 266
Stability and Negotiation Linkages 268
Policy Recommendations 271

VI. EFFICIENCY AND DISTRIBUTION ISSUES

14. **Environmental Taxation and the "Double Dividend:" A Reader's Guide** 277
Lawrence H. Goulder
Introduction 277
Three Double Dividend Propositions 280
Does the Strong Form Hold? Theoretical Considerations 286
Numerical Findings 300
Relationship to Optimal Environmental Taxation 304
Summary and Conclusions 308

15. **Energy Levies and Endogenous Technology in an Empirical Simulation Model for the Netherlands** 315
Frank den Butter, Rob Dellink and *Marjan Hofkes*
Introduction 315
Literature 317
The Entech Model 319
Simulation Results 324
Concluding Remarks 332
Appendix: Dynamic Baseline Simulations 335

16. **Welfare and the Environment: Implications of a Recent Tax Reform in Norway** 337
Haakon Vennemo
Introduction 337
The Model 338
Data 341
Baseline Estimates of Welfare Effects 344
Untangling the Reform: The Effects of Different Components 349
The Importance of Being Elastic 352
Conclusions 354

Name Index 359

Subject Index 365

PART I

INTRODUCTION

Chapter 1

PUBLIC ECONOMICS AND THE ENVIRONMENT IN AN IMPERFECT WORLD: AN INTRODUCTORY SUMMARY

Lans Bovenberg (CentER, Tilburg University)

and

Sijbren Cnossen (OCFEB, Erasmus University Rotterdam)

INTRODUCTION

In less than a decade, environmental economics has matured from little more than a special application of Pigovian taxation to a full-fledged branch of the economics discipline. Various developments have contributed to this. Growing populations and economies, for instance, have increased people's awareness that the world's environmental resources are limited. Also, the depletion of the ozone layer and the possibility of global warming have given universal significance to what were hitherto mostly local and regional pollution problems. The environment has emerged as a truly public good. Generally, Coasian negotiations fail to internalize the costs of environmental degradation. Thus, public intervention, if possible through the market mechanism, is called for. The intervention measures, however, are designed and implemented by imperfect people using imperfect instruments applied in imperfect markets.

Organization of Volume

This Chapter summarizes 15 contributions on a variety of issues relating to the interface between public economics and environmental economics. The contributions are grouped under six headings: Introduction (Part I), An Imperfect World (Part II), Assessment Problems (Part III), Institutional Aspects (Part IV), The Need for Coordination (Part V), and Efficiency and Distribution Issues (Part VI). The volume, which deals mainly with empirical and policy issues, should be read in conjunction with the mostly theoretical papers published in a Special Issue of *International Tax and Public Finance* (ITPF) issued in 1995. To facilitate cross-referencing, this Chapter also summarizes the ITPF papers under the most appropriate heading. Four papers included in the Special ITPF Issue are also published in this volume -- adding to its comprehensiveness.

Setting the Stage

Agnar Sandmo (Chapter 2) sets the stage for the subsequent chapters of this volume by surveying the role of environmental considerations in the theory of public economics. He identifies the environment as a set of public goods in the Samuelson sense and establishes the analogy between environmental protection and public goods production. Sandmo sets out the conditions of Pareto optimality in an economy with externalities, emphasizing the welfare theoretic foundation for government intervention through Pigovian taxes. As he showed in his seminal 1975 article in the *Swedish Journal of Economics*, the adoption of Pigovian taxes for only those goods that are detrimental to the environment is still defensible in the presence of distortionary taxation. He goes on to argue that under conditions of distortionary taxation society cannot afford to levy Pigovian taxes 'in full,' since that would imply an excessive supply of the environmental public good. Sandmo also discusses issues of tax incidence and employment as well as the problems raised by transnational pollution. He emphasizes the inherent tendency to underestimate (global) benefits from environmental improvement and the need for collective action on a supranational level. Finally, he turns to a brief consideration of the political economy of environmental policy.

Minding Political Aspects

Environmental priorities compete with other priorities regarding employment, investment, trade, education and other worthy goals. Therefore, the benefits (and costs) of a cleaner environment must be made apparent. This is not as easy as some environmentalists would like it to be. *Mark Pearson* (Chapter 3), drawing on work of the OECD Joint Sessions on Taxation and the Environment, lists three major areas in which political and administrative problems have hindered the implementation of environmental taxes. First, a host of administrative complications, due partly to the desire to accommodate special interest groups, have reduced the scope for implementing environmental taxes. In particular, it has been difficult to define what is to be taxed and what not, the choice of an upstream tax point has been found to be unsatisfactory, and an appropriate administrative framework has not been easy to design. Second, gainers and losers have not been properly identified. In fact, more work should have been done to reduce industry opposition to environmental taxes. Also, a countervailing constituency should have been built around the use to which the revenues from environmental taxes can be put. Most importantly, perhaps, the aims of the environmental taxes should have been specified more explicitly. According to Pearson, if policymakers are not lucid in their policy aims, taxes may lose their attractiveness as a cost-effective environmental policy instrument.

AN IMPERFECT WORLD

The textbook solution to the problem of pollution externalities is simple: levy an emissions fee on the polluter equal to the difference between the marginal social and private costs of production. This should give the polluter an incentive to adopt a combination of cleaner technology and reduced output that is optimal for himself as well as society. In the real world, however, individual emissions often cannot be monitored, so that alternative, indirect approaches must be developed.

Making the Best of Things

Gunnar Eskeland and *Shantayanan Devarajan* (Chapter 4) make a strong case for relying on a mix of indirect pollution control instruments -- those which tax or regulate activities associated with emissions -- rather than taxing the emissions themselves. They show that indirect instruments that reduce the scale of output (such as a tax on output or on polluting inputs) can be important complementary measures to emissions standards that reduce the level of emissions per unit of output. In this way, the effects of an optimal emission fee can be mimicked fairly well. The optimal mix of indirect instruments, however, requires knowledge of the "cleaner" technologies (the ease with which emissions per unit of output can be reduced) as well as the sensitivity of demand to prices (the ease with which the scale of output can be reduced). This contrasts with the optimal emission fee, which relies only on information about emissions.

Eskeland and Devarajan present empirically-based case studies of Mexico City, Chile and Indonesia to illustrate the consequences of employing a combination of presumptive charges and emissions standards. A recurring theme throughout their contribution is that the taxation of fuel use, due to the interaction between fuel use and emissions, can serve as a powerful indirect instrument to supplement pollution standards in controlling air pollution. In the case of automobiles, for example, they show that failing to employ gasoline taxes (which ensure that emissions are cut through not only cleaner cars but also fewer trips) in Mexico City would significantly harm welfare, even when regulatory standards (catalytic converters) are in place. In the case of point-source pollution, they calculate that significant potential exists for altering the fuel mix of industries in Indonesia and Chile by taxing "dirtier" fuels. Furthermore, they show that, in the case of Indonesia, the general-equilibrium consequences of such a change in the tax structure are similar, though somewhat dampened, compared to what is indicated by partial-equilibrium models.

In sum, Eskeland and Devarajan advocate taxing a bad (pollution) by taxing a good (fuels) if monitoring emissions is prohibitively expensive. Finally, they note that indirect instruments for pollution control should be viewed as improvements, and reform steps, rather than final solutions. As monitoring technologies and capacity improve, policies should shift towards instruments based on actual monitoring of emissions.

Removing Fossil-fuel Subsidies and Imposing a CO_2 Tax

As the previous chapter indicates, fossil-fuel burning is one of the main environmental culprits. Nevertheless, many countries continue to subsidize fossil fuels. *Bjørn Larsen* and *Anwar Shah* (Chapter 5) provide estimates of subsidies to energy and energy complements in OECD and non-OECD countries. They conclude that the removal of energy subsidies in OECD countries on the order of US$30 billion annually (primarily in the US and Germany) and subsidies to complements on the order of US$50-90 (United States) are likely to have only little impact on CO_2 emissions. In contrast, the removal of energy subsidies of US$270-330 billion in non-OECD countries could substantially curb the growth of global CO_2 emissions, equivalent to the impact of a carbon tax on the order of US$60-70 per ton in the OECD countries. Nonetheless, even with the removal of energy subsidies, the growth in CO_2 emissions in non-OECD countries is projected to increase by 80% from the year 1990 to 2010. Furthermore, the authors show that the introduction of a revenue-neutral national carbon tax, in addition to energy subsidy removal, can yield significant health benefits from the reduction in local pollution. Larsen and Shah note that carbon taxes are considerably less regressive relative to lifetime income or annual consumption expenditures than to annual income.

Who Bears the Incidence of a CO_2 Tax?

The imposition of a global carbon tax raises the interesting question of who in fact bears the burden: the buyers/consumers of fossil fuels or the sellers/producers? *Olli Tahvonen* (1995) in his article in the Special ITPF Issue focuses on the strategic interaction between buyers of fossil fuels (who want to reduce air pollution) and sellers of such fuels (who consider the pollution problem to be more or less irrelevant). If buyers have formed a coalition for the imposition of a CO_2 tax, the equilibrium path of prices depends on whether the sellers are competitive or constitute a resource cartel. If sellers are competitive, buyers can impose a monopsonistic import tariff and sellers lose their resource rent. If, perhaps more realistically, sellers form a resource cartel, the sellers' export fee (imposed to obtain part of the CO_2 tax revenue) reduces the buyers' pollution tax. With monopoly power reducing emissions, the buyer's CO_2 tax may fall below the Pigovian level.

ASSESSMENT PROBLEMS

Administrative feasibility of environmental taxes cannot be taken for granted. The environmentally unfriendly activity that is to be controlled must be susceptible to close definition and its perpetrators should find no easy way of evading the tax. (In contrast to traditional revenue-raising taxes, avoidance is, of course, a prime objective of the tax.)

Charging for Garbage Collection

Curbside garbage collection is perhaps one of the most common examples of pollution control. In the US nearly 2000 communities require households to pay a user fee for each bag of garbage presented at the curb for collection. The magnitudes of both the benefits (the social value of less garbage) and the costs (the social costs of additional litter and the administrative costs) of the garbage collection programs depend on the waste removal choice of individual households. *Thomas Kinnaman* and *Don Fullerton* (Chapter 6) explore these choices in a model with heterogeneous households, diverse commodities, and fixed costs associated with illegal dumping. Their model suggests that a higher user-fee decreases aggregate garbage, increases aggregate litter and, perhaps surprisingly, may decrease aggregate recycling. The magnitude of each of these changes can differ across communities. In contrast to what most empirical studies assume, each community's choice of user fee is endogenous and depends on the behavioral responses of households within the community.

Applying Correct Border Tax Adjustments

Assessment issues also arise when a domestic government wishes to apply border tax adjustments to an externality-creating intermediate good and the final goods produced with it. *Jim Poterba* and *Julio Rotemberg* (Chapter 7) examine this situation if the objective of the domestic government is to preserve the competitive position of the domestic producers (the relative marginal costs of domestic and foreign producers) of the final goods. As the authors show, the calculation of border tax adjustments is straightforward if joint production is excluded -- the border tax on the imported final good should equal the domestic marginal input-output coefficient of the intermediate good. However, with joint production, the problem becomes basically insoluble. The petrochemical and petroleum refining industries exemplify cases in which joint final outputs produced from a polluting intermediate good (namely, crude oil) are common. In these important sectors, therefore, no simple, general rules for border tax adjustments are available.

INSTITUTIONAL ASPECTS

Economists tend to be oblivious to the importance of attitudes, institutions, and administrative arrangements in determining the success of environmental policies. Environmental taxes and regulations can be enforced only with a large degree of voluntary compliance. To be effective, environmental policies should be embedded in a country's social order; people must subscribe to the government's environmental objectives. Economists also tend to pay little attention to the place of environmental policies in intergovernmental relationships. Clearly, central and subordinate levels of government are not always equal in terms of their ability to address environmental problems. Furthermore, knowledge of bureaucratic practices and arrangements is important in getting environmental policies implemented. Accordingly, institutional aspects, broadly defined, deserve a central place in the analysis of environmental policy.

The Importance of Environmental Ethics

Against this background, *Hannelore Weck-Hannemann* and *Bruno Frey* (Chapter 8) examine the relationship between incentive instruments to combat environmental problems and environmental ethics. They observe that environmental instruments based on pricing, including effluent charges and tradeable licenses, are rarely used in practice. A possible explanation of this phenomenon, Weck-Hannemann and Frey argue, is that the use of incentive-based environmental instruments may, under identifiable conditions, be counterproductive. In particular, these instruments may be viewed as a license to pollute, thereby crowding out environmental ethics. The analysis suggests ways in which environmental incentive instruments should be modified in order to protect the intrinsic motivation of environmental ethics. Moral appeals are partially rehabilitated as a means of environmental policy -- without giving up the support of incentive instruments.

The Case for a Decentralized Approach

Environmental policies should not only aim at changing markets and minds, but should also address the government's role in the decisionmaking process. Governments in developing countries, in particular, may not be sufficiently equipped to analyze and deal with environmental problems. Moreover, these governments feature a strong proclivity for direct bureaucratic intervention. Drawing on the experience in Colombia, a middle-income developing country, *Eduardo Wiesner* (Chapter 9) underscores the importance of the government's role in creating the conditions that enchance the ability of the market to deal with pollution externalities. He emphasizes that policy advice must have regard for real-world restrictions and that policy reform needs a strategy to circumvent resistance in initiating a process of change. He proposes a strategy with three interdependent components: (a) environmental federalism, relying on interjurisdictional competition to induce better local environmental public choices and increase local responsibility and participation; (b) the dissemination of the results of independent evaluations of specific local and regional pollution problems; and (c) the reduction of subsidies and a more intensive use of pollution charges. In short, Wiesner favors a stronger role for the "invisible hand," broadly interpreted to include decentralized public choice.

Poland's Example

In environmental policy, as in other fields, the best is often the enemy of the good. Poland's pollution tax system, as described by *Thomas Żylicz* (Chapter 10), is an example of this dictum. In violation of the usual tenets of public economics (Pigovian taxes, no earmarking), the country imposes a wide variety of pollution taxes (at lower than Pigovian levels) whose revenues feed various environmental funds that finance abatement, conservation projects, and the clean-up of past environmental neglect. The earmarking mechanism shields the revenues from being diverted to other "worthy" purposes. Although the funds have not brought about an ecological miracle, the pollution tax system has become an effective mechanism for funding environmental investment and ecological recovery -- despite some doubts as to the system's efficiency. The system accounts for some 1.6% of the state budget, too little to entertain the idea of making it a source of general budgetary revenue and substantially reducing the traditional distortionary taxes on labor and capital. For the time being, the case for continuing the present arrangements, although not perfect, seems self-evident.

Do Governments Face the Correct Incentives?

An aspect with institutional overtones is whether governments have the appropriate incentives to implement efficient environmental policy. In his paper for the Special ITPF Issue, *Dietmar Wellish* (1995) examines this issue using a model that incorporates elements of "costly migration" in the form of "attachment to place." Wellish shows that if households are perfectly mobile internationally, governments face the correct incentives to set environmental policy, even if pollution crosses interjurisdictional frontiers. Intuitively, household mobility induces each government to internalize interjurisdictional externalities, even though it is concerned with the well-being of only its own citizens. On an international scale, however, households are less than perfectly mobile. In that case, interjurisdictional externalities distort decentralized environmental policy. The analysis of Wellish thus suggests that environmental competition among decentralized authorities becomes more inefficient, the larger the difference is between the interjurisdictional mobility of households and the mobility of production factors and pollution.

THE NEED FOR COORDINATION

Many environmental externalities have reached global proportions. For example, the depletion of the ozone layer and the possibility of global warming call for global coordination. Without an international agency with enforcement powers, individual countries over-extend damage-generating activities, because they disregard the external cost imposed on other countries.

The EU Carbon Tax

In this context, *Stephen Smith* (Chapter 11) examines the role of the European Union (EU) in environmental taxation. As is well-known, the EU has proposed a carbon/energy tax to contribute to the stabilization of CO_2 emissions at 1990 levels by the year 2000. The discussion is organized around the "subsidiarity principle," which involves a presumption in favor of decentralizing policy functions to the member states. Exceptions are allowed only if economies of scale or cross-country spillover effects warrant EU involvement. Smith persuasively argues that environmental policy is such an exception. In particular, in a single market such as the EU, the enforcement of taxes and regulations gives rise to important international externalities. Independent policy and administration would either interfere with free trade or lead to widespread evasion. Hence, the EU should have a substantial and legitimate interest in not only environmental targets and objectives, but also in the form of environmental policy (e.g. in the choice between environmental taxes and regulation). Furthermore, Smith argues that environmental tax policies should be coordinated both with other environmental tax policies and with other fiscal policies. This implies that the decisions about environmental tax rates should be taken by the level of government responsible for other, related, environmental policies, whilst the revenues from environmental taxes should accrue in general to member states, in preference to the EU.

Balancing Equity and Efficiency

Peggy Musgrave (Chapter 12) extends the argument for a carbon tax to reduce CO_2 emissions to the global arena and examines the distribution as well as efficiency issues posed by a global agreement on carbon taxation. She distinguishes four alternative international arrangements in addition to the 'no cooperation' case. The arrangements allow for various equity rules to reflect what are considered to be equitable changes from the baseline position. Although the results of Musgrave's exercise lay no strong claims to validity as empirical findings, the relative patterns that emerge, nonetheless, contribute to an understanding of the issues that must be faced in arriving at acceptable international forms of cooperation.

Building an Environmental Coalition

The need for multilateral environmental agreements is firmly on the political agenda. *Carlo Carraro* and *Domenico Siniscalco* (Chapter 13) provide some of the tools to understand the incentives facing individual countries. Not surprisingly, they find that free-riding undermines the stability of negotiations. Moreover, cooperation can hardly be achieved through the simultaneous acceptance of the environmental agreement by all countries; instead, sequential bargaining is called for. As the commitment of the members of the coalition to the agreement increases, other countries can be induced to join. Another promising course of action involves the explicit linkage of negotiations on international environmental agreements to other international issues, including free trade and R&D cooperation.

A Scheme of Side Payments

Cooperative solutions to global environmental problems are also examined by *Parkash Chander* and *Henry Tulkens* (1995) in their contribution to the Special ITPF Issue. They proceed from the premise that, in the absence of a supranational authority, the environmental treaty must be agreed to voluntarily. Within a game-theoretic framework, the authors show that a scheme of side payments exists that makes each country better off under a global agreement compared to a non-cooperative solution or a setting in which a country would cooperate with only a subset of countries. In particular, countries with high preferences for the global public environmental good should transfer resources to the countries that bear high abatement costs.

Undesirable Governmental Competition

At a less abstract but perhaps more realistic level, countries imposing environmental taxes are concerned that the affected polluting industry will pack up and move elsewhere. In this context, *Michael Rauscher* (1995), in his contribution to the Special ITPF Issue, examines the question whether the threat of delocation of polluting industries may lead to undesirable competition among countries in the field of environmental regulation (so-called environmental dumping). Using a non-competitive partial-equilibrium framework, Rauscher shows that a wide variety of solutions is possible, ranging from a rat race (even if you win, you are still a rat) towards zero taxes to the chicken game and the case of "not-in-my-backyard." Incentives for rent shifting abound and deviations from jointly optimal policies can be substantial. Therefore, optimistic views of interjurisdictional environmental competition based on competitive general-equilibrium models may turn out to be seriously misleading if markets for environmentally-intensive goods are non-competitive.

EFFICIENCY AND DISTRIBUTION ISSUES

The need for global coordination is rooted, in particular, in the threat of global warming. As indicated above, this has led to the call for a carbon tax. In turn, the carbon tax issue has made economists aware of the enormous revenue potential of some environmental taxes. As a result, the usual assumption that the revenue from "corrective taxes" would be returned in a nondistortionary lump-sum fashion was dropped and the use to which the revenue would be put became as important as the way in which it was raised. If the revenue, it was argued, could be used to lower other distortionary taxes, wouldn't that be the nearest thing to a free lunch? And soon, the notion of "double dividend" was coined to give expression to the simultaneous attainment of enchanced environmental quality and a more efficient tax system as a result of the revenue recycling.

Bursting the Double-Dividend Bubble

Much of the early euphoria evaporated, however, when it was pointed out that in the second-best world with which we have to contend environmental taxes can exacerbate the distortions of the existing tax system through general-equilibrium interactions. Consequently, the burden of proof was reversed and it became an empirical question whether this tax-interaction effect, as *Larry Goulder* calls it in his IIPF-award-winning contribution (Chapter 14), is smaller or larger than the revenue-recycling effect. Specifically, Goulder surveys and evaluates the theoretical and empirical evidence of various double-dividend notions. He finds that the weak claim (returning tax revenues through cuts in distortionary taxes leads to cost savings relative to lump-sum recycling) is easily defended. The stronger version (the costs of revenue-neutral swaps involve zero or negative non-environmental costs), in contrast, is subject to serious doubt. Hence, the case for environmental tax reform depends on environmental benefits. The need for measuring these benefits raises significantly the amount of information that policymakers require to make a benefit-cost case for green tax swaps.

Revenue Recycling for Employment

The belief in beneficial green tax swaps has been particularly prevalent in Europe, where high levels of unemployment and strong preferences for a large public sector (and hence high tax levels) accentuate the desire for revenue-neutral, growth-enhancing reductions in labor income taxes. In this context, *Frank den Butter*, *Rob Dellink* and *Marjan Hofkes* (Chapter 15) develop an empirical simulation model for the Netherlands, which is especially designed to reckon with the effects of changes in prices on the level and direction of technological progress. It appears that the so-called employment double dividend, i.e. increasing employment and decreasing energy use at the same time, can occur. A general levy yields stronger effects than a levy on household use only. However, the stronger effects of a general levy on employment and energy use are accompanied by shrinking production and, in the longer run, by decreasing disposable income of workers or non-workers.

Dealing with Rigid Labor Markets

The employment double dividend is also the subject of two articles published in the Special ITPF Issue. In the first paper, *Søren Bo Nielsen*, *Lars Pedersen* and *Peter Sørensen* (1995) use a simple endogenous growth model with pollution externalities to study the links between environmental quality, employment, and long-term growth. The model features voluntary employment caused by the tax-transfer system, as well as involuntary unemployment stemming from union monopoly power. Assuming that environmental policy is optimal to begin with, they find that increased environmental concern calls for higher pollution taxes. This environmental policy response implies a fall not only in unemployment, but, unfortunately, also in economic growth. If environmental policy is sub-optimal at the outset, a properly designed switch from a command-and-control regime with grandfathering of pollution rights towards pollution charges may both improve the efficiency of environmental regulation and raise employment, growth and consumer welfare without damaging the environment.

A Distributional Fly in the Ointment

In the second ITPF article on the double-dividend issue, *Stef Proost* and *Denise van Regemorter* (1995) employ a two-period, dynamic, applied general equilibrium model to analyze the introduction of the proposed EU carbon-energy tax in a small open economy, i.e. Belgium. The model allows for heterogeneity of households so that income-distribution effects of a green tax reform can be measured. In addition, environmental benefits are incorporated in the welfare function. The authors consider two revenue-recycling strategies, i.e. an increase in welfare payments and a reduction of employers' social security contributions. Subsequently, they compare the welfare changes under alternative regimes of fixed and flexible real wages. Confronting benefits and costs, Proost and Van Regemorter find that even the weak double-dividend hypothesis can fail because distributional effects dominate. Typically, overall welfare and employment fall, especially if fixed real wages reflect a rigid labor market. This is a typical second-best result; internalizing environmental externalities worsens overall welfare by exacerbating pre-existing distortions in the labor market.

How Green is My Tax Reform?

Real rather than hypothetical tax reforms can be evaluated also with respect to their environmental impact. Thus, in the last contribution to this volume, *Haakon Vennemo* (Chapter 16) notes that the tax reforms that industrialized countries have recently enacted with the aim to increase efficiency exert side effects on the environment. If the effects on the environment are sufficiently negative, a tax reform that increases efficiency, narrowly defined, may not be worth doing after all. He then proceeds to evaluate the economic and environmental consequences of a recent Norwegian tax reform that reduced the marginal tax rates on capital and labor income, widened the capital income tax base, cut employers' social security contributions, raised the VAT, and introduced a tax on CO_2. Vennemo's extended dynamic computable general equilibrium framework accounts for important links between the environment and the Norwegian economy. As it happens, the tax reform seems to affect the environment only marginally, delivering an overall welfare gain of 0.9 percent.

INNOVATION AND SUSTAINED ECONOMIC GROWTH

Finally, attention should be drawn to the burgeoning literature on innovation and sustained economic growth. After all, the hallmark of good environmental policy is the extent to which it induces the development and adoption of new technologies for the efficient use of scarce environmental resources. Innovation is the product of human ingenuity, which must win the race against environmental degradation as well as resource depletion. Possibly, the survival of the human race is at stake. Whereas this topic is not dealt with in this volume, three articles in the Special ITPF Issue explore the issues.

Taxes, Permits, and Technological Change

Till Requate (1995) investigates a model in which emission taxes or auctioned permits regulate output-competing firms that have access to two types of technology: the conventional type with relatively high marginal abatement costs, and a new technology with low marginal abatement costs but higher fixed costs. The model produces a remarkable asymmetry between taxes and permits in providing the correct incentives to adopt the new, less polluting technology. In particular, permits allow for partial adoption, whilst taxes do not. Moreover, if partial adoption is optimal and the social pollution damage function is only moderately steep, taxes lead to overinvestment in the new technology, whilst permits cause underinvestment. The reverse holds if the damage function is relatively steep. Interestingly, permits, in contrast to taxes, never reduce welfare.

Will Human Ingenuity Win?

Sjak Smulders (1995) argues that human ingenuity is indispensable in simultaneously achieving economic growth and environmental preservation. The accumulation of human knowledge, which is non-rival and most likely unbounded, can, in principle, overcome the limits that physical and natural resources impose on economic activity. He employs an endogenous growth model in which the bounded physical dimension of economic activity is separated from the unbounded value dimension. Basically, economic value can continue to grow through the substitution of reproducible human inputs for natural inputs as 'entropy-saving' innovations dematerialize production. In this way, the increasing returns associated with the public good of knowledge can offset the diminishing returns implied by the second law of thermodynamics.

The Importance of Being Rich

Andrea Baranzini and *François Bourguigon* (1995) explore under which conditions sustainable growth coincides with optimal growth. A sustainable growth path is characterized by non-decreasing consumption and maximization of the probability of survival. Central in the employed stochastic optimal-control model is the possibility of introducing a new technology that raises the productivity of capital but increases the probability of extinction. Whether or not optimal growth is sustainable depends on the discount rate and the initial level of the capital stock. Sustainable growth turns out to be a luxury good, which only rich and patient societies can afford.

WHITHER ENVIRONMENTAL ECONOMICS?

Almost overnight, like a tropical flower, environmental economics has sprouted and come to full blossom. At the end of this introductory summary chapter, the question must be asked whether environmental economics has established itself as a separate branch of the economics discipline or whether it will go down in history as just another fad whose colors, however bright at the start, rapidly faded.

Although some of the early professional rhetoric must be judged for what it was: "hot air," the dispassionate analysis of environmental issues, we believe, will continue to occupy an important place in the economics discipline. Admittedly, environmental economics is not as high on the political agenda as it used to be. To some extent, environmental economics is the victim of its own success. Many environmental costs have been internalized. In industrialized countries, environmental aspects are on every government's and industry's check list. Plenty of environmental laws and regulations have been enacted to keep the momentum going.

This is not the case, however, in the developing world, which experiences genuine environmental catastrophes (air pollution, water contamination, deforestation). Economic growth does not necessarily bring more pollution in its train, but equally it does not, without government intervention, result in a cleaner environment. Moreover, it is the developing world that faces exponential population growth. And whichever way one looks at the problem, two persons, on average, will pollute more than one person.

Environmental economics, in our view, will not wither. Citing Sandmo (this volume) we believe that economists have "an important function to perform in the analysis of environmental policy; ... the tools of normative public finance theory should be employed from the point of view of promoting the public interest." As always, we should be "open about our assumptions as well as about our degree of ignorance." Environmental economics is here to stay.

References

Note: The following papers, summarized in this Chapter, were all published in *International Tax and Public Finance*, Special Issue 2/2 (1995), edited by Lans Bovenberg and Sijbren Cnossen.

Baranzini, A. and F. Bourguignon. (1995). "Is Sustainable Growth Optimal?", 339-354.

Chander, P. and H. Tulkens. (1995). "A Core-Theoretic Solution for the Design of Cooperative Agreements on Transfrontier Pollution," 277-292.

Nielsen, S.B., L.H. Pedersen, and P.B. Sørensen. (1995). "Environmental Policy, Pollution, Unemployment, and Endogenous Growth," 183-204.

Proost, S. and D. Van Regemorter. (1995). "The Double Dividend and the Role of Inequality Aversion and Macroeconomic Regimes," 205-218.

Rauscher, M. (1995). "Environmental Regulation and the Location of Polluting Industries," 227-242.

Requate, T. (1995). "Incentives to Adopt New Technologies under Different Pollution-Control Policies," 293-316.

Smulders, S. (1995). "Entropy, Environment, and Endogenous Economic Growth," 317-338.

Tahvonen, O. (1995). "International CO_2 Taxation and Dynamics of Fossil Fuel Markets," 259-276.

Wellish, D. (1995). "Can Household Mobility Solve Basic Environmental Problems?", 243-258.

Chapter 2

PUBLIC FINANCE AND THE ENVIRONMENT

Agnar Sandmo (Norges Handelshøyskole)*

INTRODUCTION

The scope of the present chapter, as indicated by its title, is a broad one. Indeed, since a large part of the literature of environmental economics is concerned with public tax and expenditure policies, there is a danger that a survey-type paper could become a general survey of environmental economics. I shall try to steer clear of this danger by relating the discussion to some of the main themes of the theory of public finance or public economics, such as the assessment of the benefits of public goods, the analysis of optimal taxation, incidence effects of tax and expenditure policies and the like. I shall concentrate on theory, both because of the principle of comparative advantage and because the empirical literature that might have a bearing on the main theme is so vast and heterogeneous that I could not possibly do justice to it. Moreover, I shall not try to give a complete survey of the literature, but rather attempt to survey some of the main problems that arise, illustrated by selected references.

The first attempt at economic analysis of environmental issues in relation to problems of public policy was that of Pigou (1920). In his impressive treatise *The Economics of Welfare*, he not only introduced the idea that has later come to be called Pigovian taxation, he also offered some interesting suggestions on how to assess the social benefits of environmental improvement. But it took a long time before these issues were incorporated into standard expositions of tax and expenditure policy. In Musgrave's classic 1959 book, there is only a very brief reference to Pigou's ideas with emphasis on the difficulties involved in assessing individual preferences and making interpersonal comparisons. By contrast, modern textbooks tend to devote separate chapters to the theory of externalities, typically illustrated by examples from the environmental field; see e.g. Boadway and Wildasin (1984), Stiglitz (1986) or Rosen (1992). This reflects both the increasing public awareness of environmental problems and theoretical

---

* I am indebted to Lans Bovenberg, David Wildasin and two referees for helpful comments on earlier versions of the paper.

developments that have paved the way for a much more systematic analysis of efficient tax and expenditure policies. The breakthroughs were marked by Samuelson's (1954) formulation of the theory of public goods and the revival and modernization of Ramsey's (1927) work on second-best optimal taxation by a number of authors around 1970, notably Baumol and Bradford (1970), Diamond and Mirrlees (1971) and Stiglitz and Dasgupta (1971).

GOODS, PRICES AND MARKETS

The physical environment offers the most obvious examples of pure public goods in the Samuelson sense. Because clean air and unpolluted water are public goods we would expect them to be undersupplied in a market equilibrium, given the absence of collective action. We might imagine a system of Lindahl prices that would characterize an efficient allocation between private goods, on the one hand, and environmental public goods, on the other. However, an obvious problem with a Lindahl equilibrium is that it does not solve the incentive problem; consumers and producers have no individual incentives to reveal their true preferences for public goods. A number of theorists have worked on the design of decentralized incentive mechanisms that would overcome the basic incentive problems associated with public goods (an example from the environmental field is Sinn (1993)); however, these are, as far as I can see, still a good distance away from practical implementation. Thus, we must conclude that there is no reason to expect a decentralized allocation mechanism to yield a Pareto optimal allocation of resources in an economy with environmental public goods.

This conclusion may need to be modified in one important respect. When an otherwise competitive economy is not in a Pareto optimum, it follows that there are prospective gains from trade to be made between individual decisionmakers in the economy. One reading of Coase (1960) has led some to the conclusion that externality problems can be solved -- and presumably will be solved -- by negotiations between the parties affected, and that there is no case for governmental interference. I agree that this is a sensible conclusion to draw in the case of externalities involving a small number of agents, but for the big problems involving the pollution of the national or global environment, the transactions costs involved in decentralized negotiations would simply be too high. Or, to put it another way, the efficient level of negotiations in this case would be at the national or supranational governmental level. Once an agreement at this level has been reached, there is a need to design a system of incentives that would lead individual agents to behave so as to act in accordance with the principles of the agreement. However, by arguing that the market allocation can be improved upon, I do not mean to imply that the mere existence of market failure, even when the number of agents is large, provides a firm case for collective action. Policy involves its own sources of failure, and these have to be taken into account when considering public intervention in private markets.

THE ENVIRONMENT AS A PUBLIC GOOD

I have argued that environmental goods such as clean air and water can be seen as public goods in the Samuelson sense. To fix ideas I will also follow Samuelson in assuming that they are public consumption goods and do not affect the production possibilities of firms. Of course, there are a large number of cases where this assumption does not hold; agriculture, fisheries and tourism are obvious examples of industries where environmental deterioration influences production conditions. But the main ideas of the analysis come through even without this concession to realism.

Public environmental goods are not produced in the same way that public goods such as national defence or transportation infrastructure are produced. The production of environmental goods is a case of joint production between private and public goods, but it is a case of "negative jointness"; the higher the production of transportation services is, the lower will be the production of clean air. It is this jointness in production that in the case of transportation services creates a divergence between private and social marginal cost, or, as Pigou would have put it, between private and social marginal net product. This divergence, in turn, provides an argument for corrective taxes. It also provides arguments for public expenditure designed to clean up the environment (e.g. for waste treatment plants). In order to know how optimally to employ these policy instruments, one must know something about the benefits that they create, and these benefits turn out in a formal analysis to involve the Samuelson sum of marginal rates of substitution between the environmental good and some private consumption good that serves as the numeraire commodity for calculating willingness to pay. If e is the environmental good and y is the private good numeraire, we can calculate each individual's marginal benefit from environmental improvement as

$$(1) \qquad MB^i = u_e^i / u_y^i, \qquad (i = 1, \ldots, I)$$

where superscripts denote individuals and subscripts indicate partial derivatives of the utility functions. The social benefit is then found by taking the sum over all individuals, i.e.

$$(2) \qquad MB = \sum_{i=1}^{I} MB^i.$$

This gives us a conceptual framework for thinking about social benefits of an improved environment, but it does not immediately provide us with a clue as to implementation. Here there are various possibilities that I will not go into, but it is worth noting that Pigou himself provided an example that is in fact fairly representative of much of the more recent applied work in this area, and also conceptually consistent with the theoretical prescription. He refers (1920, p. 185) to an investigation carried out by the Manchester Air Pollution Advisory Board in 1918, where the objective was to estimate the social cost of the heavy factory smoke in that city. To this end, information was collected on the cost of weekly washing in a sample of 100 working class families in the two cities of Manchester and neighboring Harrogate, the latter being assumed to be a clean city. It turned out that the average cost in Manchester exceeded that in Harrogate by an amount of 7 1/2 d. a week. Assuming the average family in the sample to be representative of the whole population of Manchester, the investigation arrived at an estimate of £290,000 a year. This amount would then be an estimate of the benefits that would accrue to the city of Manchester if the level of smoke were reduced to that of Harrogate. It would obviously be a low estimate, since additional benefits such as the aesthetic value of a clearer sky, health effects etc. should ideally be included in a full-scale cost-benefit analysis.

Another objection to the theory of social benefit estimation focuses on the informational problem at a somewhat deeper level. In the case of factory smoke, changes in the quality of the environment are visible to each individual consumer, and the time span between cause and effect is usually quite short. In recent years public attention has increasingly been caught by the less visible and more long-run and uncertain externalities, such as effects on the ozone layer, global warming and the dangers involved in the production of nuclear energy. In such cases, what meaning can we attach to individual marginal willingness to pay? Formally, the basic concept of social marginal benefit can be extended to include time discounting, probabilities for uncertain events and option values,[1] but it is difficult to associate such concepts with people's real perceptions and present behavior. In many cases individual consumers will have had no experience of the externalities involved, and scientific explanations will typically be hard to understand for most people. Moreover, there are well-known difficulties of communicating the risks involved in the externality-creating activities. For a number of reasons it would therefore appear that estimates of private benefits would have to be adjusted to take account of the best available expert information relating to technology and risk. This is not the area where the principle of consumer sovereignty has its strongest appeal.

Leaving these difficulties aside for the moment, let us consider the condition for optimal environmental goods production. Assume that in addition to the numeraire there are J private consumption goods which are produced in

[1] See the interesting discussion of this in Chichilnisky and Heal (1993).

amounts x_j (j=1,...,J). Imagine that the quality of the environment, e, is affected negatively by the production and/or consumption of one particular private good, e.g. the Jth. In addition, for a given level of x_J environmental quality depends positively on a public sector input, z, which may e.g. represent the capacity of a waste treatment plant. The level of environmental quality is then given by the function

$$(3) \qquad e = f\left(x_J, z\right) \quad \text{with} \quad f_J < 0, \ f_z > 0.$$

We may now ask what is the optimum allocation of resources to the waste treatment plant. If the appropriate convexity conditions are satisfied, which is not guaranteed, but which I shall assume, the first-best optimum is characterized by equality between social marginal benefit and marginal cost, which takes the form

$$(4) \qquad MB \cdot f_z = MC_z.$$

The marginal cost of z is simply the marginal rate of transformation between z and the numeraire good. The informational requirement implied by this condition is the obvious one: Optimal production of a public good requires knowledge both of the technology of pollution as incorporated in the function f and of the social marginal benefit. Equation (4) can be seen as the most direct expression of the view of the environment as a public good that is partially controlled by public action.

PIGOVIAN TAXES

The first main theorem of welfare economics tells us that, for any given supply of public goods, a competitive equilibrium in markets for private goods is a Pareto optimum. For this statement to be consistent with the presence of taxation, taxes must be lump sum, so that they do not interfere with the required equalities between marginal benefits and costs. In the presence of externalities, equality between private marginal costs and benefits will no longer imply that the allocation is socially efficient. In the model outlined in the previous section, we may imagine a competitive equilibrium in which all prices are expressed in units of the numeraire commodity, y. For the first J-1 x-goods prices should equate private marginal benefits and costs, so that

$$(5) \quad MB_j^i = P_j = MC_j. \quad \left(i = 1, \ldots, I; \ j = 1, \ldots, J - 1\right)$$

However, for the externality-generating commodity J, the following must hold:

(6) $$MB_J^i + MB \cdot f_J = MC_J. \qquad (i = 1, \ldots, I)$$

This condition cannot be achieved by an unregulated market system. Since producers will equate marginal cost to the producer price p_J, and consumers will adjust their private marginal benefit to the consumer price P_J, efficiency cannot be obtained when the two prices are equal. The consumer price has to exceed the producer price, and the difference is the Pigovian tax:

(7) $$t_J^* = P_J - p_J = - MB \cdot f_J.$$

Note that the information required to implement the optimal Pigovian tax is essentially the same as required for the optimal use of resources to waste treatment. In particular, the social marginal benefit of environmental improvement enters both conditions, and the practical as well as the more conceptual difficulties of estimating this magnitude will therefore be present both in tax and expenditure policies to improve the environment.

This model of the interaction of optimal tax and expenditure policy with environmental concerns is based on a number of simplifications and implicit assumptions. Let me note some of the most important.

a. When lump-sum taxes are not available to redistribute income or finance the net expenditure of the public sector, taxes will be distortionary, and we will be in a world of second best. There is still a belief among many economists that with second-best considerations there is really little or nothing of any usefulness that can be said. However, this is in general not true, and in this particular case theory has in fact had some useful suggestions to offer. When many prices in the economy are distorted by taxes, one might think that there would be all sorts of reasons to apply the Pigovian tax principle not only to the externality-generating commodities but also to related goods. Sandmo (1975) showed that this is not true if taxes are set optimally. In this case it is true that the consumer prices of all taxed goods will have to deviate from marginal cost, but the marginal social damage should only be included in the taxes of those particular goods that are detrimental to the environment. Thus, there is a sort of separability in the tax structure, which might be taken to suggest that environmental taxes should be employed "first," with Ramsey taxes being used to fill up the tax revenue requirement. There is something to be said for this interpretation, but it turns out that the optimal tax on the externality-creating commodity takes the form of a weighted average of a Ramsey-type and Pigovian term, the weights being determined by the tightness of the government budget constraint, i.e. by the marginal cost of public funds. The Pigovian tax component cannot, therefore, be set without regard to tax revenue considerations.

This result goes against the intuition that it might be sensible to tax complements and subsidize substitutes, but there are, nevertheless, cases in which this intuition is vindicated. One is where informational problems prevent the tax authorities from levying taxes on the correct tax bases; e.g. taxes must be levied on "gasoline" instead of "gasoline use in congested areas." In this case, as shown in Sandmo (1976), there is indeed a case for taxes and subsidies on related goods.

b. It has sometimes been maintained that environmental taxes yield a "double dividend" in a second-best situation.[2] The argument is that they not only lead to an improvement of the environment; they also allow us to lower other, distortionary taxes, leading to a reduction in the efficiency loss from taxation. (They might also allow us to expand the production of public goods through a lowering of the marginal cost of public funds.) The assumption underlying this kind of reasoning must be that the structure of the economy, including the generation of environmental deterioration, is given; the change described is from a situation in which environmental taxes are not allowed to one in which they are. In this context there is a sense in which the result must obviously be true; if e.g. the revenue generated by the environmental taxes happened to be exactly sufficient to meet the government's revenue requirement, all efficiency losses would have been eliminated.[3]

One might be tempted to generalize this insight to the case where the government has to use both Pigovian and distortionary taxes and conclude that in this case the optimal Pigovian tax would be higher than the first-best level, since it yields benefits beyond those of environmental improvement. However, as pointed out by Bovenberg and de Mooij (1994), this is not generally true. Suppose that the Pigovian tax is set at its first-best level in a situation where other, distortionary, taxes are also in use; in their case this is represented by a tax on labor, which implies that labor supply is too low. A reduction of the Pigovian tax will in their model lead to increased labor use, thus mitigating the

---

[2] The literature on this issue has been surveyed by Goulder (1994).

[3] However, it should be kept in mind that a *partial* reduction of distortionary taxes cannot be guaranteed to bring about an efficiency gain; this follows from the first principles of the second best. This part of the double dividend is a *potential* gain. For the gain to be realized, the tax cuts must be designed so as not to deviate too much from the principles of optimal taxation and tax reform.

effect of the initial distortion, the net effect of which is in fact to increase welfare.

c. With optimal Pigovian taxes a competitive equilibrium once more generates a Pareto optimal allocation of resources -- either a full optimum or a second-best optimum relative to the types of taxes available. This important insight has sometimes been formulated in terms of the equivalence of taxes and prices; taxes are introduced to repair the shortcomings of the price mechanism, and Pigovian taxes have exactly the same properties as competitive prices. There is something in this, but there are also dissimilarities. Perhaps the most important is that taxes do not adjust to changing cost and demand conditions in the way that market-determined prices do. When taxes need to be adjusted as the result of changes in preferences and technology, this has to be done by the bureaucracy and the political system. Whether they have the same ability as market agents to adjust to changing circumstances is a question that I do not believe we know very much about. Theoretically, we have no construction in public economics that corresponds to the price adjustments of the Walrasian auctioneer, and I know of no empirical studies of this kind of adjustment process. As experience with environmental taxes is gradually being accumulated, it would be interesting to see case studies of this and related issues.

d. Implementing optimal taxes is a daunting task. I do not myself believe that the theory of optimum taxation should be seen as a set of cookbook formulae where you simply have to plug in some numbers taken from econometric studies to get the answers to tax policy problems. Instead it should be seen as a guide to clear and consistent thinking about the issues and also as a guide for empirical research, since theory may often lead one to become aware of important connections and parameters that practitioners have tended to neglect. In thinking along these lines it is often helpful to think about tax *reform* rather than tax *design*. It may sometimes be easier to decide whether or not a particular tax should be increased, rather than to determine its optimum level.

e. I have concentrated here on the ability of optimal Pigovian taxes to bring marginal consumer benefits into line with marginal social cost. This is not the only role that taxes can play in the implementation of an efficient environmental policy. When there are many polluters, taxes on emissions lead to a cost effective distribution of emissions among polluters. Since cost-minimizing agents will relate marginal clean-up costs to their marginal tax savings, a tax rate that is the same for all will imply that the largest reductions in emissions will be made where marginal costs are lowest, and this is clearly efficient.[4] However, this conclusion only holds for the case where the quality of the environment

---

4 An early statement of this point is by Baumol and Oates (1971).

depends on the *sum* of individual emissions; if the emission levels of the individual polluters affect the environment in qualitative different ways, it will in general be optimal to let the Pigovian tax vary between agents; see Diamond (1973) for a discussion of this.

f. There is an implicit assumption in the model that there is no interaction between policy tools that are designed to correct market failures and those that attempt to redistribute income; in other words, redistribution is solely by means of lump-sum taxes. It is for this reason that Musgrave's objection about interpersonal comparisons does not apply in this case; we move from individual to social marginal willingness to pay by simply taking the sum. This is of course a very strong assumption that is not tenable in practice; redistributional instruments are not perfect, and individual benefits must therefore be weighted in order to arrive at the social measure. Such a measure would not, of course, reflect a social consensus. There would not in general be agreement on the relative weights, and the weighting scheme should be seen as a tool to be employed in an informed discussion about the most reasonable tradeoff between efficiency and equality.

When redistribution has to be carried out by distortionary income and commodity taxes, there is a strong theoretical case for the conclusion that environmental taxes (and naturally also public expenditure) must be designed with a view not only to efficiency, but also to equity considerations. As demonstrated in Sandmo (1975), the marginal social damage should still be reflected only in the tax on the externality-creating commodity and not in the taxes on other goods, so that the separability property still holds. The difference is, however, that individual benefits must now be weighted in order to take account of redistributional considerations. This conclusion raises a number of problems for the implementation of policy. Environmentalists will naturally worry that this conclusion will lead to a softening of environmental standards; students of politics may be concerned whether this kind of advice is a good strategy with respect to the political system. I will return to these issues below.

REDISTRIBUTION AND TAX INCIDENCE

Many economists -- even some public finance economists -- are somewhat sceptical about the theory of optimum taxation. They believe that the economist is most useful as a policy advisor when he limits his work to descriptive analyses of the effects of taxes and avoids making statements about optimality. My own view is that those who hold this view sometimes have great difficulties with sticking to their own principles. Notions like efficiency losses, equity etc. tend to creep into the most purist positive analyses of tax and expenditure policies, so that it seems to be preferable to bring such considerations out into the open by chosing to work with an explicit optimization framework. Whatever one's view on this issue, everyone will agree that incidence analysis is a central part of the work of the public finance economist, and so it should be also in the area of environmental policy.

The analysis of the incidence of a tax on a consumption good takes account both of the forward and backward incidence of a tax. The forward incidence analysis looks at the effects of the tax on consumer prices and ultimately on consumer welfare. The backward incidence analyzes the effects on factor prices and profits. Ultimately, of course, all costs related to the tax are borne by consumers, but the aim of incidence analysis is to analyze the distribution of the burden between various groups of consumers and factor owners in the economy. In this analysis it is usual to hold the expenditure side of the public budget constant. In the analysis of the incidence of environmental taxes, the meaning of this assumption is much less clear. The whole point of environmental taxes is to generate a more efficient production of environmental goods. In this sense, there will have to be "expenditure effects" that must be taken into account in conjunction with the more standard incidence effects. The question of who bears the tax burden should not be separated from that of who gets the benefits of an improved environment. The magnitude of the Pigovian tax is positively linked to the supply of public environmental goods, and this link has to be accounted for in studies of the incidence of environmental policy.

Most incidence studies are based on the assumption of competitive markets, in which agents are "tax takers," i.e. they adjust to changes in tax rates in the same way that they would adjust to changes in prices of commodities and factors. However, when the number of agents is small, there is no reason to expect them to behave as passively as this with respect to decisions of environmental authorities (the same point could naturally be made also for quantitative regulations). There is therefore a need for reconsideration of the theory of environmental taxation under conditions of imperfect competition, possibly utilizing game theoretic models of the interaction between the polluter and the public authorities.

In conformity with the theoretical literature, I have used the term "externality-generating commodity" with perhaps more confidence than can easily be justified. It seems to suggest that there are some -- possibly not very many -- commodities whose production and consumption generate external effects on the quality of the environment, while in most cases of market allocation of resources these issues can safely be neglected. For many problems this is certainly true, but when it comes to issues such as the effects of energy use on the global environment, it becomes much less clear that these problems can be solved by imposing a few selected Pigovian taxes. If it becomes a question of stimulating production that is less capital and energy intensive, the environmental aspects of tax policy become much more comprehensive, raising issues of the appropriate balance between taxes on labor and on capital and energy use. This wider view of environmental policy does not invalidate the welfare theoretic framework of analysis that I have outlined so far, but it assigns environmental concerns a much more central place in the design of tax policy than most public finance economists have thus far thought appropriate.

THE TRANSNATIONAL DIMENSION

Both public and academic interest in particular economic problems tend to go in waves that do not always reflect corresponding fluctuations in the real importance of the underlying problems. In environmental economics there was a big wave of interest in the early 1970s that seemed to recede somewhat toward the beginning of the '80s. Perhaps the main difference between the first wave and the one that I believe that we are on now, is in the present focus on global and transnational environmental problems. This focus presents the public finance economist with some new and challenging problems. We have always tended to assume -- perhaps more than we should have -- that the public sector is a fairly well-defined agent, with a government that exercises the power to tax in an autonomous fashion. The fact that the world is becoming more economically integrated is beginning to leave its mark on the public finance literature in general, and nowhere is this development more striking than in the environmental field.

I started by referring to the view of the environment as a set of public goods, and it is gradually being realized that the environment is a global and not simply a national public good. This has wide-reaching implications for the analysis of public policy. One important insight that has emerged during recent decades is that the victims of pollution are not necessarily the citizens of the country in which the pollution is generated. If the damage from pollution is to a large extent inflicted on foreigners, a national cost-benefit study will tend systematically to underestimate the global benefits from environmental improvement; too little will accordingly be invested in clean-up projects, and national Pigovian taxes will be set at levels that are too low. The limited authority of national governments will prevent the attainment of a globally efficient environmental policy. This has been recognized as an important policy issue for a long time, especially (perhaps) among European governments, and recently empirical documentation of the substantial magnitudes involved has been forthcoming; see the studies by Mäler (1991) and Newbery (1990).[5]

[5] A pioneering theoretical analysis of the effects of transboundary pollution on national fiscal policy was done by Markusen (1975); Sandmo and Wildasin (1994) explore the connection between transnational pollution and factor mobility in the form of international migration.

A somewhat different problem arises when polluters in many countries simultaneously pollute the global environment. The literature has emphasized incentive problems with respect to public goods production within a single country and studied the scope for collective action to overcome the individual incentive problems. In the global perspective, each country finds itself in a role that is similar to that of the individual consumer or firm in the national economy. There is therefore a need for collective action on a supranational level, either by the creation of international organizations that have the power to impose taxes or regulations on the national economies, or through international cooperative agreements. However, an important difference between the global and national settings is that participation in such agreements has to be on a voluntary basis. The problems that this involves have been brought out clearly in the articles by Mäler and Newbery. Mäler shows that a fully cooperative solution to what he calls "the European acid rain game" will reduce total emissions of SO_2 in Europe by 39 percent, but the efficient percentage reductions on a country basis will vary from two to 86 percent. Similarly, while the benefit to Europe as a whole will amount to 6,290 million DM, national benefits will be in the range from -336 to 1,505 million DM. To secure voluntary cooperation it is therefore necessary to redistribute the aggregate gains among countries in such a way that everyone comes out with a positive net benefit. This case -- and I should stress that Mäler regards his numerical calculations as being mostly illustrative -- provides a good example of the challenges faced by the public finance economist by the transnational perspective. In this case it involves the design of an international scheme of Pigovian taxes combined with international transfer payments that will ensure the voluntary consent of all countries to participate in a globally efficient environmental policy.

An important question is what the international system of Pigovian taxes should look like in cases like this; should it be uniform, or should the tax rate vary between countries? The answer to this question, as shown by Hoel (1994), bears an interesting relationship to an issue discussed above, that is, the ability of a Pigovian tax to lead to a cost-minimizing reduction of emissions to achieve a given improvement in environmental quality. Briefly, from the point of view of global efficiency, an environmental tax should be uniform if environmental quality depends only on the sum of national emission levels; otherwise, it should differ among countries. This is a first-best result and does not take account of the complexities that would arise if the marginal cost of public funds were at different levels in the countries concerned.

SHORT-RUN STABILIZATION ISSUES

The design of taxes and public expenditure policies for the protection of the environment requires politicians to take a long-run view of economic welfare, i.e.

to use low rates of public discount or assign high evaluations to environmental benefits in the distant future. The history of actual policymaking, by contrast, abounds with examples in which long-run perspectives have had to yield to short-run considerations. It seems to be a universal experience among those who follow economic policy that investments in public infrastructure, e.g. in the transportation sector, which from a cost-benefit point of view are clearly socially unprofitable, will nevertheless often be carried out because they provide employment. (I assume of course that employment effects have already been taken into account in the cost-benefit study.) Similar issues arise in the environmental area. On the one hand, investment in environmental infrastructure such as waste treatment plants may be carried too far because of their short-run employment effects. On the other hand, increasing taxes on polluting industries will make production there less profitable and may therefore be resisted because of the implied threat to employment. However, this threat is not an obvious one and deserves further consideration.

Theoretical models of environmental externalities typically assume that the externalities are generated by the production and/or consumption of some subset of all final consumption goods in the economy. Quite often, however, emission levels are not related to output levels as such, but to input proportions in the production process, e.g. the use of heavy fuel oils. In such cases the optimal Pigovian tax policy involves taxes not on output but on the inputs that create the environmental damage. These inputs can safely be assumed not to include labor, but rather the use of particular types of energy, chemicals etc. These taxes will involve factor substitution away from these factors towards non-polluting factors such as labor. The factor substitution effects will to some extent offset the output effects on labor use, although the output effects may perhaps be expected to dominate in the very short run. The crucial role of factor substitution has been brought out in a number of papers by Bovenberg and van der Ploeg, e.g. (1993).

However, the employment effects of environmental taxes cannot be considered in isolation. If such taxes are used with a view to increasing total public sector revenue, there will also be increases in public expenditure that will have positive effects on employment. And if the increased tax revenue is used to lower other taxes so as to hold revenue constant, there will be tax-induced increases in employment in other parts of the private sector of the economy. Seen in this perspective, the employment problem arises not because of a general decrease in the demand for labor, but because of the structural changes that will take place in the economy as a result in the changes in the tax structure. The issue is similar to the problems that arise as a result of tax reforms or liberalization of international trade. In both of these cases long-run gains in efficiency have to be bought at the price of structural adjustment costs. Seen in a somewhat longer perspective, it is perhaps reasonable to believe that the factor substitution effects will have a stimulating effect on the demand for labor, so that the concern for an environment-employment tradeoff is mostly of a short-run nature.

THE POLITICAL ECONOMY OF ENVIRONMENTAL POLICY

Economists in general, and public finance economists in particular, have often been accused of paying too little attention to the social and political framework in the economy. This objection has been seen as particularly pertinent with regard to the application of welfare economics to the analysis of optimal public policies. It has often been maintained that this is a sterile exercise that takes inadequate account of political realities. If this objection is claimed to have general validity, it must obviously apply also to the analysis of the environment as a public good and to the analysis of optimal Pigovian taxes and similar policy instruments.[6]

Many economists, both academic and practical, consider it as a natural part of their work to comment on and give advice about what constitutes good policies. If the receiver of the advice is a single person or firm with a well-defined objective by which to evaluate policy, the advice naturally has to be given with this particular objective in mind. The situation is less clear when the advice is given to the government or to society at large. Much of the advice given by economists in the environmental field is based on the idea that gains in efficiency are desirable because they can be reconciled with a potential Pareto improvement for society as a whole. However, we also know that this potential for improvement is not always realized, and sometimes we should perhaps be a little more cautious when presenting our ideas about efficiency gains. Let me illustrate this point by means of an example.

I have already mentioned the point that a uniform tax on polluters under certain conditions will lead to an efficient allocation of emissions between them, and this insight has led many economists to become strong advocates of the use of taxes rather than quantitative regulations to control emissions. When I myself took part in introducing the idea of environmental taxes in Norway in the early 1970s, I made two discoveries. The first was that politicians were surprisingly sympathetic to the idea. The other was that people in industry were surprisingly opposed to it and strongly preferred quantitative regulations.

There could of course be many explanations for these attitudes. It could be the case, for instance, that politicians had been more exposed to economic

---

[6] This point of view is prominent in the writings of James Buchanan. For a representative statement of his views, see Buchanan (1973).

theory than had the industrialists, but this did not seem quite convincing. An alternative explanation is the following: The politicians were really tax revenue maximizers but felt that their opportunities for increasing revenue by means of existing taxes were limited because of the substantial efficiency losses involved. When told that there existed taxes that would create both increased revenue and efficiency gains, they were naturally delighted. The industrialists, on the other hand, were against further expansion of the public sector and were therefore opposed to the introduction of taxes that would give the politicians an opportunity to do just that. I, for my part, hastened to point out that this particular policy recommendation did not imply any particular view as to the desirability of public sector expansion, and that this should be seen as a separate issue. Instead, the proposal should more naturally be regarded as a reform within the framework of a constant share of the government in GNP.

One lesson that can be drawn from this example is the modest one that economists should be careful to state the assumptions underlying their policy recommendations; we tend sometimes to think of them as being so obvious that they are not worth mentioning. But I am sure that some would draw a much stronger conclusion than this -- that is, that economists ought to stay away from giving policy advice altogether, primarily because they otherwise become mere pawns in the political game. With this I disagree. I believe that we have an important function to perform in the analysis of environmental policy, and that the tools of normative public finance theory should be employed from the point of view of promoting the public interest. We must of course recognize that there may be disagreement in society about the proper definition of the public interest, but I believe that we shall perform our function best by being open about our assumptions as well as about our degree of ignorance.

Our theoretical framework has its limitations, and these should be reflected in the recommendations that we make. As I have already pointed out, there are limitations to what can be achieved by means of environmental taxes because we are unable for administrative reasons, broadly defined, to identify and monitor the ideal tax bases that would be consistent with a first-best optimum. Environmental taxes will therefore frequently not only correct imperfections in the price system; they will also tend to create new distortions. These complications should be taken into account when considering whether or not to interfere with market prices; one possible outcome of such considerations could be that in particular cases there should in fact be no interference with the market mechanism, in spite of the presence of an externality.

The emphasis in this chapter, as in most of the literature in the area, has been on efficiency, with distributional considerations being mostly set aside. This can be justified by a reference to welfare economics and its separation of efficiency and equity by means of lump-sum transfers. But in the second-best world that we inhabit, little theoretical justification exists for the view that questions of equity can be ignored when designing environmental policy. This is especially so in the developing countries where there is less scope for redistribution via progressive taxation and social security. In the developed countries there might still be a more pragmatic argument for neglecting the distributional aspects of tax policy. Although redistributional policy instruments are not ideal, they are sufficiently efficient to carry the main burden of establishing a reasonable tradeoff between efficiency and equity. Moreover, there may be some practical and political advantages in allocating targets to instruments even though this may not be strictly justified by theoretical separation theorems. To paraphrase a remark once made by Leif Johansen, there is little to be gained by making the environment a battleground for redistribution policy.

References

Baumol, W.J. and D.F. Bradford. (1970). "Optimal Departures from Marginal Cost Pricing." *American Economic Review* 60, 265-283.

Baumol, W.J. and W.E. Oates. (1971). "The Use of Standards and Prices for the Protection of the Environment." *Swedish Journal of Economics* 73, 42-54.

Boadway, R.W. and D.E. Wildasin. (1984). *Public Sector Economics*, 2nd ed. Boston: Little, Brown and Co.

Bovenberg, A.L. and R.A. de Mooij (1994). "Environmental Levies and Distortionary Taxation." *American Economic Review* 94, 1085-1089.

Bovenberg, A.L. and F. van der Ploeg. (1993). "Does a Tougher Environmental Policy Raise Unemployment?" *Research Memorandum*. Tilburg University.

Buchanan, J.M. (1973). "The Institutional Structure of Externality." *Public Choice* 14, 69-82. Reprinted as Chapter 28 in J.M. Buchanan, *Explorations into Constitutional Economics*. College Station: Texas A&M University Press, 1989.

Chichilnisky, G. and G. Heal. (1993). "Global Environmental Risks." *Journal of Economic Perspectives* 7, 65-86.

Coase, R.H. (1960). "The Problem of Social Cost." *Journal of Law and Economics* 3, 1-44.

Diamond, P.A. (1973). "Consumption Externalities and Imperfect Corrective Pricing." *Bell Journal of Economics and Management Science* 4, 526-538.

Diamond, P.A. and J.A. Mirrlees (1971). "Optimal Taxation and Public Production I-II." *American Economic Review* 61, 8-27, 261-278.

Goulder, L.H. (1994). "Environmental Taxation and the Double Dividend: A Reader's Guide." This volume.

Hoel, M. (1994). "International Coordination of Environmental Taxes." *Unpublished*. University of Oslo.

Mäler, K.-G. (1991). "Environmental Issues in the New Europe." In A.B. Atkinson and R. Brunetta (eds.), *Economics for the New Europe*. London: Macmillan.

Markusen, J. (1975). "International Externalities and Optimal Tax Structures." *Journal of International Economics* 5, 15-29.

Musgrave, R.A. (1959). *The Theory of Public Finance*. New York: McGraw-Hill.

Newbery, D.M. (1990). "Acid Rain." *Economic Policy* 11, 297-337.

Pigou, A.C. (1920). *The Economics of Welfare*, 4th ed. London: Macmillan, 1932.

Ramsey, F.P. (1927). "A Contribution to the Theory of Taxation." *Economic Journal* 37, 47-61.

Rosen, H.S. (1992). *Public Finance*, 3rd ed. Homewood, Illinois: Irwin.

Samuelson, P.A. (1954). "The Pure Theory of Public Expenditure." *Review of Economics and Statistics* 36, 387-389.

Sandmo, A. (1975). "Optimal Taxation in the Presence of Externalities." *Swedish Journal of Economics* 77, 86-98.

Sandmo, A. (1976). "Direct Versus Indirect Pigovian Taxation." *European Economic Review* 7, 337-349.

Sandmo, A. and D.E. Wildasin (1994). "Taxation, Migration, and Pollution." Norwegian School of Economics and Business Administration: *Discussion Paper* 3/94; Vanderbilt University: *Working Paper* 94-W02.

Sinn, H.-W. (1993). "Pigou and Clarke Join Hands." *Public Choice* 75, 79-91.

Stiglitz, J.E. (1986). *Economics of the Public Sector*, 2nd ed. New York: Norton & Co.

Stiglitz, J.E. and P. Dasgupta (1971). "Differential Taxation, Public Goods and Economic Efficiency." *Review of Economic Studies* 38, 151-174.

Chapter 3

THE POLITICAL ECONOMY OF IMPLEMENTING ENVIRONMENTAL TAXES

Mark Pearson (OECD)*

INTRODUCTION

The number of taxes that have been introduced in the OECD area for environmental reasons and that have characteristics that might be described as Pigouvian has grown substantially but is yet to amount to a revolution in either fiscal or environmental policy. The number of emission charges (charges directly based on measured emissions) has increased somewhat, and the use of product charges (charges levied on products, not directly on emissions) has become popular in certain countries (for recent developments in environmental taxes, see OECD (1993a and 1993b)). Yet these charges remain trivial in terms of the revenue raised and they have only rarely replaced rather than supplemented direct regulation. Substantial taxes with environmental aims *do* exist; the carbon/energy taxes of the Nordic countries are notable. Nevertheless, even here the novelty of the taxes can be exaggerated, as pre-existing taxes on energy were reduced when the carbon taxes were introduced. Furthermore, these carbon taxes have been controversial, the tax rate recently being reduced in some countries and the tax base further eroded in others, and they were never justified as fully internalising externalities in the Pigouvian sense (which is not to suggest that the signal given by introducing the tax may not have had significant environmental effects).

The endless sequence of Presidential Commissions, Royal Commissions, Commissions of Enquiry etc., which have paraded their conclusions in countries throughout the OECD for the past five years, promised something of more importance than a few product charges on batteries and plastic bags. The opinion

* I would like to thank all the delegates to the OECD Joint Sessions on Taxation and Environment and Jean-Philippe Barde, Beatrice Fournier and Jeffrey Owens for discussions on which this paper is based, and the comments of Sijbren Cnossen and two referees. However, the views expressed are mine alone; this paper does not reflect the views of the delegates to the Joint Sessions, nor those of the Member countries they represent, nor those of the OECD Secretariat.

formers, economists prominent among them, anticipated if not the end of taxation as it was currently known, a modification of the tax base of sufficient size to count as the biggest change in the tax structures since the introduction of Value Added Taxes. That environmental policy was to be transformed by the widespread use of environmental taxes was so obvious as to be hardly worth commenting on.

The winds of political fashion change, and the environment is now well down the list of political priorities. McCloskey (1994) notes that uncertainty over the effects of environmental taxes, in particular on employment and trade, have made it very difficult to design politically attractive policy packages. No doubt the original claims for environmental taxes at the end of the 1980s were sometimes overstated, ignoring the obvious administrative problems and market failures that obscured price signals. Furthermore, there is a long lead-time in introducing new taxes; several countries have plans for environmental taxes that are still in the gestation stage but are likely at some point to be implemented.

Those involved in introducing environmental taxes in national capitals would probably concur that these are reasons why environmental taxes have not "taken-off" in the anticipated manner. They would, however, add a further reason. Questions about the desirability or otherwise of environmental taxes on efficiency grounds are well understood in both Ministries of Finance and Ministries of Environment; what is in doubt is the ability of policymakers to put together a policy package that achieves whatever environmental and/or fiscal objectives that might be set for a tax, while at the same time generating the support, consensus or even merely passive tolerance that is essential to the introduction of any new, large-scale tax. The problem, then, is one of the implementation of environmental taxes.

In retrospect, this should not be surprising. Analysts and commentators considered the introduction of ecotaxes as the introduction of a cost-effective environmental policy, with advantages over other environmental policy instruments. Where the new taxes have replaced existing policy instruments, and are on a small scale, there are indeed cases of successful implementation of the taxes. What this scenario does *not* describe, however, is the introduction of a large-scale tax, representing a new departure in both environmental and fiscal policy.

New taxes are not popular. Few new taxes are introduced without heavy political fallout. The governments of Canada and Japan did not benefit much from their introduction of general consumption taxes; nor did the UK government when reforming local taxes. Major new taxes have recently been introduced without politically damaging consequences in Switzerland and Iceland, but such cases are rare. Put in the context of the political unpopularity of new taxes, the fact that any major environmental taxes have been introduced becomes remarkable, rather than the fact that there are so few.

This chapter pulls together some themes from successful and (more commonly) unsuccessful implementation strategies in a modest attempt to draw some lessons from these experiences. Only new environmental taxes are considered; taxes on gasoline and transport in general have a powerful environmental impact, but have long existed and pose no problems as regards implementation. Nor does this discussion consider tax incentives or tax differentiation, both of which, whatever their environmental and economic advantages and drawbacks, pose fewer implementation problems.

The discussion of problems of implementation is grouped around three themes, each of which have their counterpart in the theoretical literature. The first two are that environmental taxes fail because of poor tax design or failure to balance constituencies of those who might gain from an environmental tax reform against those who lose. Problems in one of these areas often result in policy changes that affect another (e.g. exemptions are given to a group of polluters to reduce the opposition to a new tax, create administrative problems and undermine the argument that taxes assign the burden of adjustment equally at the margin). The proximate cause of a tax failure may not be the underlying reason for its failure. Hence the third theme: confused policy choices arise from confused aims; this problem seems particularly acute for environmental taxes.

TECHNICAL FLAWS IN ENVIRONMENTAL TAXES

Administrative problems that have reduced the probability of implementing environmental taxes include the definition of what to tax; discrimination in who is taxed; the choice of the tax point; and the administrative structures used to collect the tax.

The Definition of the Taxable Event

Most emissions resulting in environmental problems are not amenable to being measured, at least at reasonable administrative cost (see, for example, the 1992 Swedish Budget Bill, which highlighted the high administrative cost arising from widespread application of environmental charges). The problems are not technological -- measurement technology for most emissions does exist at reasonable prices -- but are intractable nonetheless.

The number of emissions sources is overwhelming. Administrative costs increase the greater is the number of separate emissions sources. Direct monitoring of emissions from cars is not considered administratively feasible, for example.

Monitoring is commercially burdensome. Most taxes that are widely used in OECD countries have as a tax base something that corresponds to a recognisable commercial event. Companies normally know how much they pay workers or shareholders; they account for sales and receipts; they know how many units of alcohol or tobacco they have sold. Hence, although the reporting and control systems of widely used taxes do impose additional burdens on businesses, these costs are limited by the fact that good commercial practise would require them to do something rather similar in any case. This is not the case with measuring emissions. Such measurement is therefore a new cost on business. This in turn is likely to have consequences for the extent of voluntary compliance with the tax, increasing the administrative burden on revenue authorities. These costs can be reduced by looking not at emissions themselves but something more easily monitored, such as purchase of inputs. This raises further problems, discussed below.

The potential consequences of attempting to operate a charging system based on the monitoring of emissions are well illustrated by considering the admittedly extreme case of the Russian Federation. Gofman and Gusev (1994) report that emission charges exist for over 450 substances in the Russian Federation. They note that this number is excessive if compared with the country's monitoring and enforcement capacity. The result is hardly surprising; compliance is extremely low, with companies routinely challenging the fees, and environmental impact is limited.

An alternative to emission charges is product charges. Taxes can be levied on the sale of products that are associated with environmental damage either in their production or subsequent use. Not surprisingly, given the high monitoring costs, outside of the arena of water charges, emission charges are less common in OECD countries than are product charges.

The extent to which product charges achieve a reduction in emissions depends on the "linkage" between what is taxed and the pollution which it is the aim to control (see OECD, (1993)). A product charge will give purchasers an incentive to use less of the product. It does not give a direct incentive to reduce pollution. Only if tax minimisation behaviour and pollution minimisation behaviour are contiguous will a product tax be an effective instrument of environmental policy. Where this linkage is weak, taxation becomes not merely an ineffective environmental policy but also imposes real economic costs.

Contiguity between a potential tax point and environmental problems. Charging for domestic refuse collection according to the number of sacks of rubbish to be collected might be thought to reduce domestic waste production. However, householders might over-fill each sack (tax avoidance) and (as noted by Oates (1994) in the context of various State schemes in the United States) seek to dump rubbish outside of the formal collection system (tax evasion), potentially augmenting rather than reducing the environmental costs. The only feasible tax point in such cases (collection of rubbish from households) is too far removed from the actual environmental problems for the product charge route to be a particularly felicitous one.

Stability in the link. Combustion of fossil fuels is a chemical process that eventually results in nearly all carbon atoms in the fuels being turned into carbon dioxide. Insofar as most production and purchase of fossil fuels is for combustion, purchase of fossil fuels can be seen as leading to a fixed and certain quantity of emissions. This is not the case for many other emissions. The concentration of emissions may be as important as the total quantity of emissions, and the timing of emissions (is there temperature inversion? Is there wind from a particular direction?) can also be crucial. Similarly, environmental problems that are locally concentrated may be unsuitable for remedy by environmental taxes. Attempts to differentiate fertiliser taxes by region have been overcome by the simple expedient of buying the fertiliser from a lower taxed region. The environmental effects of such taxes, if not combined with a suitably complex regulatory framework, are doubtful.

Furthermore, the link between product and pollution may be unstable if action can be taken by the purchaser of a product to minimise the final pollution by action other than reducing purchases of the taxed product. Taxing the sulphur content of fossil fuels in itself gives no incentive to fit Flue Gas Desulphurisation (FGD) units. Yet the FGDs may well be more efficient in the strict sense of the term in reducing sulphur dioxide emissions than reducing use of fossil fuels. Catalytic convertors reduce the emissions of harmful particulates, etc., but increase the amount of petrol consumed. An increase in the taxation of petrol makes such convertors less attractive; a potentially perverse economic incentive. Once again, an efficient tax cannot be introduced without preserving a regulatory structure. This "mixed" system is in effect the one used in most OECD countries (although one alternative approach, followed in Sweden, is to have a more complex set of effective tax rates, inputs being taxed in the textbook manner, but revenues being refunded according to the technologies used).

In summary, the linkage problem by itself accounts for much of the shortfall in use of environmental taxes as compared to environmental problems. Metering and measuring emissions directly is a surprisingly costly process. Finding a suitable transaction point to tax emissions indirectly through product charges can dramatically reduce administrative costs. But a wide range of pollution problems cannot be addressed by such taxes, because the link between the taxable transaction and the pollution outcome is too variable for the tax to be a sensible response to the problem.

That said, it has to be shown that taxes are necessarily a worse alternative to any other type of pollution control. Variable environmental damage according to geographic concentration, time of day, concentration, etc., also complicates regulatory approaches to pollution control. Partly as a result of this realisation, taxes on some pollutants that had previously been rejected as having insufficiently close linkage between input and damage are now under active consideration in some countries (e.g. Denmark and Switzerland are both considering taxation of volatile organic compounds).

Administration of Exemptions

For reasons discussed later in this chapter, it is common for special concessions to be given to industries in general or in particular. Such concessions can cause such severe administrative strains that the whole tax becomes inviable. This happened with the Belgian tax on pesticides that was levied on all purchasers except farmers. Not surprisingly, a black market quickly appeared, with farmers selling pesticides directly to members of the public.

A further example of a concession system that undermined a tax through its effects on the administration was in Sweden, where reductions in carbon taxes paid by industry were granted on a case by case basis. Quite apart from the likely problems with international trade rules if this system would have been continued, the administrative burden was extremely high (not least because the concessions were given by the government, and not by the administration).

The Choice of a Tax Point

Generally, duties on mineral oils are levied at the refinery gate, alcohol when it is bottled. The number of tax points is therefore kept to a minimum, and a presumption can be made that this is administratively desirable. This is not always possible. A variety of difficulties can push feasible tax points further down the distribution chain. The proposed United States energy tax ran into the problem of large differences in the energy content of different types of fuel, preventing use of broad categories such as "coal." More intractable still was the prevalence of fixed-price long-term contracts in the energy industry. Only by going down the chain towards the retail level was it possible to avoid fixed-price contracts.

A further reason for accepting the higher administrative costs of coming down the distribution chain when choosing a tax point is the problem of Border Tax Adjustments. Various GATT agreements have never been explicit about when border tax adjustments are permissible, but they have been interpreted to mean that whilst they are for taxes on products, they are not for taxes that depend on processes. If an excise-type tax is imposed on, for example, electricity, border tax adjustments are acceptable. A tax that tried to take into account the different amounts of fossil fuels that were used to generate that electricity is not.[1] The type of tax that can be introduced is therefore limited by international trade rules, and compliance with these rules may reduce to some extent the environmental effectiveness of such taxes. Certainly Denmark felt constrained to make its energy tax one based on final products, not inputs to the energy generation process (in part for GATT reasons), and it has been argued that early versions of the proposed United States energy tax and the Swedish carbon tax may not have been GATT-compatible.

An Appropriate Administrative Framework

It is widely accepted that the appropriate model to follow in administering and collecting environmental taxes is that of the excise taxes, because of the familiarity of excise departments with physical control of excisable products and the powers of control and enforcement that excise departments have built up over many centuries.

By and large, the advantages of administration by excise officials has been borne out by experience. Sweden, for example, switched from administration by the environmental department to the Ministry of Finance, precisely to ensure this tighter control of environmental taxes. Yet the administrative provisions available to Excise departments must be used fully for anomalies to be avoided. In Belgium, before the environmental tax reform was implemented, producers and consumers built up stocks of soon-to-be-taxed products, leading to arbitrary effective tax rates, far lower revenues than expected, disillusionment with the tax reform itself, and the development of a group of aggrieved taxpayers who had failed to exploit the loophole. Clearly, such problems could have been avoided by the use of a tax on stocks (and indeed the United States had such provisions in their various proposed energy taxes).

---

[1] The most recent GATT treaty might permit a different interpretation, the argument resting on the interpretation of a footnote. Equally, recent rulings by GATT arbitration panels have been interpreted by some as showing signs of more flexibility.

Belgium has also encountered problems in enforcing its environmental taxes in the context of a European Union operating without fiscal checks at the borders. Most European Union countries have provisions allowing for temporary import of products for re-exportation without imposing any tax (as otherwise the tax would have to be refunded). It is believed that quantities of such untaxed products are leaking into the Belgian economy. With traditional excises, such leakage is made difficult because of the use of "banderoles" (physical marking, or tax stamps) applied to alcohol and tobacco, or by the operation of bonded warehouses, where control is extremely tight. Such administrative systems are administratively expensive, but their cost is minor in comparison to the taxes being levied on the products concerned (most cigarettes and mineral oils in the European Union are taxed at over 100 percent of the underlying cost). This cannot be said of, for example, taxes on disposable razors or non-recyclable drinks containers: the costs of such an administrative system would be disproportionate. As a consequence, no tight enforcement of such taxes is feasible.

"GAINERS AND LOSERS"

Whilst there are cases of environmental taxes failing due to administrative problems, far more taxes never reach the stage of being designed because of an inability to get political support for the tax. In part, this reflects the fact that environmental taxes are often proposed to deal with "new" environmental problems, or as a way of tightening existing environmental regulations. Opposition to environmental taxes is often not so much against the tax instrument as against the policy aim. But even were this not to be the case, and where taxes are designed well and directed at important environmental problems, the distribution of "gainers and losers" from the policy is such that the latter have both the incentive and the ability to prevent introduction of a tax.

The distribution of gainers and losers is of importance because of the intra-marginal transfers arising from a tax. The economic costs of the tax arise from the costs of adapting behaviour, but the financial costs to those paying the tax are far higher. The financial burden will in the end be borne by workers, owners and consumers of the industry concerned; these groups are likely to act through industry organisations to oppose the tax. Hence there can be an *a priori* assumption that environmental taxes will be opposed by a coherent set of interests.[2] Policymakers can counter this opposition in two broad ways: by

[2] This is a simplification. It might be expected that "industry" should be divided over the attractions of environmental taxes when compensation is available (lower social charges, etc), polluters still being losers but other industries being gainers. That this rarely

working with the industries concerned to reduce their opposition (for example by reducing the amount of transfers involved in the policy change); they could also use the funds raised from the tax in such a way as to create or reinforce a constituency arguing in favour of the environmental tax.

The Courting of Industry

A variety of strategies have been followed in attempting to reduce industry opposition to environmental taxes: Commissions have been used to reduce the monopoly of information that industry might otherwise hold; measures have been taken to limit industrial disruption following introduction of a tax; changes in behavior have been encouraged so as to avoid the need for taxes; and revenue recycling has been implemented.

Committees, Commissions, Groups of Enquiry

Almost every country in the OECD has at some time within the past ten years set up a Committee of some sort to look at the use of economic instruments for environmental control. Sometimes this reflects the ordinary decisionmaking processes in a country, but in many cases the use of such Commissions has been particular to environmental policy. There appear to be two reasons for their widespread use in this area: information gathering and consensus forming.

Information on the marginal damage of pollution normally being absent, tax rates are often set in an essentially ad hoc manner, aiming to achieve some level of pollution abatement. But this means that the information required in order to set the appropriate tax rate is large, and is possessed only by the industry being taxed. Without detailed data on available technologies and likely response strategies, it is unlikely that authorities could correctly anticipate the economic costs of any particular tax rate. Issues such as the treatment of stocks of goods, the appropriate point of taxation, the physical definition of the tax base and its measurement and the definition of exempt activities are all areas critical to the successful operation of environmental taxes, but are dependent on the successful interpretation of industry data. One reason for the success of industry in blocking the introduction of environmental taxes is this control over necessary information.

Such data cannot be reliably collected without the good offices of the industries effected, and the administrative feasibility of a tax that is not widely regarded as just and fair can quickly be undermined. Hence, the information-

---

happens presumably reflects internal dynamics in employers' organisations.

gathering role of Committees is usually supplemented by an intended consensus-building. The consensus may not result in an outcome favourable to the development of environmental taxes. In Sweden, industry representatives on a committee set up to review the carbon tax successfully managed to create a consensus with other representatives that the tax should no longer be applied to industry. More generally, the fear of the capture by industry of the tax-setting process is widespread among authorities.

Nevertheless, Belgian experience provides an example of a Commission successfully buttressing environmental taxes. An environmental tax reform, introduced as part of a political compromise and without the benefit of extensive consultation, did not go smoothly. A "follow-up" Commission was therefore set up such that no environmental tax may be introduced or modified without the Commission giving its opinion on the desirability or otherwise of the change (although the advice of the Commission need not be followed). Each year, an annual report is produced that analyses the effects of the existing environmental taxes. The perceived impartiality of the Commission, and its apparent willingness to listen to the views of industry has quickly made it a central player in Belgian environmental policy. It is widely seen not just as the provider of technical, scientific information (the role for which it was designed), but as a "Court of Appeal." Industry goes to the Commission to argue its case, rather than directly to the administration. It is clear that the existence of the Commission has substantially increased the legitimacy of the environmental taxes in Belgium.

Pre-announcement and gradualism

One way of reducing industry opposition to ecotaxes is to keep industrial disruption to a minimum by announcing the introduction of a tax long before it is actually introduced; to introduce it at a level well below the desired final level, and to increase it only gradually. The rationale for such a strategy is to give companies time to adjust before the tax is introduced by investing in appropriate emission abatement technology without having at the same time to suffer the financial drain of tax payments.

Generally, policymakers accept that environmental taxes should be more predictable than traditional taxes. A particularly effective example of preannouncement was the German experience in preannouncing a large increase in waste-water charges in the early 1980s. Some recent examples of this strategy of preannounced and predictable tax rate changes can be seen in Switzerland (where the carbon tax and the tax on volatile organic compounds (VOC) are to be introduced in three steps, each one two years after the previous step, with the first step not applying until 1996 at the earliest. In the case of the VOC tax, the third step will only apply if the first two steps have not been sufficient in reducing emissions). In Denmark, tax changes on existing environmental taxes up until 1998 have been announced; a new tax on municipal and industrial waste water will not be introduced until 1997. In France, the present plan of the "agences de l'eau" involves a doubling of charges over the period 1992 to 1996.

Offering an alternative

The private costs to the polluting industry of ecotaxes are higher than the costs to the economy as a whole by the amount of intra-marginal transfers out of the industry. Hence, in return for the tax not being introduced, polluters have an incentive to commit themselves to reducing pollution by as much or even by more than they would if it were introduced. And industry has in fact offered to reduce pollution, or to introduce recycling schemes, or has undertaken pollution-reducing investment in order to obviate their need.

The Swedish experience of taxes on SOx and chlorine (used for bleaching in the chapter and pulp industry) is one example. Emissions of sulphur by Sweden fell by 80% during the 1980s, compared to a 25% fall in the rest of Europe. The sulphur tax was introduced only after much debate and several false-starts. While precisely identifying which of several policy instruments was responsible for the fall in output is not possible, it seems likely that the threat of introduction of the tax as much as the actual introduction had a major effect on sulphur emissions. More clearly, the Environmental Charge Commission of Sweden suggested a tax on chlorine. This was vigorously opposed by industry. However, the very threat of taxes appears to have been enough to have caused a dramatic cut in chlorine use by the paper and pulp industry. Similarly, in Norway the threat of a tax on ozone was apparently extremely effective in getting a voluntary agreement.

Exemptions and revenue recycling

If industry cannot be persuaded to accept the transfer of funds out of the industry, then one way of reducing opposition is to reduce the extent of the transfers. Exemptions and tax reductions for some industries could be introduced. Such exemptions need not reduce the effectiveness, narrowly defined, of a policy in achieving environmental goals. Governments have sought to negotiate voluntary agreements with the exempted sector to pursue environmental aims. This has been true in Denmark, for example. The original European Carbon/Energy tax proposal involved widespread exemptions for energy-intensive industries, but these were defended on the grounds that they were dependent on efforts being made to reduce energy consumption.

Nevertheless, the variety of problems caused by exemptions is extensive. Administrative costs go up sharply, as measures have to be taken to ensure that untaxed products do not leak into general distribution. The marginal cost of abatement across different sectors diverges, undermining the case for using environmental taxes. The competitive position between different industries (or even different companies within any particular industry) that are taxed at different effective rates is distorted. All these problems are well understood by policymakers. Yet they have been unable to resist calls from industry for such special treatment. The complaints about the "damage to competitivity" have proved overwhelming.

A Danish review of environmental taxes and industry made in April 1994 looked at recycling through reductions in income tax, VAT, corporate taxes and social security contributions, and increased subsidies to investment and depreciation allowances. The report favoured income tax or employer social security contribution reductions, often claimed by industry representatives themselves to be causing high labour costs. Such a framework for considering the use of the revenues points to a key policy choice: should they be used to reduce net transfers out of the taxed industry, reducing that industry's opposition to the tax, or should they be used for other popular policies, creating an alternative constituency in favour of the tax? The following section considers this issue.

Use of the Tax Revenues

In the face of industry opposition, successful introduction of an environmental tax will depend on the existence of a countervailing constituency in favour of the introduction of the tax. This might be built around the environmental benefits of the tax. For small-scale, well-defined environmental problems, this may be sufficient. Empirically, however, all large-scale ecotaxes in existence have been introduced as part of a more wide-ranging policy package, emphasising that the manner in which the tax revenues are spent will lead to "gainers" as well as the more visible and vocal "losers." This has been done by emphasising revenue and distributional neutrality.

Revenue neutrality

Countries introducing new environmental taxes like to stress that this will be done in a revenue-neutral manner. The EC proposals for a carbon/energy tax go furthest in making this explicit. Revenue neutrality may be appropriate, given existing economic conditions and state of public finances, so other taxes should be cut when ecotaxes are introduced. On the other hand, there may be good reasons to use the tax revenues to close budget deficits (a point which the Italian authorities have made in response to the proposals from the EC). But it is felt that making the revenue neutrality commitment may improve the public perception of the tax, by emphasising that these taxes are "different" from normal taxes that just have a revenue-raising function.

Support for emphasising revenue neutrality can be found when considering the cases of Norway and Sweden. Starting from the end of the 1980s, Norway and Sweden have undertaken two of the largest tax reforms in history. Marginal tax rates on labour income have been reduced substantially (from 75 to 63 percent even at incomes little above average in the case of Sweden), and average tax rates on labour have fallen by ten percentage points of total income.

The carbon taxes introduced at this time could have been justified to the public by the need to fund these other tax reductions. This happened to some extent, but generally the environmental benefits of the taxes were emphasised. Yet this proved to be a weakness of the taxes. The authorities introduced reductions and exemptions in the carbon tax base, for the understandable reason that the competitive position of Swedish and Norwegian industries would otherwise be damaged by such taxes. The result was that effective tax rates were far lower than apparent nominal energy tax rates (Haugland, (1994)). In Sweden this led to the government-appointed review commission pointing out that the tax had no environmental effect at the margin and so the tax must be for revenue-raising purposes, and one on an inappropriate tax base. The recognition that other countries were not going to follow the Norwegian lead in introducing substantial measures to reduce carbon emissions again made the environmental case for carbon taxes very unconvincing, and it now seems likely that Norwegian carbon tax rates may fall (or at least will not rise). In retrospect, there might have been advantages in explicitly making the link between the revenue from the carbon taxes in these two countries with the tax reductions which were taking place at the same time. This might have muted industrial criticism of the tax.

Switzerland, on the other hand, in its recent proposals for taxes on carbon and VOCs, did not emphasise revenue neutrality. Instead, the proposal identifies the different ways in which the funds would be used. In the case of the VOC tax, SFR50m would go to help industry make appropriate investments to reduce VOC emissions; the remaining SFr250m would be recycled to the population through reductions in health insurance premiums. For the carbon tax, up to SFr430m would be earmarked "to secure the financing of existing environmental and energy measures, and to finance new measures, including international funds for global environmental problems"; SFR220m would be given to the business community by reducing old-age pension and invalidity insurance; the remaining SFr650m being distributed to the population through reduced health insurance premiums. Similarly, in the United Kingdom, revenue from much increased landfill charges has been linked to reductions in employers' social charges, with apparent success in reducing industry opposition.

This Swiss approach aims to establish a connection between the tax and the use of the tax revenues. This logic is used to justify earmarking. Attitudes to earmarking vary widely throughout OECD countries. One extreme is that of

France, where the "Principle of Mutualisation" is given wide credence. This views environmental taxes as a form of fee-charging in a "club-good" situation. The funds raised may be used either for research and development, or to aid polluters in making investments in facilities to prevent or treat the pollution on which the tax is levied. Hence the tax is systematically returned to the polluters. The system in the Netherlands is similar.

The disadvantages of this type of earmarking are widely recognised (including in France). There is a risk that environmental spending could become too low or too high due to reluctance to change environmental tax rates. Delache and Henry (1994) argue that in the case of water pollution there is now systematic overinvestment in waste-water treatment, because these investments are funded by a charge whose rates are not changed to reflect investment needs.

However, the advantages of earmarking are equally apparent. Distrust of official motives in introducing taxes is widespread. As the "Principle of Mutualisation" suggests by its very name, industry sees that there is no net transfer of financial resources out of the industry; and environmental groups often support earmarking, apparently believing that governments are unlikely to raise taxes to a level sufficient to have appreciable incentive effects. Earmarking is often used, therefore, as a way of maximising support for the introduction of a measure which is controversial.

It is not surprising that earmarking is often found in those countries where distrust of government is strongest. In the United States, virtually all environmental charges are earmarked (e.g. the CFC charge, the oil spill charge and the Superfund charge). At the opposite extreme, in the Nordic countries public policy is strongly opposed to earmarking, an opposition which is frequently restated. However, the advantages of earmarking are sufficient such that even in countries with strong public sectors (such as France and the Netherlands) earmarking is sometimes used.

Distributional neutrality

The distributional consequences of environmental policy are generally regressive, reflecting the environmental costs of the food and energy industries. The degree of regressivity varies according to how regressivity is defined and measured. As Poterba (1992) argued, within a life-cycle approach, looking at the increase in expenditure on energy following the introduction of a carbon tax as a proportion of total expenditure gives a very different picture of the regressivity of a tax than does looking at the increased tax payment as a percentage of income. Pearson (1993) confirms that analysis using European data. The regressivity of a carbon tax would differ dramatically between different European countries, with tax payments being high and strongly regressive in the United Kingdom and Ireland, and relatively low and proportional in southern Europe.

No ecotax reform will be complete unless the distributional effects of any tax cuts that take place are taken into account. The distributional effects of the Nordic carbon taxes alone is misleading, given the policy context of reductions in marginal tax rates on earnings and other sources of income. Second-round effects may also have a distributional impact. Hence, for example, it is argued that the distributional effects of the European Carbon/Energy tax could be broadly neutralised by reducing employers' social security contributions for those on low incomes, which would lead to the double dividend of increased employment as well as reductions in carbon emissions. These offsets to the apparent regressivity of environmental taxes are often ignored by those opposed to the environmental objectives of the tax in an attempt to mobilise a broader coalition against the tax. Also, alternative environmental policy instruments need not necessarily have better distributional effects than taxation.

However, it is unlikely that tax cuts alone can ever fully offset the distributional consequences of any environmental tax reform (see OECD (1994b)). Those whose welfare receives a particularly high weight in public discussion are often reliant on public benefits: pensioners, the unemployed etc. Tax cuts are unlikely to fully compensate such groups for the loss in welfare arising from increases in the costs of expenditure items which form a large proportion of their budget. The only way to compensate such groups is through the benefit system. But as soon as this is accepted as part of the environmental reform package, various other policy aims cannot be pursued. Any compensation of those outside of the labour force for increases in ecotax payments will pre-empt the use of funds which might otherwise have gone towards reducing the cost of labour or public sector deficits. In addition, the income elasticity of demand may be high enough to substantially offset the demand changes arising from the changes in relative prices, reducing the efficacy of the tax.

The distributional effects of ecotaxes were one of the reasons why sustained increases in the price of mineral oils in the United States were politically unacceptable; mineral oil taxes there are slightly regressive (see Poterba, (1992)). More recently, the public debate which followed the imposition of VAT on domestic fuel (gas, electricity, etc) in the UK (ostensibly for environmental reasons) was focused entirely around the distributional impact. Even after (belated) direct expenditure measures were taken to protect some of the most vulnerable groups (e.g. pensioners), the full tax increase could not be introduced. When introducing its environmental tax reform of 1994, Denmark was careful to offer substantial compensation to pensioners and students.

However, some taxes (e.g., on water consumption) may result in a pattern of increased tax payments that cannot be successfully neutralised by tax cuts or increased benefit payments (see Rajah and Smith, (1994)). The Wolfson Committee (1992) for the Netherlands also argued that whilst compensation for a carbon tax could be made through the tax system with rates reductions based on average energy usage per income group, high energy consumers within income groups would still be worse off. There is wide variation in consumption of energy even within particular income brackets, with research suggesting that energy consumption is only weakly related to income and strongly to factors such as age, household composition and type of housing. It is therefore unclear whether redistribution of revenues through the tax system can be designed in such a way as to be socially and politically acceptable.

WHAT ARE ENVIRONMENTAL TAXES FOR?

The examples of administrative failures and, particularly, of difficulties in overcoming the opposition of "losers" following the introduction of an ecotax can often be traced to confusion over the aims of the environmental tax reform. Inevitably, environmental taxes have been introduced for more complicated reasons than simply modifying market incentives in order to reflect environmental costs. The question is not so much *whether* other policy issues have to be considered, but the importance they are given in the design, implementation and public justification for the introduction of ecotaxes.

The inevitability of having to consider non-environmental goals arises because environmental taxes involve intra-marginal transfers. The overall economic costs of environmental taxes are those arising from changes in behaviour in response to the new set of price signals. But the costs to the affected consumers and industries are higher, reflecting these intra-marginal transfers. These funds are redistributed, with policy goals other than those relating to the environment having a legitimate claim on the funds. In other words, even supposing that the setting of the tax rate can be considered as being determined entirely according to environmental concerns, it is inconceivable that environmental concerns can monopolise the discussion of what to do with the funds. As the tax by definition raises funds, the simplistic answer of "to help the environment" to the question posed as the heading of this section is not credible; or, more precisely, if the simplistic answer is given in a simplistic way it is unlikely to be convincing.

The revenues raised from environmental taxes could be spent on environmental issues; equally, they could be spent on improving the distribution of after-tax incomes (by tax and benefits changes); on altering the costs of different factors of production (by cutting labour taxes); on attracting mobile capital (by cutting capital taxes); on compensating those who pay the environmental tax (by redistributing the funds according to some method unrelated to tax paid); on reducing public sector deficits; or on reducing charges for health care. All of these policies have been followed when introducing ecotaxes or have been officially suggested by governments trying to introduce such taxes. Not surprisingly, the effects of spending the money in these different ways have been pushed as reasons for introducing the taxes quite as much (often more so) as the environmental effects from the changes in price signals.

Certainly some countries have benefitted from putting the use of environmental taxes in a context much broader than environmental policy. Sweden was the pioneer in this respect, reforms of the tax system in 1991 being linked to the reform of environmental taxes. Norway followed suit in a series of budgetary changes through the 1990s. In Denmark, the 1994 environmental tax reform (which introduced several new taxes and substantially increased existing taxes) was directly linked to a package of government policies, collectively entitled "A New Growth Perspective." All the policies were presented as a coherent package; the environmental policy changes were particularly linked to the income tax reform and the labour market reform.

Combining policy aims *can* reinforce the case for an environmental tax reform, but there are risks in so doing. Precise policy aims help determine the scope for flexibility in ensuring administrative and political viability. Failure to determine clearly the objectives of a tax -- and therefore what policy compromises can be made to get the tax introduced, and which should not be made -- can be fatal to the chances of implementation.

Difficulties caused by confusion over aims have been found in a wide variety of OECD countries. In Belgium a major difficulty in introducing the environmental taxes has been the definition of the environmental objectives themselves. Pesticides have been divided into three categories in Belgium; toxic, fairly toxic and very toxic. Disputes over in which category to place particular pesticides have been more controversial than the introduction of the tax and the setting of tax rates. A clear statement of the precise environmental aims could have avoided this problem. Similarly, the lack of agreement on the potential costs of continued carbon emissions has led to attempts to present carbon taxes as no-regrets policies with other beneficial impacts on energy-related pollution and congestion. This can backfire, with opponents of the tax able to credibly argue that carbon taxes are not well targeted at these secondary aims.

Even where an environmental concern is accepted, the issues regarding the use of a tax to reduce emissions and the use of a tax to eliminate emissions are rather different; the former type is more vulnerable to the charge that the tax is being introduced simply for revenue-raising purposes, the latter to the charge that a tax is an inappropriate instrument. Further confusion has arisen over the role of voluntary agreements to reduce emissions, which polluters' have proposed as an alternative to environmental taxes. Where the aim of the tax is purely to meet a given emission target, it may be that such an offer should be accepted. Where aims go further than meeting a given target, or revenue requirements are important, the tax might still be preferable. Once again, confused responses have been given to such offers, reflecting official confusion about whether they meet the original (often unarticulated) aim or not.

Swedish experience represents a cautionary tale about the consequences of policy confusion. The Swedish carbon tax was introduced in 1991, so that total Swedish energy taxes were the highest in the OECD. However, environmental aims were diluted by a concern over the competitive position of Swedish industry. Until the end of 1991, all energy and carbon taxes exceeding 1.7 percent of the value of sales were refunded. In 1992 this threshold was reduced to 1.2 percent. A Committee was set up with the brief of ensuring that the energy tax system was compatible with international trade rules, but also to ensure that the competitiveness of Swedish industry would not be damaged by the energy tax. The committee concluded that as the energy taxes were having no impact at the margin on energy consumption, the fundamental purpose of the tax must have been fiscal not environmental. Following this conclusion, the committee considered several policy options. The first (removing the concessions benefitting heavy users) was rejected. If the purpose of the charge was fiscal, then better tax bases (labour) could be found that would not lead to relocation of industry. If the tax were made Pigouvian, then it would have to be levied on global CO_2 emissions. In the absence of that, there was no point in a tax that simply caused relocation of emissions. A general tax reduction for certain industries was considered, but was considered excessively arbitrary, potentially causing trade problems. The Committee therefore suggested harmonising energy taxation on industry with those of the EU. As the EU has no energy tax, that amounted to saying that the carbon and energy taxes should be abandoned. The tax reform of 1993 therefore exempted all industry from the energy tax and from 75 percent of the carbon tax.

CONCLUSIONS

This chapter has discussed some failures in the strategies which have been followed by governments when attempting to introduce environmental taxes. Inevitably, therefore, the tone has been negative, pointing out problems and difficulties. This should not be taken as implying that environmental taxes are, and always will be, minor in importance and disappointing in effect. There have been policy successes from introducing environmental taxes; environmental problems have been reduced and controlled, and in a more efficient way than would have been possible using conventional command and control techniques.

But these successes have been fewer than was expected a short time ago. It has now become apparent that in most countries the discussion of the introduction of environmental taxes will be against a background of hostility. New taxes are seen as imposing new burdens. "Pensioners and the poor will be worse off; industries will lose export orders and jobs; the tax will not work anyway; it would be far more effective to use regulations; the administrative costs will be huge" -- the arguments that can be used against new environmental taxes are politically very sensitive. This chapter has not discussed the obvious demerits of these views. But any successful implementation strategy must contain a credible response to these worries.

Several ways have been suggested for putting together a policy package that might help the implementation of environmental taxes. The optimal package will differ from country to country, reflecting different national preferences regarding the environment, distributional effects, the role of government and the importance of various non-environmental objectives. Nevertheless, some common themes seem to be encountered in the successful implementation of environmental taxes.

a. The environmental case for taking some sort of action must be unassailable. If there is doubt over the basic policy goal, it will be easy for those who will suffer from pursuit of that goal to point up the costs without the environmental benefits being apparent. In practise, this suggests that polluters must be persuaded that the environmental case against them is so strong that the argument becomes about policy instruments rather than goals. The possibility of successfully implementing a carbon/energy tax to reduce global warming is, it is increasingly clear, heavily dependent on a wider consensus on the existence of global warming than currently exists.

b. Administrative feasibility cannot be taken for granted. The case for environmental taxation rests on marginal costs of abatement being equalised across emissions; if there is evasion or avoidance of the tax, its justification is undermined and the tax is perceived as unfair. Administrative feasibility imposes constraints not just on which emissions might be controlled by taxes, but also on the level and structure of the tax. If taxes are imposed on consumption of products associated with emissions, countries with open borders must take account of cross-border shopping or imports of uncontrolled products. The granting of exemptions to a tax creates borderlines of taxed and untaxed goods which must be policed, increasing the administrative complexity of the tax.

c. An environmental tax involves some economic agents suffering a welfare loss. The core constituency opposing an environmental tax will therefore be well-defined. A major part of a strategy to implement large-scale environmental taxes must consist of measures to define a group of "gainers" who will support the introduction of the tax. The assurances of revenue neutrality, or the earmarking of environmental tax revenues for particular (usually environmental) purposes are among the strategies that are followed to emphasise the existence of winners, but the experience of several countries suggests that less binding commitments may be effective.

d. The "spending" of the environmental tax revenues will inevitably involve assessment of policy aims other than those relating to the environment. But clarity is required in weighing the demands of the different policy goals. The "selling" of an environmental tax on the grounds that reductions in direct taxes and hence labour costs can promote employment may be politically attractive. But an answer is then required to the complaints of those industries that ask for exemptions from the tax because they employ many people and are particularly affected by a tax. What also should be the response to offers to meet some environmental aims voluntarily in order to avoid the financial transfers out of the polluting industry? The "blurring" of policy objectives can be seen as a great opportunity to promote a package involving environmental taxes as pursuing a range of desirable objectives, but if policymakers are not lucid in their policy aims, the pursuit of multiple policy objectives can fatally reduce the attractions of taxes as a cost-effective environmental policy instrument.

References

Delache, X. and C. Henry. (1994). "La Fiscalité et l'Environnement: Le Cas de la France." Paris: OECD.

Gofman, K.G. and A.A. Gusev. (1994). "Taxation and Environment in the Russian Federation." In *Taxation and the Environment in European Economies in Transition*. Paris: OECD.

Haugland, G. (1994). "A Comparison of Carbon Taxes in Selected OECD Countries." *Environment Monographs* 78. Paris: OECD.

McCloskey, W. (1994). "Implementation Strategies for Environmentally-friendly Tax Reforms." Paper presented at the 50th Congress of the International Institute of Public Finance. Boston, MA.

Oates, W. (1994). "Environment and Taxation: The Case of the United States." In *Environment and Taxation: The Cases of the Netherlands, Sweden and the United States*. Paris: OECD.

OECD. (1993a). "Environmental Taxes in OECD Countries: A Survey." *Environment Monographs* 71. Paris.

OECD. (1993b). *Taxation and the Environment: Complementary Policies*. Paris.

OECD. (1994a). *Managing the Environment: The Role of Economic Instruments.* Paris.

OECD. (1994b). *The Distributive Effects of Economic Instruments for Environmental Policy.* Paris.

Pearson, M. (1993). "Equity Issues and Carbon Taxes." In *Climate Change: Designing a Practical Tax System.* Paris: OECD.

Poterba, J. (1992). "Lifetime Incidence and the Distributional Burden of Excise Taxes." *American Economic Review* 79, 325-330.

Rajah, N. and S. Smith. (1994). "Distributional Aspects of Household Water Charges." *Fiscal Studies* 14/3.

Wolfson Committee. (1992). *Stuurgroep Regulerende Energieheffingen: Final Report on Regulatory Taxes on Energy.* The Hague.

PART II

AN IMPERFECT WORLD

Chapter 4

TAXING BADS BY TAXING GOODS: TOWARDS EFFICIENT POLLUTION CONTROL WITH PRESUMPTIVE CHARGES

Gunnar S. Eskeland (World Bank)

and

Shantayanan Devarajan (World Bank)*

INTRODUCTION

Anyone who has visited Mexico City or Jakarta or Santiago knows that pollution is part of everyday life in the developing world. Yet economics textbooks have been prescribing solutions to the problem of pollution for at least half a century. One reason why the textbook solution -- levying an emissions fee on the polluter -- has not been applied is that it requires the government to monitor emissions, something that may not be feasible in many countries. In fact, most pollution control policies in place employ *indirect* instruments -- those which tax or regulate activities *associated* with emissions -- rather than neutrally putting a brake on emissions. In this chapter, we argue that such policies economize on scarce capacity in monitoring and enforcement, but generally leave room for improvement.

An important feature of the textbook emissions fee is that it gives the polluter the incentive to choose the optimal mix between cleaner technologies and reductions in the scale of output supply or input use. With indirect instruments, this optimal combination does not come by itself. For instance, emissions standards make cars cleaner and have modest monitoring requirements (annual inspections), but they do not provide incentives to reduce the number of miles driven. The purpose of this paper is to show that indirect instruments that reduce the scale of output (such as a tax on output or on polluting inputs) can be important complementary measures in a cost-effective pollution control program. A combination of the two types of instruments -- those that reduce the scale of output and those that reduce the level of emissions per unit of output -- can mimic fairly well the effect of an optimal

* The authors thank C. Ballard, M. Cropper, G. Hughes, conference and seminar participants, as well as two anonymous referees.

emissions fee. Furthermore, when regulations to adopt cleaner technologies are in place, the addition of a tax on an input associated with emissions can reap substantial benefits for a pollution control program. These complementary instruments require very little or no additional resources for monitoring and enforcement. Finally, schemes that use presumptive charges have the advantage of facilitating refinement of incentives as capacity allows. Thus, they are not in conflict with the recommendation of "first-best" instruments.

Unlike the situation with emission fees, however, the optimal use of indirect instruments requires knowledge of the "cleaner" technologies (the ease with which emissions per unit of output can be reduced) as well as the sensitivity of demand to prices (the ease with which the scale of output can be reduced). We rely on empirically based case studies of Mexico City, Chile and Indonesia to illustrate the choice between the two types of instruments -- the interaction between the two and the consequences of employing them.

A recurring theme throughout the chapter is that taxation of fuel use, due to the association between fuel use and emissions, can serve as a powerful indirect instrument to control air pollution. In the case of automobiles, we show that failing to employ gasoline taxes in Mexico City would significantly reduce welfare, even when regulatory standards are in place. In the case of point-source pollution, we calculate that there is significant potential for altering the fuel mix of industries in Indonesia and Chile by taxing the "dirtier" fuels. Furthermore, we show that in the case of Indonesia the general-equilibrium consequences of such a tax are similar, though somewhat dampened, compared to what is indicated in partial equilibrium models. In sum, we advocate taxing a bad (pollution) by taxing a good (fuels) as part of a program to address air pollution when monitoring of emissions is prohibitively expensive.

The next section spells out our basic analytical framework, including its relationship with the popular "market-based" approach to pollution control. Next, we treat the case of mobile-source pollution with an examination of gasoline taxes and regulatory policies in Mexico City. Then, we address point-source pollution and the potential for altering the fuel mix in industries in Indonesia and Chile, based on firm-level data. Subsequently, a general-equilibrium model of Indonesia portrays the economy-wide consequences of changes in fuel taxes. The last section contains our concluding remarks.

THE ECONOMICS OF POLLUTION CONTROL: A BASIC ANALYTICAL FRAMEWORK

To illustrate some important principles in the economics of pollution control, we begin with the textbook example of a polluting factory or an automobile -- let us call it a plant. The plant produces a good -- say bread or passenger-kilometers -- whose marginal social cost of production initially exceeds its marginal private cost, since the pollution will harm people other than the plant manager. The textbook solution to this problem is to levy a fee on pollutant emissions equal to the difference between the two costs. The imposition of this emissions fee gives the plant manager an incentive to reduce emissions in two ways: by adopting cleaner technologies and reducing output. Moreover, he will choose a combination of cleaner technologies and reduced output that is optimal, not only for himself, but for society as a whole. Note that the policymaker need not know what this optimal combination is -- just as the purchaser of bread need not know the optimal combination of inputs required to produce it.

However, the imposition of an emission fee requires the government to *monitor* the plant's emissions. For many pollution problems, especially those that are caused by a large number of polluters, this is virtually impossible. In this case, the government has two alternatives: tax the good produced (or used) by the polluter or ask the polluter to adopt a cleaner technology. In the next section, we show that each of these contributes towards achieving the benefits of an emissions fee, and that a combination of the two can mimic the effects of an emission fee reasonably well. In this situation, knowledge about control technologies and elasticities of demand for the polluting good is valuable to the policymaker (such information is redundant if emissions can be monitored, because of self selection). We also discuss the problems with applying 'presumptive charges' and other indirect instruments, and how they may be overcome.

The Textbook Case: An Emissions Fee

The case for taxation of emissions is made in most texts on welfare- and environmental economics, such as Tietenberg (1992) and Baumol and Oates (1988). We add here a more detailed treatment of how emissions are generated (and can be abated), thereby highlighting some results of practical relevance.

Sandmo (1975) showed how a polluting sector would be taxed optimally in a setting with distortionary taxes. A distinction between emission taxes and input or output taxes could not be made, however, as emissions were related merely to the scale of output in the polluting sector. Others, such as Greenwald and Stiglitz (1986), Sandmo (1976), Balcer (1980) and Wijkander (1985), have treated taxation of quantities *associated* with emissions when an emission tax is not possible or not perfect in providing incentives for emissions reductions. They can be seen as developing further an insight provided by Corlett and Hague (1953). Corlett and Hague's problem was one of optimal taxation with one untaxable good, and they showed that the distortionary costs of revenue generation are minimized by taxing goods complementary with the untaxable good (leisure, rather than emissions, in their case) at a higher rate. None of these authors, however, spells out a model in which emissions from a sector depend both on its scale of operation and on *how* it operates (we will describe the latter by the application of pollution abatement technology).

Figure 1.1 (see Figures 1.1-1.4 on next page) illustrates the case of a polluting firm that produces a good whose demand curve is given by $p=d(q)$. *PMC* and *SMC* are the private and social marginal costs of production, respectively. The gap between *PMC* and *SMC* is due to emissions associated with production, and the fact that these affect the general public, not just each user according to his or her own emissions. The classic examples are a factory that emits toxic wastes, or an automobile that sends fumes into the atmosphere when burning gasoline. Since it will be familiar to so many, we will use the automobile example.

The problem with the situation in Figure 1.1 is that the consumer is in equilibrium at point q^*, where $p=PMC$, which is not a *social* optimum. This can be seen by observing that consumer surplus, which we will take as the measure of social welfare, is the area under the demand curve that lies above the *SMC* curve, or the shaded area *A* less area *B*. By reducing production to the left of q^*, we can increase consumer surplus, by reducing area B. Therefore, q^* is not a social optimum. As noted earlier, the standard solution to this problem is to levy a fee on emissions. When facing such a fee, the motorist has two choices: he can drive less and reduce total emissions, or he can choose technical solutions, such as a catalytic converter, which will reduce emissions per liter of gasoline (in both cases, the polluter will pay fees for the emissions that remain). For any given emission fee, we can be sure the motorist will choose the socially optimal mix of these two ways of reducing emissions. To the extent that he reduces his emissions per liter of gasoline consumed, the external costs (the excess of *SMC* over *PMC*) are lowered. However, the catalytic converter would also raise his *PMC*. If the emissions fee is set so that it reflects the difference between the new *PMC* and *SMC*, the resulting equilibrium will be that in Figure 1.2 (we assume here that tax revenues are transferred to consumers as lump-sum income). The shaded area *A* has grown, while *B* has disappeared, so that social welfare, *A* - *B*, has increased.[1]

---

[1] The assumption essential to this argument is that the shadow price of public revenue is equal to that of private incomes. Assuming (costless) lump-sum transfers is one way of ensuring this, and an assumption of marginal utilities of income independent of the price of the polluting good allows the use of consumer surplus as a measure of welfare costs. Our discussion, thus, is only tangentially relevant to the "double-dividend" literature (see, for instance, Goulder (this volume)); in a nondistortionary environment, there is only a single dividend -- that from internalizing the environmental externality. For analysis in a context with (optimal) distortionary taxes, see Sandmo (1975).

When Monitoring is Costly: Demand Reduction and Technical Controls

Consider now the case when it is difficult to monitor individual emissions, and therefore difficult to levy an emission fee. Is it possible to achieve some or all of the benefits of an emissions fee, using other means to curtail demand or induce technical controls?

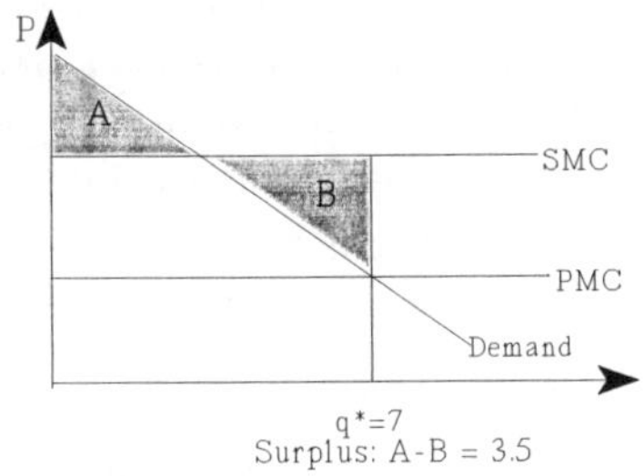

Figure 1.1 No intervention

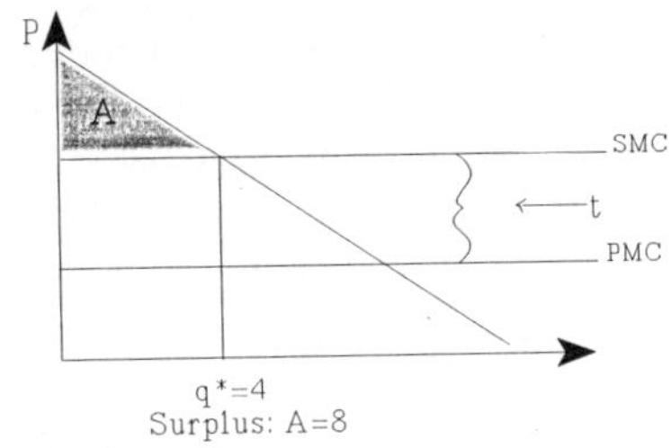

Figure 1.3 Pollution tax on output

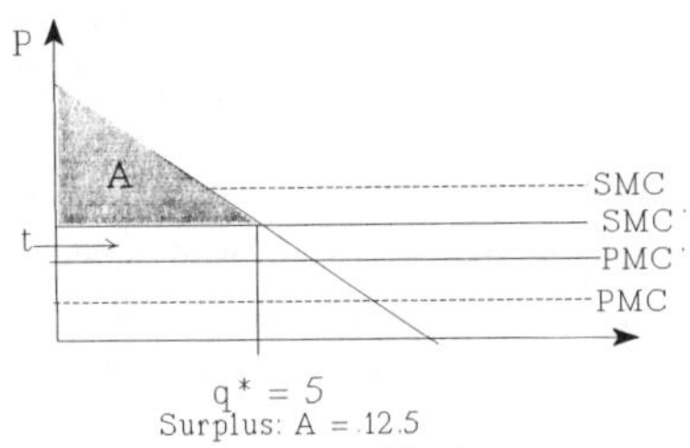

Figure 1.2 A pollution tax (or output tax and abatement)

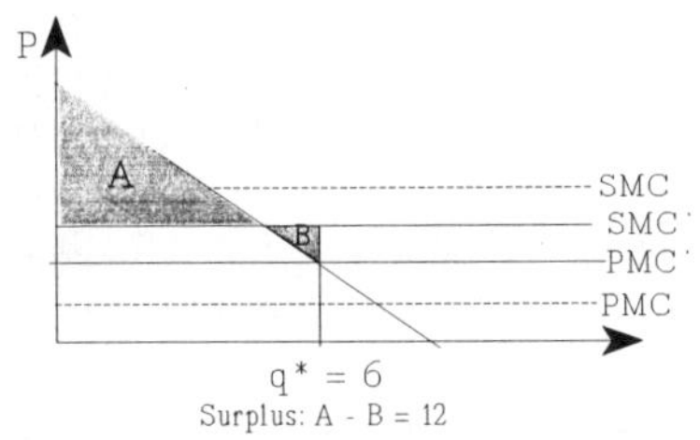

Figure 1.4 Mandated abatement

The first possibility is illustrated in Figure 1.3, where we have imposed a tax on output or a variable input, such as gasoline, equal to the difference between social and private marginal costs. The tax induces consumers to reduce their consumption by manipulating the price that they face.[2] Consumer surplus, in this case, will be the area A. Society has made net gains equal to *B* in Figure 1.1 by eliminating the consumption for which marginal social costs exceeded marginal social benefits. Notice that the social savings made from such a strategy, area *B* in Figure 1.1, depends crucially on the slope of the demand curve: the steeper the curve, the smaller the gains from demand-management instruments such as input and output taxes. However, the size of the optimal tax is independent of the slope of the demand curve whenever the difference between *SMC* and *PMC* is constant.

---

2 In case an input tax is envisaged, the argument in this section assumes fixed coefficients of that input to output, i.e. that the gasoline tax can reduce the use of the car, but not make car use less fuel intensive. This assumption, and its relaxation, are discussed in detail in subsequent sections.

In Figure 1.3, we assumed that a tax on a commodity such as gasoline would suppress demand in a "neutral" fashion, not changing the amount of pollution per liter of gasoline consumed. In Figure 1.2, we assume that an emission control technology is available. The application of the technology, such as a catalytic converter, would increase the private marginal costs from *PMC* to *PMC'*, but also would reduce pollution per unit so much as to position the new social marginal cost curve, *SMC' below* the original one. Figure 1.4 displays what happens if such a technology is induced directly, for instance by regulation. First, as social marginal costs are shifted downward, the area *A* expands, and the area *B* shrinks. Second, since consumers now face a private marginal cost which includes the costs of catalytic converters, consumption is also reduced (in comparison with Figure 1.1), so the area *B* is further reduced. Consumer surplus (*A* - *B*), will necessarily be greater than in the uncontrolled case.

Finally, consider what happens when you combine mandated abatement with a gasoline tax that is optimal given the abatement technology. We introduce a tax to suppress the consumption that was excessive given the reduced emission coefficient. The new equilibrium is identical to that achieved by an emission tax in Figure 1.2. In comparison with Figure 1.3, mandated abatement has been introduced to make consumption cleaner, and the consumer can therefore face lower taxes on the polluting good, and benefit from additional consumption. Notice that the area *A* is bigger, and the gasoline tax, per liter of gasoline, smaller. However, the tax rate *per unit of pollution* is the same in Figures 1.2 and 1.3: the difference in tax rates per liter merely reflects the difference in emissions per liter.

We have emphasized here that emission reductions can be provided by a combination of instruments, some of which make the polluting activity cleaner and others which reduce its scale. Which combination of instruments will be used depends, as Figures 1.1 - 1.4 illustrate, on the elasticity of demand for the polluting good and on the ease with which the activity can be made cleaner. The reduction in emissions will be provided mostly by technical controls if these cost little per unit of emission reductions *and* demand elasticities are low.[3] By contrast, emission reductions will be provided mostly by demand reductions if demand elasticities are high and control effectiveness is low (see Table 1.1). A typical example of the former could be reductions of the waste-load in domestic sewage: it would be hard to argue that the generation of human waste loads is sensitive to prices, so all

[3] *Technical controls*, *abatement*, and *cleaner consumption* are here used synonymously. What is essential is that emissions are reduced per unit of input or output. Technical controls show high effectiveness if they can shift emissions downward so that the SMC curve is shifted downward sharply. The SMC curve denotes the sum of PMC and the external costs due to emissions, so the SMC curve can be shifted downward sharply only if emissions can be reduced sharply *without* a major increase in PMC.

pollution reductions would have to be provided by technical controls. An example of the latter could be the control of carbon dioxide to prevent climate change. There are no control technologies available to reduce carbon dioxide emissions from fossil fuel burning, so all emission reductions will have to be provided through changes in demand for fuels (including shifts towards fuels with less CO_2 per calorie, such as natural gas). Both demand reductions and technical controls will be important in providing reductions in the two remaining cases: in one case, polluters have two important ways of responding to emission taxes, and emission reductions will be cheap. A good example in this category could be air pollution from urban transport. In the other case, polluters can do little to reduce emissions, and emission reductions will be costly. Initially, ozone depleting substances such as CFCs appeared to be in this category, although technical alternatives have been developed for most important uses. In general, the costs of obtaining emission reductions increase as one goes towards the south-east of the table.

Table 1.1. How Most Emission Reductions Are Delivered, When . . .

Control efficiency is		Demand elasticities are: High	Demand elasticities are: Low
	High	Both (control costs are low)	Technical controls
	Low	Demand reduction	Both (control costs are high)

Table 1.1 characterizes responses that deliver emission reductions under instruments such as emission taxes. With indirect instruments, the policymaker will generally need separate instruments to induce technical controls and demand reductions. However, to know which can deliver the bulk of emission reductions, the policymaker needs to know the elasticity of demand and the efficiency of technical controls. This is an important difference between the world of direct instruments, based on perfect monitoring, and indirect strategies. Simply put, if the pollution control agency knows little about individual emissions, then it should know a lot about other things -- what characterizes polluters, what their options are, and more. Table 1.2 summarizes which indirect instrument will be most important in providing emission reduction, based on the same variation in underlying conditions as in Table 1.1.

Table 1.2. Indirect Instruments: How Emission Reductions are Stimulated, When . . .

		Demand elasticities are	
		High	Low
Control efficiency is	High	Do both	Stimulate technical controls
	Low	Tax polluting goods	Do both

The lower left-hand cell, in which demand elasticities are high and control effectiveness low, is the case where it is most important to tax goods and services used in polluting activities. We call it a strategy of taxing bads by taxing goods. To follow the examples above, taxes levied on fuels will be a good strategy to reduce emissions of carbon dioxide. Similarly, to reduce the discharges from domestic sewage, technical solutions that treat or redirect the waste-loads are necessary, and indirect instruments mandating either these technologies or connection to collective systems can stimulate such responses. For cars, as we shall show later, it is important both to make trips cleaner and to discourage trips, and instruments that do both are necessary unless emission taxes can be levied on individual emissions. A simple and balanced, although blunt solution is to apply emission standards with periodic testing to ensure that cars are reasonably clean, and gasoline taxes or other demand management instruments to discourage their use. A more sophisticated solution is to read both the odometer and the emission test annually, and tax the owner according to presumed accumulated emissions. The latter proposal has yet to be applied in practice.

It should be noted that combinations of instruments are useful in all the cells; the emphasis here is on which instruments will be most important. Another way of reading the table is to ask when it is costly to apply only a narrow range of instruments. For instance, a program consisting only of instruments stimulating emission control technologies will be a costly one, and unnecessarily so, if demand elasticities are high (such narrow programs are the rule, rather than the exception, in practice). Similarly, a program consisting only of input taxes will be costly if emission control technologies would be effective in reducing emissions per unit of input.

Presumptive Charges: Problems and Promises

A Pigovian tax equal to external damages, levied on emissions, would generally be the first-best policy prescription. We have made the argument that taxes can also be levied on polluting goods and services, in presumption of emissions associated with their use. However, the link between external damages and the use of polluting goods and services may be weak, so that marginal emissions vary among polluters paying the same presumptive charges. In brief, the incentives provided by presumptive taxes will inherit any weakness in the association between the tax base and damages.

To illustrate, let emissions from polluter *I*, e^i, be determined by the quantity consumed of a polluting good, x^i, and also by a parameter describing the type of technology, a^i: $e^i = e(x^i, a^i)$. The cost to the polluter of a one dollar price increase for the polluting good is equal to the number of units used, x^i. The change in tax proceeds received by the government, assuming that other goods are untaxed,[4] is $x^i + tx^i_t$, where a subscript stands for partial derivatives, so that x^i_t denotes the marginal change in demand in response to the tax increase. Thus, if we assume that income is worth the same when accruing to the government as when accruing to the polluter, the social cost of raising the tax rate is tx^i_t. The marginal change in emissions from polluter *I* in response to a change in the tax rate will be $e^i_x x^i_t + e^i_a a^i_t$. Assuming, perhaps conservatively, that technology does not change in response to a price increase for the polluting input, $a^i_t = 0$, the marginal social cost of emission reductions produced by a price change for the polluting good is

$$\frac{t}{e_x(x^i, a^i)} \tag{1}$$

or simply the tax rate divided by the marginal emission factor (the latter may depend on technology as well as on consumption).

There are two problems with the presumptive charge on the polluting good, *t*. One is related to the failure of *t* effectively to stimulate other initiatives

4 The argument also holds for another setting for nondistortionary taxation: when all goods are otherwise optimally taxed at uniform *ad valorem* rates (i.e., no distortionary taxation). In that case, the tax rate here referred to as the tax rate on the polluting good, *t*, should be interpreted as a special additional unit rate. Note that we assume that the production cost of the polluting good is given, so that the change in the user cost and in the tax rate is the same.

that can reduce emission, here represented with the variable a^i.[5] As noted in the previous section, one may address this problem by using separate instruments (incentives or regulation) to induce the application of cleaner technologies, depending on how attractive these are, and the monitoring and enforcement requirements. As shown in the following section, however, at given abatement levels marginal cost expressions like (1) apply, as long as the emission factor reflects the abatement level. Thus, when taking into account the estimated marginal emission factors, appropriately averaged if there are many polluters, one can reduce emissions by raising the price of the polluting input at a marginal cost indicated by equation (1).

The other problem with the presumptive tax levied on a polluting input is that it is "unfair" in a way that is (also) inefficient: it discourages the use of the polluting good with the same strength (tax rate) for polluters who may have different marginal emission factors. Thus, while a polluter with a high emission factor could reduce emissions cheaply by reducing the use of the polluting good, she would only face average inducements for doing so, as long as she faces the same tax on the polluting good as others. Importantly, this may be a problem even if separate strategies can be pursued to induce cleaner technology, since emission factors may vary even when separate instruments push abatement optimally.

Thus, if one recommends presumptive taxation of polluting inputs and goods in lieu of high monitoring and enforcement costs, one needs to keep these weaknesses in mind. Importantly, an incentive scheme that is based mostly on presumptive charges can be complemented and refined as technical and institutional capacity allows.

It may be argued that presumptive taxes, while not themselves pointing towards ideal incentive schemes (which would be based on emission monitoring, when monitoring costs come down), provide a setting conducive to institutional and policy development in that direction. The argument rests on the observation that presumptive charges are both "unfair" and "transfer-intensive." "Unfairness" will induce the polluters that are or could be cleaner than average to self-report, referring to their lower emission factors to ask for refunds of parts of the taxes paid, or simply lower tax rates. Incentives to do so are high, since the taxpaying polluter considers the transfer she makes to the government as a cost, and any savings as 100 percent genuine savings. The pollution control agency can announce, when it is institutionally and technically capable, that polluters may obtain specific treatment, such as partial refunds, for those who prove cleaner than

---

5 It will often be the case that replacement of older, dirtier machinery by newer, cleaner machinery will be accelerated by increased fuel prices, but this does not mean that fuel prices are *effective* in stimulating cleaner technologies.

average. It would then be up to the polluter to take the initiative and make the case that she is "cleaner" than the average. She can support the claim by self-reporting, credible monitoring, and by submitting to random audits. Such a policy refinement would provide incentives both for pollution abatement technologies and for monitoring and auditing technologies. Importantly, as polluters with the lower emission coefficients opt out of the "basket case" treatment, the tax rate for the remaining polluters will increase (since their weighted average emission factors increase). Starting from a basis of crude presumptive instruments, such a policy stance could lead in the direction of a very sophisticated emission fee system.

The risks with this "policy trajectory" argument may be that the incentives provided by blunt presumptive charges (and those by environmental regulation, which will also be based on averages), will remain rigid and blunt, rather than continuously refined. The pollution control agency (and the treasury), infatuated by proceeds from taxes on polluting goods, may resist calls for special treatment, since they come singularly from polluters who would like to pay less (and also because rigidity and uniformity may be seen as virtues in vulnerable bureaucracies). Polluters, if they expect a negative response, may not press for special treatment, and will in such cases not face the additional incentive to become cleaner than the average. The latter risk is higher if the taxes are perceived as ordinary indirect taxes, rather than as a specific charge that they are invited to avoid through specific avenues. It should be remembered, however, that presumptive charges serve a purpose in pollution control even if they remain blunt instruments, as long as they do not block developments of the more refined incentive schemes.

POLLUTION FROM MOBILE SOURCES

We have mentioned that a pollution control agency that could levy taxes on each individual's emissions need not know anything about the technical and behavioral changes that could produce pollution reductions. The agency would be in a position quite analogous to a buyer of fruit who is confident of his ability to assess the produce upon arrival. Such a buyer pays whoever brings good produce and has no interest in the background of the suppliers or the history of their shipments. Indeed, had a social planner possessed data on how much pollution each individual caused through the year, then a year-end tax bill based on related pollution damages would provide appropriate incentives for pollution reductions. For pollution caused by motor vehicles (as well as for other important polluter categories) continuous monitoring of emissions is not applied, however. Then, the planner needs to investigate which sectors are polluting, what options exist within each sector, and how he can best stimulate each of these options.

Monitoring Technologies and the Need for Detailed Program Evaluation

Although continuous monitoring emissions at the source (say, by a running meter, as for water or electricity consumption) is not yet possible, several "windows" give approximate information about *rates* of emissions. *New car models* are tested through a "model driving cycle," which is designed to mimic typical driving. New-car standards allow certification if the estimated *average* emissions *per mile* do not exceed a predefined limit. Periodic tests, used in mandatory inspection and maintenance programs, measure the percentage of various gases in the exhaust, and thus provide a measure of emissions per liter of gasoline for *vehicles in use.* More recently, technology for remote, roadside measurement has been developed and tested. As of yet, however, it can only provide the percentages of HC and CO in the exhaust, and has not been used for enforcement purposes. The technology has provided useful corrections to the picture of the fleet given by mandatory inspection and maintenance programs -- and reduced the confidence in periodic nonrandom tests.

The two important limitations of these approaches to monitoring are that they give emission rates, rather than cumulative flows, and that the measurement may not be representative of the vehicle's in-use emissions. To build an analogy, consider a tea-plantation where we are not measuring how much each tea-plucker has collected through the day, but we spot-check the speed with which she appears to pick in a one-minute observation every day. The problem of representativeness is clear: if the pluckers could sense when they were being observed, the monitoring would be of little value (being a test of who can best speed up for a limited time, rather than who brings in most leaves through a day). We would have another problem of representativeness if the speed measurements were subject to random fluctuations and other disturbances, i.e., if cloud-cover and bushes vary greatly or if the inspector can easily be corrupted and distracted. The problem that we monitor the picking rate rather than cumulative pickings would not be important if we (and the pluckers) felt the measure was representative, *and* all the workers worked the same hours. But what if some worked half-day or less, and we did not know who? Effectively, we would end up rewarding part-time workers with full-time pay, and end up with only part-timers.

The recent demonstration of remote-sensing techniques has emphasized the problem of representativeness. Lawson et al. (1990) found that the length of time since the last periodic test had little influence on whether a car's emission rate complied with the standard.[6] Thus, it appears, some "repairs" to make

---

[6] Harrington and McConnell (1993) find that using remote (and unscheduled) sensing techniques is cost-effective compared to mandated periodic testing under plausible assumptions. It would also provide additional emission reductions cost-effectively in areas

polluting cars pass the mandatory tests have no lasting effect on emission rates. The fact that one measures emission rates rather than cumulative emissions is quite important with cars, since one car will be traveling eight thousand and another seventy thousand kilometers per year. By only measuring their emission rates, one gives equal incentives to clean up both, although the social returns from reducing emission *rates* are about nine times greater for the latter.

Periodic tests and the institutional and technical systems that support them are improving, reducing the scope for both poor measurement and weaknesses due to corruption.[7] The representativeness of the tests is gradually improved as better technology and more resources are put into the tests (loaded mode tests, longer cycle tests and a closer vehicle inspection). Other, less likely developments that could improve representativeness are surprise roadside tests, or a technology reading cumulative emissions since the last test.[8]

Given present monitoring capacity, an obvious way to improve the applied systems would be to read the emission rate from the car's biannual test, multiply it by the cumulative mileage (on a tamper-proof odometer), and apply an emission fee to the result. The fee could be paid upon testing, or it could be paid as a presumptive tax uniformly at the gas-station, to be refunded in part to the owners of the vehicles found "cleaner" than presumed in the prepaid tax. The efficiency gains from such a reform would come through several channels: First, all owners would have a continuous incentive to drive less, and owners of more polluting cars more so. Second, all owners would have incentives to make their cars cleaner, but owners who use their cars rarely would be under less of this pressure than owners with average mileage. As a consequence, society would waste few resources cleaning or scrapping cars that are rarely used. Third, the car market would facilitate exchange of vehicles, to make sure the households that use theirs intensively would end up with the cleaner ones.

---

with established inspection and maintenance programs.

7 For instance, computerized testing equipment may print, automatically, a certification sticker based on the test result. Such advances may be of little help, however, if random or other disturbances may allow the gas sensor to be held away from the tailpipe.

8 Of course, the development of such a technology could be stimulated by offering car owners a rebate on the gasoline tax if their cumulative emission meter should indicate that this is justified (it would have some interesting incentive effects). Other illustrations of monitoring techniques, and problems, are: Surprise road-side tests were, perhaps laudably, resisted as an enforcement measure in Mexico City. It was felt the public should be spared from police interruption unless upon reasoned suspicion. In Jakarta, policemen have been authorized to fine smoking vehicles based on visual assessment. In Los Angeles, the public may call authorities to report smoking buses.

To summarize, current monitoring methods allow inducements to make cars and fuels cleaner, but not to reduce demand for the car's services. We now consider a control program for Mexico City, where we analyze the cost effectiveness of various policy measures that can make the car fleet cleaner, in terms of emission rates. We then ask whether and how such a program can be combined with a *uniform* gasoline tax. The role of the gasoline tax is to provide polluters with an incentive to reduce their demand for the polluting good. The reform has the principal weakness of discouraging vehicle use uniformly, ignoring differences in emission rates. The practical reason for suggesting such a modest reform is the general suspicion that administrative and technical systems for emission testing are still vulnerable. Thus, it is doubtful whether they can be used as major tax-collecting devices. The beauty of the fuel tax is administrative simplicity. All countries interfere in fuel markets (whether with a tax or a subsidy), so their ability to manipulate this price is proven. Thus, the collection of a pollution motivated charge requires no new monitoring or administrative set-up. When monitoring and administrative capacity so allow, a system using a uniform charge may be further refined. The most immediate direction would be to use emission test results *and* utilization rates to collect the emission fee, in order to increase, in relative terms, the pressure for cleaner cars on those who use their cars a lot.

Using Indirect Instruments: Cleaner Cars, Fewer Trips

The pollution control agency can be seen as one who purchases emission reductions on behalf of the public. Viewing cleaner cars and fewer trips as alternative suppliers of emission reductions, the agency should buy from each up to the point where emission reductions, at the margin, cost the same from each.[9]

Since each liter of gasoline consumed produces a certain amount of pollution, the gasoline demand curve shows the price at which consumers can reduce pollution by means of reduced gasoline consumption. If we use consumer surplus as a measure of the welfare costs of reducing consumption, the part of the demand curve that is above the gasoline supply curve (marginal production costs) also shows the welfare costs of pollution reductions provided through demand reductions. Formally, let welfare depend on gasoline consumption, x, and other consumption, $y{:}w=u(y,x)$. Let emissions depend on gasoline consumption and the application of pollution abatement technologies, $a{:}e=e_x(a)\,x$. Further, assume a fixed coefficient cost structure (abatement affects the cost of the polluting good,

[9] This perspective is examined in detail in Eskeland (1994), from which much of this material is selected. We use the term gasoline tax, but the same principles apply to other polluting inputs, such as diesel.

though), so that society's resource constraint is $y+f_x(a)\,x=0$. Also, gasoline revenues are redistributed as a lump-sum subsidy $S=R=t_x x$. We shall here assume, for simplicity, that there is one representative consumer. These assumptions ensure that the individual budget constraints, $y+p_x x=S$, where $p_x=f_x(a)+t_x$, sum to society's resource constraint, so that compliance with the latter is ensured by individual demand functions (notice that taxing gasoline does not influence a, so that $\partial p_x/\partial t_x=1$). Assuming that the individual considers the effect of her consumption decisions on emissions and tax revenue to be zero, the marginal welfare costs of adjusting the gasoline tax rate, per unit of associated emission reduction, is the following:[10]

(2)
$$\frac{dw/dt_x}{de/dt_x}=\frac{\frac{\partial v}{\partial p_x}\frac{\partial p_x}{\partial t_x}+\frac{\partial v}{\partial I}\frac{\partial S}{\partial R}\frac{dR}{dt_x}}{e_x\cdot\frac{dx}{dt_x}}=\frac{\frac{\partial v}{\partial I}t_x\frac{dx}{dt_x}}{e_x\cdot\frac{dx}{dt_x}}=\frac{\frac{\partial v}{\partial I}t_x}{e_x}$$

where $v(p,I)=u(y(p,I),x(p,I))$ is the indirect utility function (utility does not depend on emissions, for simplicity).

The formula is most easily explained in terms of Figure 2. The additional welfare cost of a tax change "dt_x" is the trapezoid *abcd*, approximated by the rectangle $t_x\cdot dx$. Emission reductions, "de", will equal $e_x dx$, and "dx" cancels out in the expression for marginal costs, which is the ratio between the effect on welfare and the effect on emissions. We may notice that the term "dx" is determined by the slope of the demand curve, or by the demand elasticity. Thus, when this term cancels out, it means *that the marginal welfare cost of providing emission reductions through tax rate changes, at a given tax rate, is independent of the demand elasticity.* Of course, the emission reductions provided by a given tax change are not independent of the demand elasticity, but the marginal cost at any given tax rate is. As an example, if the elasticity is large, then the emission reductions are large, but so would be the costs from sacrificed consumption, so the demand elasticity would not affect the costs at which the emission reductions are delivered.

---

[10] Detailed calculations are shown in Eskeland (1994). Importantly, income is here valued equally when accruing to the public sector and to private households. Thus, income distribution effects, and revenue effects, are abstracted from.

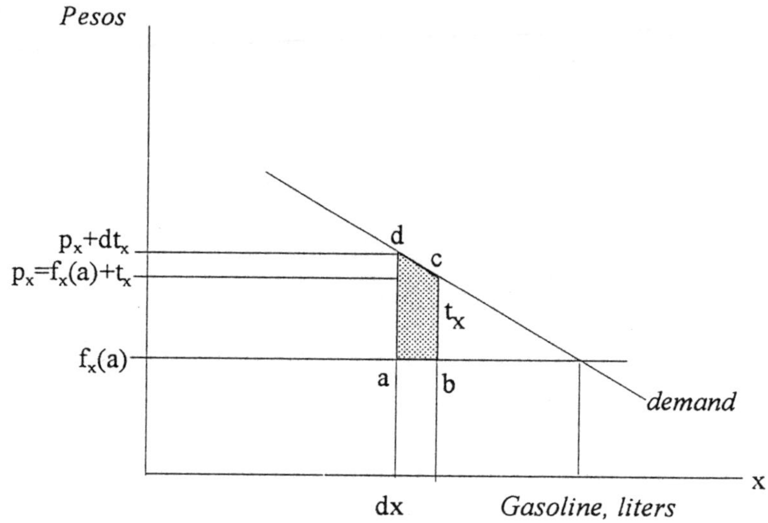

Figure 2: The cost of an increase in tax, dt_x.

A similar expression for the marginal welfare cost of emission reductions provided by an adjustment in required abatement, ā, is:

$$(3) \qquad \frac{dw/d\bar{a}}{de/d\bar{a}} = \frac{-\frac{\partial v}{\partial I}\left(x\frac{\partial f_x}{\partial a} - t_x\frac{dx}{da}\right)}{e_x\frac{dx}{da} + \frac{\partial e_x}{\partial a}\cdot x}$$

A cost-effective program would seek emission reductions from cleaner cars and fewer trips such that the marginal costs are the same for both "providers." Setting (2) equal to (3) and rearranging, a cost effective program is characterized by:

$$(4) \qquad \frac{t_x}{e_x} = -\frac{\partial f_x/\partial a}{\partial e_x/\partial a}$$

where $-(\partial f/\partial a)/(\partial e/\partial a)$ is the marginal cost of emission reductions via pollution abatement ("cleaner cars and fuels").

This formula quantifies the principle that cleaner cars and fewer trips should deliver emission reductions in such a way as to equalize the marginal costs. The framework is easily amended to characterize an *optimal* pollution control program, which is a special case amongst cost-effective programs. Then, one would let welfare depend on emission as well $w=u(y,x,ne)=v(p,ne,I)$, where e is an indicator of total pollution, and n aggregates the damages from pollution over all consumers that are harmed. Optimality requires

(5) $$\frac{t_x}{e_x} = -\frac{\partial f_x/\partial a}{\partial e_x/\partial a} = -n\frac{\partial v/\partial e}{\partial v/\partial I}$$

Equation (5) is the Samuelson condition for optimal provision of public goods ($-(\partial v/\partial e)/(\partial v/\partial I)$, marginal individual willingness to pay, equals $-u_e/u_y$, the marginal rate of substitution between pollution reductions and the numeraire good). The framework can also be used to show that an optimal combination of abatement and demand reduction can result from the use of one instrument -- an emission fee.

What if a pollution control program did not follow this principle, for instance by pursuing technical controls, but not taxing gasoline (as in the United States)? Consumers would be loaded with some technical control requirements that were quite costly at the margin, compared with the costs at which they could have sacrificed some nonessential trips. They could be made better off if asked to abstain from some of their least essential trips, in return being freed from some of the costliest controls, holding total emissions at the same level.

A technical control-cost curve was constructed for a study on air pollution in Mexico City (Eskeland 1993 and 1994, and World Bank, 1992). Table 2 shows the cost-effectiveness ranking of the proposed interventions to make cars and fuels cleaner, together with the rates for a gasoline tax that would optimally match these (equation (4)).

It may be noticed that the gasoline tax rate, in cents per liter, does not increase proportionately with the technical control costs. The reason is that the emission factor per liter of gasoline (e_x), is declining as one climbs further along the control-cost curve (because abatement technologies are applied, making cars progressively cleaner). Thus, while the gasoline tax follows the control-cost measures in terms of dollars per ton emitted (as stated in equation 4), it increases less than proportionately when measured in terms of cents per liter.

Table 2 Mexico City: Abatement Measures and Matching Gasoline Tax Rates

	Cost ($1,000 USD/ wt'd ton)	Cumulative emission reduction (Mn wt'd ton)	Cumulative costs abatement only (Mn USD)	Matching gasoline tax (cents/liter)
Gas truck LPG retrofit	-379	90	0	-4.4
Minibus CNG retrofit	-248	148	0	-2.8
Gas truck CNG retrofit	-225	231	0	-2.4
Gasoline vapor recovery	-80	275	0	-0.8
Re-engine buses	140	299	3	1.4
Minibus 1992 standards	181	391	20	1.7
I&M High-use vehicles	209	545	52	1.8
Gasoline truck 1993 standards	264	632	75	2.1
Taxis tier 1 standards	322	641	78	2.5
Re-engine R-100 buses	482	651	83	3.7
Taxi replacement, 1993 standards	510	714	115	3.7
Centr. insp./ maintenance, cars	651	771	152	4.4
Passenger cars to 1993 standards	669	883	227	4.0
Diesel especial	699	893	234	4.2
Lower regular R. vapor pressure	836	904	243	4.9
Provide regular unleaded	923	954	289	5.1
Decentral. insp. & maint., cars	1034	1018	356	5.3
Replace Gasoline trucks	1114	1096	442	5.0
5% MTBE in regular unleaded	1201	1116	467	5.3
Lower RVP in premium unleaded	1313	1128	482	5.6
Road paving (1000 km)	1335	1136	498	5.7
Passenger car 1991 standards	1367	1180	508	5.4
Reduce sulphur to .1% in diesel	1371	1187	569	5.3
Passenger cars tier 1 standards	1629	1201	578	6.2
US specifications for diesel	2097	1207	601	7.9
11% MTBE in regular unleaded	2447	1219	613	9.0
5% MTBE in premium unleaded	13487	1222	643	49.0
11% MTBE in premium unleaded	14728	1226	686	53.2

Note: Columns 1 and 2 give, respectively, the height and length of the steps on the technical control-cost curve (Figure 3). The matching gasoline tax (column 4) is monotonously increasing per gram of pollutant (equation (4)), but not per liter of gasoline, since pollutants per liter (e_x) are brought down progressively by controls.

Whether the optimal discouragement of trips matter much, quantitatively spoken, depends on the slope of the demand curve, in comparison with the slope of the technical control cost curve. Eskeland and Feyzioglu (1994) used panel data from 31 Mexican states over seven years, to estimate the relationships to income and prices of optimal car stocks, gasoline consumption per car, and total gasoline consumption. The results are in line with a rich body of empirical literature in terms of long-term elasticities, which are the most important from the perspective of efficiency (in the range of around one for income elasticities, and -0.5 to -1 for price elasticities). They stand out, however, in displaying quite rapid adjustment in the short term, so short-term elasticities are larger, in absolute value, than often found. The results are summarized in Table 3.

Table 3 Elasticities for Total Gasoline Consumption

	Short run	Long run
Gasoline price	-0.79	-0.80
New-car prices	-0.03	-0.03
Income	0.98	1.02

Source: Eskeland and Feyzioglu (1994).

Figure 3 shows two supply curves for emission reductions: the lower is a curve that results from cost minimization when an adjustable gasoline tax is included in the tool box of the control agency, the higher would result if the control agency were prevented from using a gasoline tax. The gasoline tax rates reflected in the curves are given by equation (4), and listed in Table 2.

An otherwise well-designed control program would, without the gasoline tax in the toolbox, become 23 percent costlier, at an additional cost to the citizens of Mexico City of US$110 million annually. Thus, since demand for the polluting good is quite sensitive to pricing, demand management is important for the solution to this problem.

The proposed program removes about 70 percent of weighted emissions from the vehicle fleet. The gasoline tax that is optimal under such a target is about 25 percent, and removes 20 percent of emissions, under the assumption of a price elasticity of -0.8.[11]

Market-Based Versus Regulatory-Demand Management

Once it is realized that the demand for polluting goods and services may be manipulated directly to reduce pollution, what is the best way of doing so? Initiatives to economize on polluting trips may be gasoline taxes, driving bans, parking fees, toll rings and subsidies to public transport. As a demand management instrument, a gasoline tax has the shortcoming that it allows only a limited differentiation

[11] In Mexico City alone, such a gasoline tax would collect US$300-350 million in tax revenues annually. A gasoline tax motivated by pollution-control objectives should, ideally, be higher in urban areas, but the geographical differences may have to be limited for administrative reasons. Gasoline may also be taxed for general revenue and other purposes, of course.

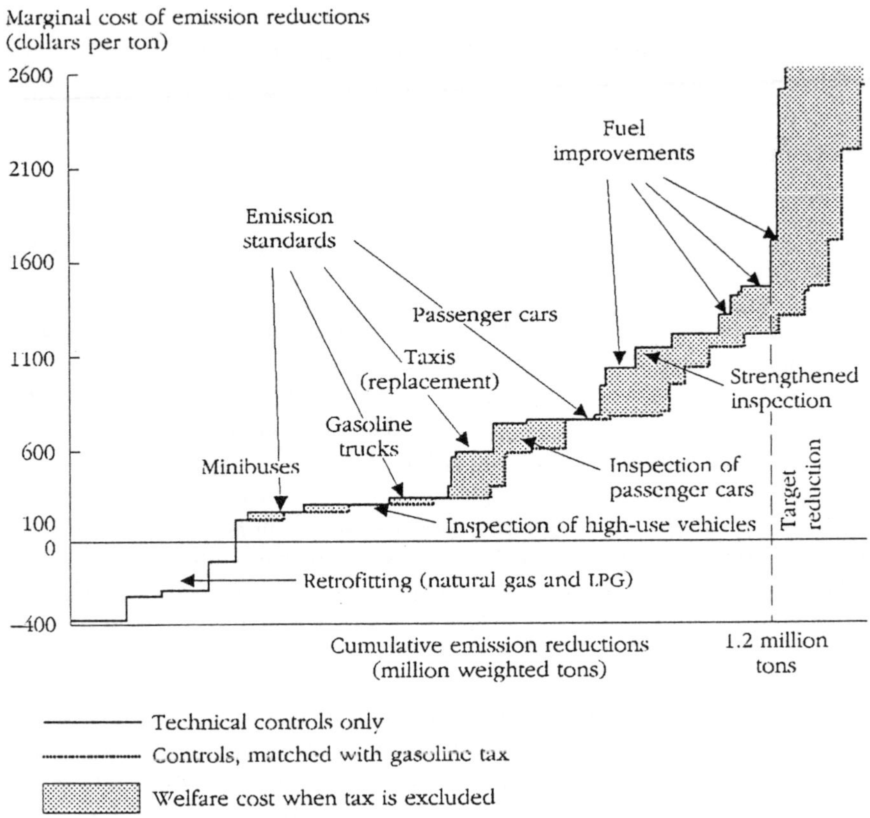

Note: Calculations are based on –0.8 elasticity of demand for gasoline.

Figure 3 Supply curves for transport emission reduction in Mexico City, with and without a gasoline tax.

geographical area, and no time-of-day-variation. Apart from that, it has many advantages. It may be used alone if sharp geographical and temporal differentiation is not important, and may be supplemented or substituted by toll rings, peak-load charges, parking fees and public transport policies when such variation is necessary.

When consumers sacrifice trips in response to demand-management instruments, they incur welfare costs of doing so.[12] Importantly, when a trip is sacrificed due to a marginal increase in the gasoline price, the value of the sacrificed unit to the consumer is the retail price of gasoline. Thus, while there are inframarginal units of gasoline (and trips) that are worth more to consumers, a gasoline price increase will screen out, systematically, the trips that are worth the least. This property of the gasoline tax allows it to reduce trips at the lowest possible welfare cost.

In Mexico, while gasoline is now taxed, a driving ban is also used in an attempt to curtail driving. The capital's "Day without a car program" is based on the vehicle's license plate number, and bans each car from driving on a specific workday. Thus, the maximal potential effect on demand that the regulation could have would be to remove 20 percent of workday driving, and the effect would be less if people adjusted in ways other than simply by canceling trips.

To quantify the costs and benefits of the regulation, one needs to quantify the demand reduction that it causes (if any), and the welfare costs of obtaining these reductions. When demand falls because of a marginal price increase, we know that the sacrificed consumption is worth the commodity's market price to the consumer. Thus, referring to Figure 2, we know that it is a marginal quantity, or the "trip least essential" to consumers, that is sacrificed. As one consumer reduces his demand by one liter, we know that there is not another who uses gasoline, but who has a lower marginal willingness to pay. This property with price increases, that they select the "least essential units" if and when sacrifices occur, guarantees that the welfare costs of the demand reductions are minimized.[13]

A demand reduction resulting from a regulation will rarely have this selection quality. The "Day without a car program" may curtail trips in households with a very high willingness to pay, and it may block a household's Tuesday-driving, say, even if the household could more easily have sacrificed other trips. Both of these effects result because the regulation does not allow 'trading' of the rationed commodity: trips. With reference to Figure 2, the regulation may curtail inframarginal as well as marginal trips. If we compare it with a gasoline tax that

---

12 We use trips and gasoline as metaphors for polluting goods and inputs. Consequently, if gasoline is saved by using more fuel-efficient cars, the argument applies as if trips were saved. Other fuels should be taxed according to emission as well.

13 When calculating the welfare costs of manipulating consumption, we need to emphasize again that we employ the "compensation criterion": we assume no value (nor penalty) to the transfer of money from private to public hands, nor from poor to rich. Thus, the analysis ignores any concern for income distribution, as would be correct if transfers could be arranged costlessly with other policy instruments.

would have yielded the same demand reduction as a regulation, the unit costs of the latter will be at least as high, and possibly much higher. A comparison of the welfare cost of a regulation and a tax increase calibrated to give the same demand reductions is shown in Figure 4, below. A key assumption in this argument is that the regulation, if providing emission reductions at all, would provide these through its impact on *aggregate* gasoline consumption. Then, using market forces to allocate *any* reductions (in gasoline consumption, this time, rather than in emissions) will assist in containing the total costs of the reductions.

In order to estimate the demand reduction caused by the regulation, a demand function was estimated based on data from Mexico City in the periods before the regulation (Eskeland and Feyzioglu, 1994b). Since the demand function is estimated in periods without regulation, it can be used to simulate counterfactual demand developments as they would have been in subsequent periods, had no regulation been implemented. The result, which was indeed surprising, is shown in Figure 5 below. One would expect actual consumption to be lower than the estimate in the counterfactual simulation, but this occurs only in the first two quarters! After that, actual consumption exceeds and persists above the levels that the simulation model forecasts.[14] Moreover, the indicated confidence intervals (obtained by shifting forecast values up and down by one standard deviation) show the extremely low probability of observing the actual string of consumption levels if demand had not been shifted upwards by the regulation (the probability of observing one value outside the confidence interval is about three percent-- observing many is much less likely). Thus, we can safely conclude that, after an initial period of adjustment, gasoline consumption was not curtailed by the regulation, but rather boosted, or at best held constant.[15]

Since the result is not intuitive, one may speculate what produces it. One way is to think through how households can respond to the regulation. One household complies as expected -- it manages without its car on Tuesday -- ride-sharing with friends, using the metro, postponing or canceling the trip. Of course, if the trip is only moved to another day, then total demand reductions would be zero, but otherwise one would expect demand reductions to be positive. Another household may find that way of complying too costly, and decides to buy an additional car. If the household had more driver's licenses than cars, then its total driving is likely to increase, so the household provides negative demand reductions in response to the regulation. Thus, for the market as a whole, one might see total

[14] Simulations are performed inserting actual price and income developments in the estimated demand model. All calculations take seasonality and stationarity into account.

[15] This result is also supported by the appropriate Chow test.

demand reductions close to zero or even negative, if there are many households in the latter group.[16]

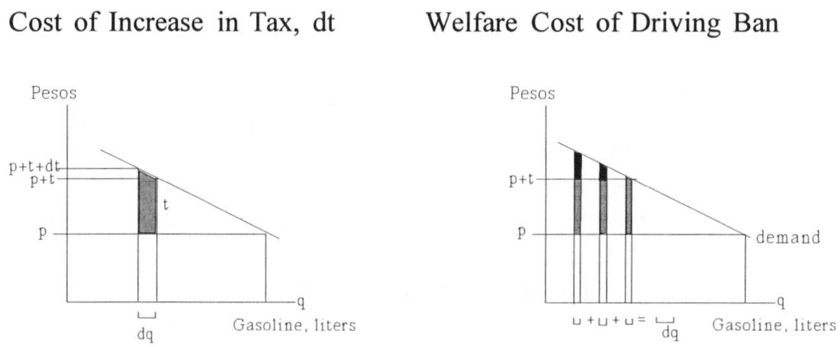

Figure 4 The double shaded area in the right figure illustrates the extra costs when the demand reduction is found on different parts of the demand curve, rather than selecting the least essential trips, as a tax increase does.

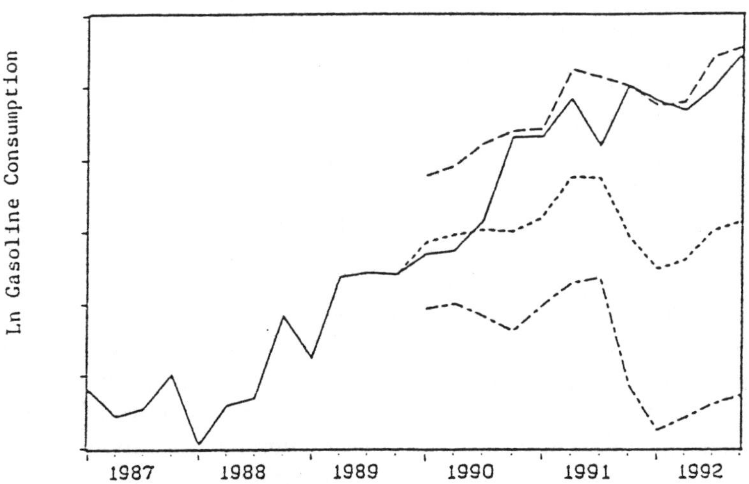

Figure 5 Day without a car, gasoline consumption. Simulated without regulation versus actual under regulation.

---

16 There is anecdotal evidence that many households bought *used* cars, so that Mexico City imported old cars from the rest of the country. If this happened, then the additional cars could be dirtier *per liter* than the average car otherwise would have been, so that the consequences for emissions are worse than indicated by aggregate consumption.

Of course, having found that the demand reductions, and thus the emission reductions, offered by the regulation were zero or negative, there is no need to attempt the difficult task of calculating the welfare costs -- welfare costs are definitely positive, and if they do not produce benefits, then the regulation is not an attractive proposal (rather, removing it is).

Similar regulatory measures are used in some other cities. In Santiago, Chile, it is used as an emergency measure in high-pollution periods, and its effect will thus less likely be jeopardized by households buying additional vehicles. Also, the Santiago regulation applies only to the old, uncontrolled cars, so it has an additional effect in making cleaner cars more attractive. We may notice, however, that this particular regulation has more weaknesses than do most rationing schemes. One is that it allows households to purchase "implicit driving permits" by buying cars, thus inducing the country to tie up capital unproductively. Another is that the limits involved relate to certain trips, rather than to a household's total driving -- thereby introducing more limitations than necessary, even for within-household reallocations.[17]

If pollution (or congestion) problems have sharp variations spatially or by time of day, demand management instruments such as mass transport provision and subsidies, parking policies etc., may be used to complement the effects of gasoline prices. In transportation projects, in contrast, time savings are the main benefits produced. Krupnick (1993) found that the inclusion of emission benefits in analysis of projects aimed to reduce congestion and produce time saving is not difficult in principle, and should be done. He also found, however, that congestion- and time-saving benefits would likely dwarf pollution benefits in such projects. Thus, the inclusion of emission benefits will unlikely tilt the balance in favor of large transportation infrastructure investments, if the projects would otherwise be difficult to justify.

Among the stylized facts cited by Krupnick is that developing country cities with air pollution problems invariably are growing fast. Thus, he advises against using modelling frameworks that do not incorporate long-term phenomena such as residential developments. This is, of course, relevant if one considers the empirical findings of short-term studies, such as the discrete-choice models of travel mode. These studies often conclude that travel-mode choice responds very little to parameters such as travel costs and travel times (elasticities usually in the range between zero and 0.2 in absolute value), and thus that better or cheaper mass transport will do little in reducing the use of private cars. Swait and Eskeland (1994) estimated a mode-choice model based on travel survey data from São

[17] Rationing nontradeables, in this case trips, thus has higher welfare costs than rationing tradeables with low resale costs, such as alcohol and tobacco.

Paulo. The results are quite in line with others from developed as well as developing countries, and can be read as a warning against the belief that mass transport can perform wonders in curtailing private transport. It is important to remember, however, as indicated above, that these models represent household behavior in a very restricted setting. Thus, in the long run, when households choose where to live, where to work, how many cars to own, etc., behavior may indeed be more sensitive to gasoline prices, fares and travel times than what these models indicate.

Market-Based and Other Inducements to Get Cleaner Cars and Fuels

Even for measures making cars (and fuels) cleaner, we may pause to emphasize the distinction between the "physical" changes that can deliver emission reductions, such as a tune-up for that purpose, and the policy instruments that may be used to induce them. For cars in use, standards and mandatory inspection and maintenance programs are ordinarily among the most important policy instruments. Higher gasoline prices will also cause some additional attention to maintenance and tuning, and may even increase the market share for smaller cars. However, gasoline price increases can only trigger those options that are driven by fuel efficiency considerations. Emission standards and gasoline taxes are indirect instruments, the first a regulation addressing a characteristic of the equipment (the car's "emission rate"), the second a market-based instrument addressing a proxy for its utilization.

A problem with emission standards is that they have to be designed to stimulate those physical solutions that give cost-effective emission improvements for the *average* car.[18] Other instruments could possibly trigger options that were attractive only for selected cars. Further, while cars that are in noncompliance are pushed to become cleaner, certain weaknesses follow from the binary nature of compliance/noncompliance: no impetus for emission reductions is given to owners of cars in compliance, even though some of them may face cheap options. Similarly, standards will, inadvertently, require costly changes for vehicles that are close to, but not in compliance with regulations -- even though those changes may yield very low benefits compared to costs.

---

[18] Regulations, or their enforcement, can allow for some differentiation, which may be desirable for efficiency or equity reasons, for instance to be applied only in polluted areas (see Harrison, Jr. (1977) who made this point early), or differentiated by model year, weight class, etc. Cars that would not pass the emission test in Mexico City might still be useable in other areas of the country: a desirable feature for a program in a poor country where ambient air pollution is the least of problems in rural villages.

Could there exist many vehicles in use for which there are economic ways to reduce emissions, ways that are not stimulated by a standard for the average vehicle? Clearly, there could.[19] We may use a feature frequently recurring in the economics of technical controls to illustrate it: fixed costs. A tune-up, a retrofit with a catalytic converter and a conversion to use natural gas all have one thing in common: a fixed cost is paid up-front, while emission reductions will come as a flow through subsequent periods, proportional to the utilization rate and equipment lifetime. Inadvertently, such vehicle modifications are more cost effective the higher the vehicle's utilization. Thus, if the standard is set to induce measures that are cost effective for a vehicle travelling the average, eight thousand kilometers per year, it will fail to induce many of the control options that are costlier, but more cost-effective, for cars used more intensively.

Of course, standards for vehicles in use can be complemented with a wide range of instruments, and these could target some of these options directly. However, technical details set limits for how finely these instruments may be designed. Two possible vehicle modifications -- retrofitting with a catalytic converter, and the conversion of a vehicle to use natural gas -- may serve as illustrations. In the case of natural gas, the conversion gives the owner access to a fuel that can be priced independently. Thus, adjustments in the price for conversions (which may be taxed) and the price of natural gas (which should, then, be cheaper to use than other fuels), may be used to induce high-use vehicles to self-select for conversion. A similar scheme to induce retrofits with catalytic converters is not feasible, however. The reasons is that a lower price on unleaded gasoline cannot be used to induce such conversions, since usage of unleaded gasolines does not require a catalytic converter (The emission reductions provided by thc catalytic converter will be damaged by the presence of lead, however).

---

[19] Anderson (1990) provides an interesting, practical review of strategies applied to reduce emissions from in-use vehicles. Evaluation of strategies to detect high-emitting vehicles is the topic of McConnell and Harrington (1992 and 1993). The presence of high-emitting vehicles, despite standards and scheduled testing, is partly due to monitoring problems, but also due to caps allowing a vehicle to operate when a specified tune-up cost has been expended. The efficiency of programs to make cars and fuels cleaner depends on how well they screen vehicles with respect to relevant parameters. As an example, Alberini, Harrington and McConnell (1993) analyzed a program aiming to buy off the streets cars with higher emissions. They found the program to select for some desirable characteristics (vehicles in poor operative conditions), but also some undesirable ones (vehicles with a low expected remaining lifetime mileage). Comparing various instruments, Krupnick, Walls and Collins (1993) does not find a tougher corporate average fuel economy (CAFE, imposed on vehicle manufacturers, allowing averaging between the producer's models) to be cost effective, even when the emission control objective is one for greenhouse gases.

Also, these fuel prices may need to follow each other for other reasons).[20] One could look for other ways of selecting high-use cars to be retrofitted with catalytic converters, for instance by introducing separate requirements for taxis and delivery trucks. Such classifications of vehicles are surely rough proxies for utilization -- taxis in Mexico City travel about nine times the annual average mileage for personal vehicles, and may thus be useful as bases for indirect instruments, if it is feasible administratively.[21] In this case, since taxis need licenses issued by the city, one could make special requirements for them.

The important lesson here, of course, is not that taxis and delivery trucks should be cleaner than most cars (though they most probably should). The lesson is that the absence of comprehensive information based on continuous emission monitoring leads the agency into a detailed examination of where emissions can be squeezed. The agency then considers the economics of the reductions themselves, under alternative inducement mechanisms. A good agency then tries to allow maximum flexibility for polluters, leaving room for solutions it has not imagined, or thought uneconomic. It may still be the case, however, with indirect instruments, that different instruments must be used to pursue separate groups of polluters, or separate avenues. In that case, much responsibility for analysis rests with the agency.

CONTROL OF POLLUTION FROM INDUSTRIAL SOURCES

Heterogenous Polluters, Quantitative Importance of a Few

In what way should the economics of pollution control differ by type of polluter? Perhaps the most important feature distinguishing industrial sources from those of urban transport is the much greater heterogeneity. While motor vehicles certainly differ in important respects (age, size, habits, fuel, combustion and emission-control system, maintenance and utilization), they may easily be placed in a small

---

[20] It is difficult to prevent mis-fueling if the fuel prices are very different (mis-fueling *either way* being undesirable, but in particular the use of leaded gasoline in emission controlled vehicles, since the catalytic converter is poisoned, and its emission-reducing capability irreversibly wiped out). Other differences may prevent mis-fueling (nozzle size regulations, enforcement, octane differences), but these easily fail if the leaded/unleaded price difference is large.

[21] As a matter of illustration, then, emission reductions cost nine times as much for the average vehicle as for the taxi (per gram of emissions saved) if we assume equal expected lifetimes for the equipment, or five to six times as much if we assume the taxi equipment lives half as long, allowing for some discounting.

number of categories. In each of these categories (two might be new and used personal gasoline vehicles), there will be a high number of polluters that are reasonably homogenous (almost all use an internal combustion engine to produce transportation services), and we may know each group reasonably well by sampling a small subset.

Industrial sources will necessarily be a group with greater heterogeneity. Cars will mostly pollute the air, while firms pollute all media, and with a wider range of pollutants. Firms' technologies, equipment and input use vary much more, mainly because they are dedicated to producing a much wider range of outputs. Thus, *a priori*, it would appear more dangerous for the planner to assume he knows the emission control opportunities open to firms than it would be in the case of vehicles.

Another important difference is that firms' quantitative impact will vary much more. While there will generally be thousands of polluting firms in a polluted city, just as there will be hundreds of thousands of vehicles, it may be the case that 50 percent of the vehicular emissions in Mexico City originate from the 15 percent of vehicles with highest emissions. For industries, in contrast, the distribution can be much more skewed, in the sense that just a few polluters play a major role: two pulp factories represent 37 percent of discharges in one Java river, and two others represent 92 percent in another.[22] According to a registry of the 90 largest polluters in the Valley of Mexico, two firms emit 95 percent of total dust emissions, and 10 percent of firms emit 99.5 percent.

These two features distinguishing industrial polluters, or fixed point sources as they are often called in the literature, have implications for pollution-control policy. The greater heterogeneity of sources is a problem for the planner, since he will have a poor idea about individual emissions and control options for the great majority of polluters. Thus, if he tried, on the basis of sampling and a rough categorization of polluters, to make prescriptions for each -- either in terms of allowable emissions or in terms of abatement technologies to be applied -- he would do so on the basis of grossly inadequate information. The resulting emission reduction would, in consequence, come at a high cost. The greater dominance of a few polluters will, in contrast, make it easier for the planner to devise strategies for cost-effective emission reduction. He might, for instance, commission engineering studies to understand a firm's or a sector's opportunities to reduce emissions, with

[22] Java is, naturally, characterized by short rivers, so the predominance of few polluters will be greater than in greater river systems. The reporting system is still weak in many respects, but the tendency that a few dominate will likely remain, assuming that nonreporting and under-reporting is greater among truly smaller dischargers (the figures are for biological oxygen demand).

which he could bridge part of the gap in knowledge between the firm (or the sector) and himself.

In actual pollution-control policies in industrialized as well as poorer countries, pollution control agencies issue standards for industrial-pollution control, whereby a firm's "allowable" emissions or practices are determined on the basis of rough categories (such as whether you report that you produce "textiles" or not). Standards and regulations in developing countries will, at best, be determined on the basis of a study that has sampled a few firms in the category. More often, the standards will be based not on local studies, but on studies from other countries, without much questioning of their adequacy or relevance to the local conditions. Also, one will see plant-specific studies made for the large, dominant polluters, attempting to take into account factory-specific and recipient-specific conditions. Such studies will underpin plant-specific emission permits, allowable time to obtain compliance, etc. Thus, there is ample evidence that the agency recognizes the costs associated with the information gap. We shall argue in the following, however, that there are policy instruments available to deal with such problems that are not exploited to their full potential.

Expanding Society's Interest: Output Supply, Input Use and Residual Generation

When setting out to protect the environment from pollution, society expands its interest in firm behavior to include its generation of residuals. An undisturbed market can, as we know, under certain conditions be relied upon to induce firms to use inputs and supply outputs efficiently. Under what conditions will a market, undisturbed by authoritative intervention, also make firms efficient in containing residual generation? Indeed, those conditions are rare, mostly because the benefits of reduced residuals will be public, rather than private.

The Coase theorem states that a firm may take efficient action to control pollution optimally if it negotiates with the beneficiaries of pollution control, irrespective of whether the firm initially holds the right to pollute or, alternatively, the victims hold the right to a clean environment. However, let us first examine the situation where polluters hold the right to pollute, and victims should negotiate bribes to polluters so that they reduce pollution to optimal levels. If victims are many (as in Mexico City: 19 million, or in greater Jakarta, 15 million) a voluntary *club* to purchase emission reductions would easily suffer under the *free-rider* problem: a nonmember would enjoy the benefits of the emission reductions as well, so the purchases made by a voluntary club would be backed by few. The situation would not be much better if polluters initially held no rights to pollute, but would have to buy emission permits from the victims of pollution. An individual holding such rights would sell his with a light heart (i.e., even if he were to be offered a low price), since he would suffer only a minuscule share of the consequences in terms of pollution increases (most would be suffered by others), while increasing his private income. In both cases, it is the *publicness* of the environmental good that makes voluntary negotiations so impotent, and in both cases, application of *authority* can solve it. Thus, if we view municipalities and provinces as citizen-minded clubs in which membership is *compulsory*, then the pollution-control agency could be seen as its representative, with powers to negotiate with polluters. Then, the Coase theorem could ensure efficient pollution control and pollution levels.

Often having broad powers, the planner may choose to make his demands without negotiations with polluters. When issuing emission standards for a pulp manufacturer, he would usually allow him to emit an amount proportional to his pulp production. Emission standards for cars, similarly, are emission permits conditional on (and proportional to) car use.[23] As noted in the section providing the analytical framework, such individual permits, or quotas, will fail to minimize the total costs of the emission reductions provided, as long as they are not based on all the information relevant for the costs of reducing individual emissions. The problem may be repaired if trades in emission permits are allowed, or if fees rather than quotas are used to induce pollution abatement. These approaches rely on good information about individual emissions, information we may refer to as continuous monitoring of individual emissions. With such information, emission taxes or tradeable permits, defined in such a way as to take into account the damages caused by emissions, will be all that is needed to induce efficient pollution control, even when sources are very heterogenous. We thus proceed to the case where the planner, in the absence of information on individual emissions, works with proxies for such emissions.

General Economic Policies, Input and Output Taxes, Regulation

If the planner, or society, has an interest in total emissions, but does not possess information about individual emissions, what solutions exist? Inevitably, the planner will find himself seeking proxies not only for individual emissions, but also for individual abatement opportunities and costs.[24] Thus, he is immediately brought far away from a situation with full information, in which he would include only information on individual emissions, or damages, in the incentives scheme which he

---

[23] Both standards will, effectively, set a limit for the *rate* of emissions (grams of NOx per vehicle mile, kilograms of BOD per ton of pulp), implicitly allowing emissions proportional to the utilization of the "plant."

[24] We may recall from Table 1.2 that the proxies addressed by indirect instruments are chosen not only to be related to emissions, but also to emphasize *choices* the polluter may make which actually change emissions.

designs for polluters. In the case of vehicles, we may recall, the planner conducted a study to learn about the determinants of emissions, before deciding to work with manufacturers to have new vehicles cleaner, to intervene in fuel markets to tax fuels and make them cleaner, and to use periodic testing to give incentives to the "dirtier" among vehicles in use. Similarly, for firms, the planner needs to have an idea about how emissions are generated in order to know where and how to squeeze emissions.

Kopp (1992) describes the planner's needs as an expanded -- though traditional -- model of the firm, by including residuals or emission among the variables for which demand and supply functions are specified. As an example, if an expanded matrix of substitution elasticities shows that emissions are highly sensitive to the characteristics of a firm's capital equipment, then this may give useful indications about the design of the incentive scheme, both in terms of the variables to address and in terms of a strategy for monitoring and enforcement. Similarly, should the elasticity matrix indicate that emissions, say of sulfur, are sensitive to certain input or output prices, such as prices of sulfur-rich fuels, then taxation of these may provide emission reductions in a way that greatly economizes on monitoring and enforcement.

In economic studies that comprise pollution, quite superficial models of emission generation are often used. For instance, marginal emission factors may be assumed to be fixed by sector to either output or to fuel use, abstracting from the variation between plants according to their characteristics and behavior. They will, thus, not be adequate in the analysis of policies aimed at modifying the determinants of emission intensities at the firm level. They have credibility, in contrast, in the analysis of policies that would mainly change incentives at broader levels. In the case of output-based emission factors by sector, the models simulate well the effects of sector-specific taxes, of trade policy and industrial policy. In the case of emission factors linked to fuel use, the models may simulate well the effects of energy pricing policy or conservation initiatives.

A major research effort in the World Bank (Hettige, et al., 1995) has estimated sector-specific emission coefficients based on US manufacturing data. Using such emission coefficients, Ten Kate (1993) studied how industrial pollution had developed in Mexico in the period 1950 to 1989. Growth in manufacturing over the period was remarkable -- a tenfold increase in output. But Ten Kate observed that pollution had also been increased through an uninterrupted change in output composition towards the more polluting sectors. Surprisingly, Ten Kate found no breaks in this trend that could be associated with distinct changes in economic policy in the period -- from import substitution, through petroleum -- (and public-sector) led growth, and finally to stabilization and liberalization in the 1980s. He found evidence that Mexico, through low domestic energy pricing, had "escaped" the environmental benefits associated with rising world energy prices after 1973. Mexican industry increased its average energy intensity by six percent from 1970 through 1990, a period in which OECD reduced its energy intensity by 38 percent. Apart from this observation, Ten Kate found that the rapid expansion in a few highly polluting sectors, such as agro- and petro-chemicals, had been driving much of the increase in average emission coefficients. Since those are sectors with predominant public-sector influence, and domestic energy pricing has relied on public subsidies, (equivalent to four to seven percent of GDP, for different years) Ten Kate left open the possibility that domestic policy had a major influence over the increased emissions.

Energy Conservation and Interfuel Substitution

While the differences between sectors in terms of their emission intensities are important, it is highly desirable to provide more sophisticated incentives, in order to induce responses along a greater range of possibilities than simply to sectoral output levels.[25] One way of doing that, stopping short of monitoring emissions from each firm, is to exploit the link to energy use, which is particularly strong for air pollution.

For industrial as well as for other polluting activities (see transport, previous section) -- it is cost-effective to pursue emission reductions through technical controls as well as demand reductions. The marginal costs of emission reductions through these two means will be equalized if emission taxes (or tradeable permits) are used. Otherwise, however, presumptive taxes on polluting goods should be used, complementing mandated technical controls to equalize marginal costs:

$$\frac{t_x}{e_x} = c_e \tag{6}$$

Thus, cost-effective programs (and, among these, optimal programs) are characterized by tax levels on polluting goods (t_x) which, when normalized by marginal emission coefficients (e_x), equal marginal costs of emission reductions through technical controls. The formula emphasizes that a polluting good should be taxed in proportion to its emission coefficient as long as emissions themselves are not taxed.

---

[25] Indeed, general equilibrium simulations have shown surprisingly little responsiveness in economies' sectoral shares, though a common conjecture is that this impression will change as models improve in the ability to capture dynamic effects. (See, for instance Dahl, Devarajan and van Wijnbergen (1986), Bell and Srinavasan (1984) and Go (1994) on the literature on tax and tariff reform, Jorgenson and Wilcoxen (1990) and Hazilla and Kopp (1990) on the impact of pollution-control policies).

For firms, we shall not here develop technical control-cost curves, as we did for automobiles. We shall, however, investigate the scope for "purchasing" emission reductions via taxation of a particularly important category of polluting goods: fuels. With a rather simple model of how air pollution emissions from manufacturing respond to price changes[26]

$$(7) \qquad \frac{de_m}{dp_j} = \sum_i \frac{\partial e_m}{\partial x_i}(\bar{a}_i) \cdot \frac{\partial x_i}{\partial p_j}(p_1,..,p_j,.\cdot)$$

where e_m is emissions of pollutant m, summation is over a set of inputs categorized as polluting at the margin (fuels), $\partial e_m/\partial x_i(\bar{a}_i)$ is the marginal emission coefficient for fuel i when $\bar{a}_i$ characterizes the technology, and $\partial x_i/\partial p_j$ is the price responsiveness of demand of fuel i with respect to a price, p_j. We thus turn to the topic of emission coefficients and demand elasticities for fuels in manufacturing.

On the link between fuel use and emissions, it is worth noticing that the technically based literature (summarized, for instance in the US EPA's AP-42) usually will project emissions by recommending (constant) emission factors, specified in terms of grams of each pollutant per unit of fuel. These will, generally, vary by measurement, in part according to identified characteristics (for burners, according to configuration and firing conditions, for coal, according to ash content, etc.) but will be averaged when applied to a population of polluting sources.[27]

Basing emission models on fuel use has traditionally been motivated mostly by technical reasoning. Experience about emissions rates from different types of equipment and processes will generally be based on sample testing or theory (repeated sampling is needed for pollutants such as particulates, carbon monoxide and nitrogen oxides, while for SOx and CO_2, theoretical reasoning is sufficient) and fuel throughput may be viewed as a proxy for the utilization of the equipment. An alternative check, to our knowledge previously untried, is to regress reported emissions intensities against information on energy use and other possible

---

[26] The simplifying assumption is that the marginal emission coefficient itself is not changed by the price change, i.e., $\partial a_j/\partial p_j = \partial a_i/\partial p_j = 0$. See the introduction to equation (1).

[27] A fuel may have different emission factors for different users, and the marginal emission coefficient for a population of users -- when these are exposed to a fuel price change -- would then be a weighted sum of their individual emission factors, with their share in total fuel-demand change as weights. To our knowledge, studies allowing comparison of individual demand elasticities do not exist, so an averaged emission factor and an aggregate demand elasticity is applied.

determinants. For PM-10, or small dust particles, arguably the air pollutant of greatest concern in many third world cities (see second section and Ostro, 1994), the energy-cost share is highly significant, and with a positive sign, in explaining variations in emission intensities.[28] This holds whether or not other variables, such as dummies for SIC industry code, cost shares for material inputs, etc., are included among the explanatory variables. Moreover, many SIC industry dummies that are not significant in explaining variations in PM-10 emission intensities when the energy-cost share is not included in the model, are significant when energy is included. Those industries can be said to represent "clusters" in terms of PM-10 emission intensities, but only if the variation that is due to differences in energy intensity is taken care of. A similar test has yet to be performed with data on the different *types* of fuel among the explanatory variables, since such data have not been merged with emission data.

Given that different fuels pollute differently, the scope for interfuel substitution and energy conservation will be an important determinant of the effectiveness with which fuel taxes can be used to reduce emissions. What do we know about the substitutability of different fuels? In trying to answer the question, we have resorted to econometric as well as engineering studies. This is necessary because econometric studies provide no information on emissions, and, partly for the same reason, because econometric techniques often require abstractions where specificity would be desirable for the purposes of this study (see, for example, Kopp, 1992, on the eclectic use of estimates based partly on econometric studies, partly on technical studies and expertise).

The experience of the OECD countries since the first OPEC oil price shock indicates that the possibilities for energy conservation and interfuel substitution are quite high. Between 1971 and 1988, manufacturing output in OECD countries rose 62 percent, while energy use remained unchanged (Bacon, 1992). In Mexico, as mentioned earlier, where energy price increases were suppressed domestically, manufacturing energy intensity *increased* in the same period. Using separability

---

[28] We are grateful to David Wheeler and associates, who, with the collaboration of the US Bureau of the Census and the EPA for the first time have merged data on firms' costs, toxic releases, air emissions and water-pollutant discharges (see Hettige et al. 1995). We also thank Manjula Singh, performing the regressions, for the opportunity to present these results. With emission intensities, we understand a measure of emissions divided by output. We have not performed the test for other air pollutants, but the energy cost share is an important determinant also for some other important pollutant measures. Based on technical reasoning, among the main air pollutants, one would expect the link between PM-10 and energy use to be of "intermediate" strength: weaker than for sulfur and carbon dioxide, for which strict proportionality by fuel should hold (atoms are simply passed through), stronger than for volatile organic compounds, nitrogen oxides and carbon monoxide, for which the transformations are very sensitive to firing conditions.

assumptions and flexible functional forms, Fuss (1977) and Pindyck (1979) estimated own price elasticities for individual fuels in manufacturing to be in the range -0.7 to -2.2, and own-price elasticities for aggregate energy of -0.5 (Fuss) and -0.7 (Pindyck). The whole set of own- and cross-price elasticities is relevant for the responsiveness of emissions of each pollutant to a specific fuel price, but the price elasticities for aggregate energy has a more immediate interpretation: A price elasticity for aggregate energy of -0.5 implies that an equiproportionate price increase of 20 percent for all fuels (and electricity) would reduce demand for all fuels by ten percent, and consequently emissions of all pollutants by ten percent as well.[29]

The evidence in developing countries has been more limited. An important study is Pitt (1985), for Indonesian manufacturing. Using plant-level data over three years (1976-78), he estimated the own-price elasticity for aggregate energy to be between -0.07 and -0.8, i.e., overlapping with the Fuss-Pindyck range. Pitt had greater challenges in handling plant-level data, but estimated models which distinguished between subsectors of manufacturing. In general, he concluded with higher elasticities for the more energy-intensive sectors.

A study of Chile asks these questions: how much of the change in manufacturing's energy intensity is due to shifts in output levels among subsectors, and how much is due to changes in input use at the plant level (Tybout and Moss, 1994)? The answer is, perhaps surprisingly, that much of the change comes from shifts in input use at the plant level. This is an important finding, since many econometric studies based on aggregation over firms and sectors are unable to distinguish flexibility within firms and sectors from shifts between energy-intensive and other sectors. A subsequent study (Guo and Tybout, 1994) estimated interfuel substitutability for selected subsectors based on firm-level observations, to indicate considerable substitutability, but varying by subsectors and across plants (the estimates are also sensitive to the choice of model for estimation).

Eskeland, Jimenez and Liu (1994) estimate, for manufacturing, own- and cross-price elasticities for different fuels in Chile and Indonesia, with a view towards assessing the potential of fuel taxes for reducing emissions. Using a model with separability between an energy aggregate and other inputs, various results may be presented, depending on what is assumed constant in a policy experiment (the methodology follows that of Fuss, 1977). One possibility is to assume aggregate energy is held constant; another is that output is constant, but substitutability between energy, materials and labor occurs. When aggregate energy is held constant

---

[29] These models assume that different fuels form a separable homothetic aggregate called energy, so that fuel shares in the aggregate do not change unless relative prices between fuels change.

(the "energy submodel"), the relative price of one energy source (a fuel, or electricity) is increased, while the price and use of aggregate energy is constant. When only output is constant, the price of the energy aggregate changes when an individual fuel price changes. For Chile, they report on estimates based on both assumptions, while for Indonesia, only the energy submodel was estimated.

Table 4 presents estimates of Chilean price elasticities of demand for the first case -- holding aggregate energy constant. The six energy sources in the original study have been aggregated into three to provide a simpler display of policy tools. Notice that all the own-price elasticities are negative, and fairly large in absolute value. At such a general level, the results are in line with earlier research, for instance by Pindyck (1979). In the full elasticity matrix, twenty-four of the thirty-six elasticity estimates are significantly different from zero at the five percent level, and an additional two at the ten percent level. The fossil fuels are sorted according to their TSP emission intensities, with coal, fuel oil and "grouped fuels" (the latter includes wood and coke) comprising "heavy fuels," while diesel and natural gas comprise "light fuels."

Table 4 Chilean Manufacturing: Energy Group Elasticities, Energy Constant[a]

	Electricity price	Heavy fuels' price	Light fuels' price
Electricity demand	-0.98	0.77	0.22
Heavy fuels' demand	0.61	-0.89	0.28
Light fuels' demand	0.20	0.53	-0.74

Note: a. See text. Heavy fuels are coal, fuel oil and grouped fuels, light fuels are natural gas and diesel. Elasticities are combinations of elasticity estimates in a six-fuel matrix, most of which are significant at a five percent level.
Source: Eskeland, Jimenez and Liu (1994).

The elasticities in Table 4 are calculated under the assumption that prices for various fuels in a group are increased proportionally.[30] As an example, if prices for heavy fuels increase by ten percent, demand for electricity and light fuels will increase by 7.7 and 5.3 percent respectively, while demand for heavy fuels will fall by 8.9 percent.

[30] Aggregation is as follows: when several prices increase in the same proportion (say, prices for "heavy fuels"), a fuel's demand elasticity with respect to this aggregate price change is the sum of its elasticities with respect to the individual prices. The elasticity for an aggregate fuel (say, "heavy fuel demand") is the quantity weighted sum of the elasticities for individual fuels in the aggregate.

What are the implications of these elasticities for the responsiveness of emissions to price changes? The answer depends to a great extent on what happens in the electricity sector, and there are good reasons to give the sector a separate treatment. One reason is that the electricity sector is reasonably easy to subject to separate policy instruments, including those based on emission monitoring. Another is that electric power plants may consist of tall-stack units located far away from the most polluted and populated areas, so one may not want to treat emissions from the sector in the same way as one treats emissions generated directly in the manufacturing sector (the source of electricity is not given, since the data is from a manufacturing survey). Finally, the extent to which the electricity sector is polluting at the margin will depend not only on the composition of plants but also on their specific order of dispatch. Table 5, for these reasons, provides emission elasticities under alternative assumptions. The assumption "Electricity Polluting" assumes that electricity at the margin is produced 50 percent by coal-fired, uncontrolled plants, and 50 percent by unpolluting technologies. The alternative assumption is that the electric power sector is not polluting at the margin.

Table 5 Chilean Manufacturing: Emission Elasticities, Energy Constant

	Electricity price	Heavy fuels' price	Light fuels' price
SOx Emissions:			
Electricity polluting	0.07	-0.19	0.11
Electricity not polluting	0.50	-0.58	0.07
TSP Emissions:			
Electricity polluting	-0.14	-0.09	0.24
Electricity not polluting	0.48	-0.73	0.25

Source: Eskeland, Jimenez and Liu (1994).

Again using a ten percent increase in prices for heavy fuels as example, SOx emissions would be reduced by 5.8 percent and TSP emissions by 7.3 percent under the assumption that the electricity sector is not polluting at the margin. Price increases for light fuels would, as expected, increase emissions under all assumptions, since they would lead to substitution towards more polluting energy sources. Electricity price increases would increase emissions significantly if electricity is assumed to be nonpolluting. If electricity is assumed to be polluting at the margin, electricity price increases would lead to a small increase in SOx emissions, but to a small reduction in TSP emissions. As a general summary, price increases for heavy fuels are quite powerful in reducing emissions if and only if the electricity sector is not polluting at the margin. The reason is, as can be seen in Table 4, that much of the induced demand response is towards electricity.

Table 6 Indonesian Manufacturing: Energy Price Elasticities, Energy Constant

	Electricity price	Heavy fuels' price	Light fuels' price
Electricity demand	-1.02	0.26	0.76
Heavy fuels demand	0.24	-1.41	1.17
Light fuels demand	-0.01	0.37	-0.35

Note: Heavy fuels are coal and a group which includes fuel oil. Elasticities are combinations of elasticity estimates in a six-fuel matrix, most of which are significant at a five percent level.
Source: Eskeland, Jimenez and Liu (1994).

Table 6 shows the calculated group-demand elasticities for Indonesia. Comparing them with those of Chile, there are differences, but the dominance of negative own-price elasticities and positive or small cross-price elasticities is repeated. The own-price elasticity for heavy fuels is larger in absolute value, a fact that will make tax instruments more powerful.

Emission elasticities are given in Table 7. As is the case with Chile, price increases for heavy fuels will reduce emissions for both SOx and TSP. The magnitudes of the effects are greater for Indonesia. Referring back to Tables 4 and 6, Indonesia has a higher own-price elasticity for heavy fuels (in absolute value) and a smaller cross-price elasticity with electricity. Thus, emissions respond more to an increase in heavy fuels' prices. Under the assumption that the electricity sector is not polluting at the margin, emission elasticities with respect to the price of heavy fuels are -1.02 and -1.52 for TSP. Even under the assumption that electricity is polluting, emission elasticities with respect to heavy fuels' prices for Indonesian manufacturing are quite large, in the range of -0.8 for SOx and TSP.

These estimates are based on the assumption that the prices of the aggregate "bundle" of energy does not change when a fuel price rises.[31] In reality, unless compensatory price changes are made for other fuels, firms will also alter their use of aggregate energy by substituting labor and materials (still holding

[31] Technically, the experiment entails making compensatory price changes for other fuels, so that the price of aggregate energy is held constant. With aggregate energy price constant, aggregate energy use is held constant.

output constant). For Chile, the necessary deflators for outputs and other inputs could be constructed, and a short-run model with three aggregate inputs -- energy, labor and material could be estimated.

Table 7 Indonesian Manufacturing: Emission Elasticities, Energy Constant

	Electricity price	Heavy fuels' price	Light fuels' price
SOx Emissions:			
Electricity polluting	0.02	-0.81	0.79
Electricity not polluting	0.22	-1.02	0.80
TSP Emissions:			
Electricity polluting	-0.21	-0.86	1.07
Electricity not polluting	0.26	-1.52	1.25

Source: Eskeland, Jimenez and Liu (1994).

Table 8 displays the own- and cross-price elasticities in this aggregate model. The own-price elasticity of aggregate energy is considerable, at -0.63, just between Pindyck's -0.7 and Fuss' -0.5. This indicates that demand for all fuels -- and emissions of *all air pollutants* -- would fall by 6.3 percent if all energy prices increased by ten percent. We may notice that the cross-price elasticity between energy and labor is positive and substantial, indicating that subsidies to energy would be a poor strategy if one pursued an employment objective.

Table 8 Chilean Manufacturing: Elasticities For Aggregate Inputs[a]

	Aggregate energy price	Wage	Material input price
Energy demand	-.63 (.36)	1.57 (.49)	.95 (.31)
Labor demand	.52 (.16)	-1.89 (.34)	1.37 (.30)
Material demand	-.07 (.02)	.32 (.07)	-.25 (.08)

Note: a. Standard errors in parentheses. All coefficients significant at five percent level.
Source: Eskeland, Jimenez and Liu (1994).

With aggregate energy adjusting, price elasticities for individual fuels will be modified -- own-price elasticities are higher in absolute value, while the cross-price elasticities are somewhat dampened -- to be smaller, and perhaps will switch sign if they are positive in the energy submodel. The intuition is that the presence of other variable factors makes it easier to reduce demand for the fuel being taxed.

Importantly, it also reduces the tendency that demand spills over to other fuels. The implications for policy are seen in the emissions elasticities in Table 9. The greater own-price elasticities and the lower cross-price elasticities have increased the elasticities of emissions with respect to the price of "heavy fuels." Under these assumptions, a tax increase on heavy fuels can be a potent instrument for pollution control -- whether or not electricity is polluting at the margin.

Exploration of General Equilibrium Effects

Up to now, we have been discussing the effects of fuel taxes on emissions in a partial equilibrium context. For transportation, we emphasized the role fuel taxes could play in reducing the output in the polluting activity, while other instruments could be used to make each "polluting plant" less polluting per unit of output. In the applications to industrial emissions, in contrast, the emphasis has been on the extent to which the same amount of aggregate output from manufacturing could be delivered with less use of energy, or with less polluting types of energy used. In this section, we use a computable general-equilibrium model of the Indonesian economy to address, in an exploratory fashion, some extensions.

Table 9 Chile: Emission Elasticities, Energy Adjusting

	Electricity price	Heavy fuels' price	Light fuels' price	All energy prices
SOx Emissions:				
Electricity polluting	-0.15	-0.52	0.04	-0.63
Electricity not polluting	0.28	-0.91	-0.01	-0.63
TSP Emissions:				
Electricity polluting	-0.37	-0.42	0.16	-0.63
Electricity not polluting	0.25	-1.06	0.18	-0.63

Source: Eskeland, Jimenez and Liu (1994).

Increased input taxes, if uncompensated, will alter output levels, since they change production costs. In the case of a fuel tax, fuel-intensive producers and sectors will initially experience the greatest increase in costs, and thus have more of a pressure either to change their input combination or to raise output prices. The latter will occur to a great extent if fuel-intensive producers have little flexibility in input choice, and one would expect to see reduced equilibrium output levels in fuel- intensive sectors. The new equilibrium, with a different output combination and different emission levels, will reflect the opportunities for change by all agents in the economy.

To analyze these changes, we employ a computable general-equilibrium (CGE) model of the Indonesian economy that is designed to highlight the features of importance for energy use (Devarajan and Lewis, 1994). In particular, the model distinguishes among three types of primary energy -- oil, gas and coal -- and three energy products -- electricity, gas and refined oil products. The model pays close attention to the existing system of taxes and subsidies in the economy. The rest of the model is a fairly standard CGE model, with 26 productive sectors, four households, five labor groups and fixed, sector-specific capital.

The model enables us to go deeper than the rather abstract "aggregate manufacturing output" in the previous section. Table 10 shows the greater degree of specificity in terms of outputs. Eight sectors constitute manufacturing in a traditional sense, while six "sectors" convert petroleum resources to refined products and LNG. Four sectors are engaged in extraction of natural resources, two sectors represent agricultural activities. The first column in the table shows emissions of total suspended particles (TSP, or dust) from each sector, as modelled on the basis of fuel use. In manufacturing, three sectors stand out: chemicals and fertilizer, nonmetallic minerals and basic metals. Importantly, two "miscellaneous" sectors (electricity and transport), are the two largest in terms of TSP emissions, while construction would rank equally with a manufacturing sector such as food processing.

The second column in the table shows the "direct emission coefficients," or total emissions from the sector divided by the sector's gross output.

The CGE model allows us to evaluate the importance of intermediate deliveries between sectors. The third column presents the sum of the direct and indirect emission coefficients. That is, it includes not just the direct effect on emissions of increasing output in a sector by one unit, but also the effects due to the inter-industry demand stimulated by this increase in output. These are important if one looks at the economy with a blunt planning perspective. The column informs us, for instance, that an expansion of a sector such as metal products and machinery would increase output from quite polluting sectors, and thus increase pollution, even though the sector itself does not have a very high emission coefficient. This may be relevant if major industrial expansion plans are considered and other, adequate, policy instruments are not in place to influence the levels of pollution. If polluting suppliers to the industry in question are facing appropriate tax levels, however, then expansion plans for a downstream industry should only pass the test of paying for the effects of its *own* emissions (i.e., according to direct coefficients).

Table 10 TSP Emissions by Sector in the Computable General Equilibrium Model

Sector	Total emissions	Direct coefficient	Total coefficient	CGE coefficient
Manufacturing				
Food processing	44.7	0.003	0.015	-0.006
Textile & leather	9.8	0.003	0.039	-0.003
Wood & furniture	10.2	0.004	0.024	-0.012
Paper & other industries	3.5	0.003	0.042	-0.002
Chemicals & fertilizer	157.5	0.041	0.095	0.016
Nonmetallic minerals	220.0	0.136	0.200	0.076
Basic metals	164.5	0.090	0.166	0.028
Metal products & machinery	14.8	0.002	0.056	0.002
Refining				
Gasoline	0.1	0.000	0.013	-0.095
High-speed diesel (HSD)	0.1	0.000	0.013	-0.017
Industrial diesel	0.0	0.000	0.013	-0.211
Kerosene	0.1	0.000	0.013	-0.171
Fuel oil	0.1	0.000	0.013	-0.570
LNG	2.8	0.001	0.004	-0.003
Agriculture & Natural Resources				
Food agriculture	0.9	0.000	0.007	-0.010
Traded agriculture	4.1	0.000	0.010	-0.010
Oil mining	6.4	0.001	0.004	-0.000
Natural gas	1.1	0.001	0.004	-0.017
Coal mining	39.1	0.515	0.603	3.484
Other mining	5.5	0.004	0.003	-0.020
Miscellaneous				
Electricity & gas	1668.9	0.999	1.225	0.586
Water supply	0.2	0.002	0.058	0.006
Construction	42.2	0.002	0.043	-0.005
Transport	692.2	0.073	0.088	0.020
Trade & storage	5.0	0.000	0.015	-0.007
Services	6.0	0.000	0.025	0.000
TOTAL	3037.3			

Source: Devarajan and Lewis (1994).

The calculation of the total coefficient in the third column assumes implicitly that resources are available in perfectly elastic supply to expand output in this economy. This is why all the total coefficients are larger than the direct coefficients. In reality, however, when a sector expands output, it will draw resources from other sectors by bidding-up factor prices. As a result, some othersectors will contract, and the overall effect on pollution may be positive or negative. The fourth column displays this overall effect when these overall resource limitations are brought into effect.[32] Note that some of the coefficients are negative, because the higher factor costs due to the output expansion in that sector lead to a contraction in more polluting sectors. In particular, some of the petroleum product sectors, such as kerosene and industrial diesel, which are "light" fuels to begin with, actually have a negative coefficient because an increase in their output crowds out other, more polluting products.

In general, of course, taxing the polluting activities themselves according to their pollution and damage levels is a good idea. In the case of fossil fuels, since pollution is related to the *use* of these fuels (the outputs from the refining sectors, strictly speaking), taxation of fuels will be a tax on polluters and have much of the desired effect in terms of inducing sectors using these fuels (and users of these using sectors' outputs) to reduce polluting inputs.

We turn therefore to the impact of changes in tax rates on energy use and emissions. As noted earlier, Indonesia's energy sector already has a complex system of taxes and subsidies. A question one could ask is therefore: What would be the effect on emissions of eliminating these? The answer, not reported here in detail, is that emissions of most pollutants would increase. The main reason is that many of the less polluting fuels are subsidized at the outset (kerosene and two types of diesel, in particular) and some of the heavier fuels are taxed. Thus, an elimination of taxes and subsidies makes the more polluting fuels more attractive, contributing to emissions.

To evaluate the impact of changes in tax rates, we display in Table 11 general equilibrium arc elasticities for grouped energy sources. As before, own-price elasticities are negative, and cross-price elasticities are positive or small. Compared to the partial-equilibrium elasticities for the manufacturing sector (Table 6), the general equilibrium (tax) elasticities are all smaller in absolute value. We may obtain

[32] These coefficients are calculated as follows. Government demand for a sector's output is increased exogenously (financed through lump-sum taxation). The model solves for the new set of output- and factor prices that equate demand and supply in the new equilibrium (i.e., the one with the exogenously increased demand for the output of one of the sectors). We then calculate the emissions associated with the new production structure and compare it with emissions in the old production structure. The difference is labelled the "CGE coefficient" in Table 10.

a better understanding of this by turning our attention to the differences between the models.

Table 11 Indonesia: General-Equilibrium Arc Elasticities

	Electricity price	Heavy fuels' price	Light fuels' price
Electricity demand	-0.41	-0.04	0.04
Heavy fuels' demand	-0.08	-0.43	0.45
Light fuels' demand	0.00	0.02	-0.14

Source: Devarajan and Lewis (1994).

The general-equilibrium model distinguishes itself from the partial- equilibrium model in three important ways. First, the model attempts to describe the Indonesian economy as a whole, rather than covering only the input side of the manufacturing sector. This implies coverage of additional sectors, but also that manufacturing industries may adjust their output levels in addition to their energy use. Second, the supply of fuels is not perfectly elastic, implying that users will face a price increase smaller than the increase in the fuel tax. Refining is modelled as one sector transforming crude oil into a range of petroleum products, with a joint output, constant elasticity of transformation technology. Third, the model incorporates budget constraints for the various "households." Importantly, when energy tax rates are changed, the model has adjusted income-tax rates, so that government revenues are preserved.

How would one expect each of these differences to influence the responsiveness reflected in the elasticities? The added flexibility in aggregate energy use and in output composition should make the CGE model display greater flexibility than the partial equilibrium model, but the result would also depend on the flexibility within the additional, nonmanufacturing sectors.[33] Fuel supply that is imperfectly

[33] The manufacturing sectors of the CGE is calibrated with a view to the estimated parameters of the econometric model, but no close mapping can be achieved, due to the differences in structure. Most importantly, since the manufacturing sector in the estimated partial-equilibrium model (previous section) is one for aggregate manufacturing, it should display greater input-demand flexibility than what each manufacturing sector displays in the general-equilibrium model, if the two are to approximate the same technology (since some changes in output mix will be recorded as such in the CGE, but will be masked -- as changes in input mix holding "output constant" or "energy constant" in the partial equilibrium model).

elastic, in contrast, would tend to make quantities less responsive to fuel tax changes in the CGE model than in the partial-equilibrium model. In particular, imperfect elasticity of transformation between petroleum products limits the substitution between fuels, since suppliers will prevent user prices for fuels from moving too far apart. A similar effect hurts the effective flexibility that users have in their choice between heavy fuels and electricity: CGE (cross-tax) elasticities are close to zero, as opposed to 0.25 in partial equilibrium. The reason is that electricity producers also face higher costs when heavy fuels' prices increase. They will raise electricity rates, and thus dampen the resulting change in relative user prices between heavy fuels and electricity. Finally, the compensatory changes in income tax rates is an assumption chosen to give specific meaning to the allowed flexibility in output composition. As compared to an alternative -- e.g., let the government keep the proceeds from a fuel tax increase -- the result may not be very different: it would depend on whether the marginal consumption basket among households is more or less energy intensive than that of the government.

In sum, the CGE model displays notably less flexibility than the partial-equilibrium model for the manufacturing sector. In accordance with the above discussion, the added sectors are as flexible in their energy use as manufacturing; thus the imperfectly elastic supply of fuels, the joint production in refining, and the cost-pass-through in electricity production are responsible for this lower flexibility.

Table 12 displays the resulting elasticities for emissions of TSP and SOx. In this model, the electricity sector chooses its inputs endogenously, and the line "electricity not polluting" simply assumes that tight emission controls apply when electricity is based on fossil fuels.

Table 12 General Equilibrium Arc Emission Elasticities

	Electricity price	Heavy fuels' price	Light fuels' price
SOx Emissions:			
Electricity polluting	-0.04	-0.20	0.14
Electricity not polluting	0.06	-0.19	0.07
TSP Emissions:			
Electricity polluting	-0.08	-0.41	0.41
Electricity not polluting	0.06	-0.55	0.50

Source: Devarajan and Lewis (1994).

As Table 12 shows, the general-equilibrium model nevertheless displays a significant response of emissions to changes in fuel and energy prices. For example, assuming that electricity is polluting, a ten percent price change for heavy fuels will reduce emissions of TSP and SOx by four and two percent, respectively. For larger price increases, the emission reductions can be expected to increase in the same proportion (or more), though such estimates are made on a weaker empirical basis. The elasticities are, however, noticeably smaller than for the partial-equilibrium model. An important contributing factor is that the sectors supplying fuels are modelled as imperfectly elastic in their supply -- thereby limiting the changes in user prices for fuels that follow from changes in fuel tax rates. Importantly, this reflects the modelers' assumptions about technology and about the trade regime for petroleum products. If Indonesia were to trade crude oil and products without barriers, users would face almost perfectly elastic fuel supplies. Another lesson in instrument choice is seen by observing that the cross-price elasticities between heavy fuels and electricity would be positive if taxes on heavy fuels were not applied to the electricity sector's purchases. Such an exemption, or refunds, would be justified if the electricity sector were applying tight emission controls. Thus, if the electricity sector is "not polluting" and can purchase heavy fuels without paying the emission-presumptive taxes, then the SOx and TSP emission elasticities with respect to heavy fuels' prices would be larger (in absolute value) than the indicated 0.19 and 0.55.

In sum, the evidence points to significant potential for using fuel taxes to combat air pollution. Partial-equilibrium analysis and general-equilibrium simulation show significant responsiveness of pollutant emissions with respect to changes in tax rates for fuels and electricity. The general-equilibrium modelling serves to remind us of the importance of sectors other than transport and manufacturing, and also of the importance of the technical and policy conditions governing supply of fuels and electricity. Clearly, a restrictive trade regime for fuels can restrict an economy's responsiveness to fuel- and energy-tax changes. Further, if the electricity sector can be emission controlled, and (then) exempted from presumptive fuel taxes, then the scope for substitution in using sectors will be enhanced, and emissions will respond more strongly to fuel taxes. Apart from these insights, the general-equilibrium simulations showed us much the same types of effects as those observed in the partial-equilibrium models for manufacturing (and for transport, though they were from another country).

It is worth noticing that much of the relevant flexibility lies in substitution between heavy fuels (coal and fuel oil) and electricity. This flexibility renders taxation of heavy fuels a more powerful instrument if pollution from the electricity industry is low or is taken care of by other instruments. Thus, if the costs of monitoring and controlling emissions from electricity plants are such that that industry will not be too polluting at the margin, then it would lend additional strength to simple tax instruments in the pursuit of emission reductions from other users of energy.

In terms of which fuels should be taxed, our analytical framework indicates that each fuel should be taxed according to its direct emission coefficient, with damage estimates providing weights by pollutant. Whether emphasis should be placed on sulphurous emissions, dust emissions, or other measures depends on local conditions. However, many studies now point towards emphasis on dust (in particular the smaller particles, PM10) in urban air pollution. However, taxation of sulphur and dust will often single out the same fuels (called heavy fuels here), at least until the taxation scheme and the market operators become sophisticated enough to distinguish more finely and encourage fuels with lower ash and sulphur content. When tax rates are levied according to each fuel's characteristics, the responsiveness of emissions to price changes will be greater than displayed in this section, where uniform price changes for a broader range of fuels were considered.

While the general lesson from this section is similar to the title of the chapter -- tax bads by taxing bad goods more -- one needs again to emphasize the many possible avenues towards emission reductions that cannot be stimulated with fuel taxes. Thus, we highlight the value of complementing this instrument with others, and to refine the instruments as institutional and technical capacity develops. An important question will always be when the agency finds itself capable of allowing partial refunds of fuel taxes to users who prove that they pollute less than the average user of the fuel. Only at that point will the fuel taxes, in themselves, add encouragement for control technologies to the encouragement they provide for fuel economy, demand reduction and fuel switching. At earlier stages, and perhaps a for long time, other instruments that stimulate cleaner technology (such as emission standards, popular in the real world) will be valuable supplements in a program of emission-presumptive fuel taxes.

CONCLUSION

The literature on pollution control has debated the merits of market-based instruments versus command-and-control approaches, concluding that, in general, the former is preferred to the latter. This chapter has looked at the problem from a different angle, namely, as one of choosing a mix of instruments when continuous monitoring of emissions is impossible. When viewed from this perspective, some of the command-and-control policies appear more sensible because they focus on processes, equipment and machinery, giving the government valuable information about where it can intervene. For example, if an agency cannot monitor emissions continuously, but can inspect machinery periodically and tax fuel use, then emission standards and presumptive fuel taxes would be used.

However, a focus on processes, input use and equipment is only a partial solution to the problem, and this chapter has addressed the implications of recognizing this. First, when continuous emissions monitoring is feasible (and not costly), the government should monitor emissions only, and not evaluate equipment, processes and input use. Second, in the more realistic case (when continuous monitoring is costly), the government should focus on factors causing emissions: the equipment and processes, and the scale of the activity. Pollution policy consists of incentives to make all activities cleaner (for instance by addressing equipment and processes), and incentives to reduce the scale of the more polluting activities. Furthermore, when only one set of policies are in place -- for example, only standards requiring catalytic converters in cars -- the introduction of demand-management instruments (such as a gasoline tax) can significantly improve welfare.

These principles were illustrated using empirical information gleaned from case studies of Mexico City, Chile and Indonesia. In analyzing the options for controlling vehicle-induced air pollution in Mexico City, we showed that "cleaner cars" and "fewer trips" are two ways of achieving emission reductions, and therefore the contribution of each should be equated at the margin. Specifically, we observed that a gasoline tax, which will induce fewer trips, is part of a cost-effective control strategy. We calculated the size of such a tax, as well as the costs of not adopting it.

Noting that industrial pollution sources are much more heterogeneous, we examined the potential for fuel-demand management in the industrial sector by estimating fuel-demand elasticities based on firm-level data from Chile and Indonesia. The results indicated that firms have the flexibility to respond to fuel-price increases. They can switch from dirty to clean fuels, and they can reduce overall energy intensity. Finally, simulations with a general-equilibrium model of Indonesia revealed that these fuel taxes will not lead to perverse or unintended effects when all of the adjustments in the economy take place: in fact, the economy shifts towards cleaner fuels and becomes less energy intensive, both of which benefit the environment.

The mobile-source example and the fixed-source example both illustrated how effectiveness depends on the quality of monitoring. In Mexico City, while fuel taxes provide 20 percent of emission reductions, 80 percent is provided by technical controls, most depending on testing and periodic inspections. For fixed sources, fuel and energy taxes work much better if the electricity sector can be kept tightly controlled by other means, and then be exempted from taxes on heavy fuels.

To be sure, these conclusions about indirect instruments for pollution control (when emission monitoring is a problem) should be viewed as improvements, and reform steps, rather than final solutions. As monitoring technologies and capacity improve, the instruments should shift towards policies based on actual monitoring of emissions. In the meantime, however, instruments aimed at increasing the marginal cost of polluting activities, either by taxes on marketed outputs and inputs, or by standards on equipment and processes, can reduce pollution in a cost-effective manner with modest monitoring requirements. Furthermore, these same instruments can complement existing policies (such as the existing standards), as well as new instruments based on emissions monitoring, which may be phased in as new technologies and information become available.

References[34]

Anderson, R. (1990). "Reducing Emissions From Older Vehicles." *Research Study* 053. Washington, DC: American Petroleum Institute.

Alberini, A., W. Harrington and V. McConnell. (1993). "Determinants of Participation in Accelerated Vehicle Retirement Programs." *Discussion Paper* QE93-18. Washington, DC: Resources for the Future.

*Bacon, R. (1992). "Measuring the Possibilities of Interfuel Substitution." Policy Research Department *Working Paper* 1031. Washington, DC: World Bank.

Baumol, W.J. and W.E. Oates. (1988). *The Theory of Environmental Policy.* Englewood Cliffs, NJ: Prentice-Hall.

Balcer, Y. (1980). "Taxation of Externalities, Direct versus Indirect." *Journal of Public Economics* 13, 121-129.

Bell, C. and T.N. Srinivasan. (1984). "On the Uses and Abuses of Economy-wide Models in Development Policy Analysis." In M. Syrquin, L. Taylor, and L. Westphal (eds.), *Economic Structure and Performance.* Orlando, FL: Academic Press.

Corlett, W.J. and D.C. Hague. (1953). "Complementarity and the Excess Burden of Taxation." *Review of Economic Studies* 21, 21-30.

Dahl, H., S. Devarajan and S. van Wijnbergen. (1986). "Revenue-neutral Tariff Reform: Theory and an Application to Cameroon." *CPD Paper* 1986-25. Washington, DC: World Bank.

Devarajan, S. and J.D. Lewis. (1994). "Energy Pricing, Economic Distortions and Air Pollution in Indonesia." HIID and World Bank (in process).

*Eskeland, G.S. (1994). "A Presumptive Pigovian Tax on Gasoline: Complementing Regulation to Mimic an Emission Fee." *World Bank Economic Review* 8/3, 373-394.

*Eskeland, G.S. and T. Feyzioglu. (1994a). "Is Demand for Polluting Goods Manageable? An Econometric Study of Car Ownership and Use in Mexico." Policy Research Department *Working Paper* 1309. Washington, DC: World Bank.

[34] A star denotes a study performed as part of the World Bank research project.

*Eskeland, G.S. (1994b). "Rationing in Demand Management -- the 'Day Without a Car' Program in Mexico City." Policy Research Department (in process). Washington, DC: World Bank.

*Eskeland, G.S., E. Jimenez and L. Liu. (1994). "Energy Pricing and Air Pollution: Econometric Evidence from Manufacturing in Chile and Indonesia." Policy Research *Working Paper* 1323. Washington, DC: World Bank.

*Eskeland, G.S. and C. Kong. (1994). "Environmental Policies and the Poor: A Welfare Function Approach to Air Pollution Control on Java." Paper presented at the 1995 Annual Meeting of the American Association of Environmental and Resource Economists, Washington, DC. Also, in process for publication by the Policy Research Department. Washington, DC: World Bank.

*Eskeland, G.S. (1992). "Attacking Air Pollution In Mexico City." *Finance and Development* 29/4, 28-30.

*Eskeland, G.S. and E. Jimenez. (1992). "Policy Instruments for Pollution Control in Developing Countries." *The World Bank Research Observer* 7/2, 145-169.

*Eskeland, G.S. and E. Jimenez. (1991). "Curbing Pollution in Developing Countries." *Finance and Development* 11/1, 15-18.

Eskeland, G.S. (1994). "The Costs and Benefits of an Air Pollution Control Strategy for Santiago, Chile." Chapter 5 in *Chile: Managing Environmental Problems: Economic Analysis of Selected Issues,* Report No. 13061-CH. Washington, DC: World Bank.

Fuss, M.A. (1977). "The Demand for Energy in Canadian Manufacturing." *Journal of Econometrics* 5, 89-116.

Go, D. (1994). "External Shocks, Adjustment Policies, and Investment in a Developing Economy: Illustrations from a Forward-looking CGE Model of the Philippines." *Journal of Development Economics* 44, 229-261.

Goulder, L.H. (1995). "Environmental Taxation and the "Double Dividend:" A Reader's Guide." *This volume.*

Greenwald, B.C. and J.E. Stiglitz. (1986). "Externalities in Economies with Imperfect Information and Incomplete Markets." *Quarterly Journal of Economics* 101, 229-264.

*Guo, C.C. and J.R. Tybout. (1994). "How Relative Prices Affect Fuel Use Patterns in Manufacturing: Plant-Level Evidence from Chile." Policy Research Department *Working Paper* 1297. Washington, DC: World Bank.

Harrington, W. and V. McConnell. (1993). "Modeling In-Use Vehicle Emissions and the Effects of Inspection and Maintenance Programs." *Discussion Paper* QE93-02. Washington, DC: Resources for the Future.

Harrison, Jr., D. (1977). "Controlling Automotive Emissions: How to Save More Than $1 Billion Per Year and Help the Poor Too." *Public Policy* 25/4.

Hazilla, M. and R. Kopp. (1990). "Social Costs of Environmental Quality Regulation: A General Equilibrium Analysis." *Journal of Political Economy* 98/4, 853-873.

Hettige, M., P. Martin, M. Singh and D. Wheeler. (1995). "The Industrial Pollution Projection System." Policy Research Department *Working Paper* 1431. Washington DC: World Bank.

Jorgenson, D.W. and P.J. Wilcoxen. (1990). "Environmental Regulation and U.S. Economic Growth." *Rand Journal of Economics* 21/2, 314-340.

*Kopp, R.J. (1992). "Economic Incentives and Point Source Emissions: Choice of Modeling Platform." Policy Research Department *Working Paper* 920. Washington, DC: World Bank.

*Krupnick, A.J. (1993). "Measuring the Effects of Urban Transportation Policies on the Environment: A Survey of Models?" Policy Research Department *Working Paper* 1030. Washington, DC: World Bank.

Krupnick, A. and M.A. Walls. (1992). "The Cost Effectiveness of Methanol for Reducing Motor Vehicle Emissions and Urban Ozone." *Journal of Policy Analysis and Management* 11/3, 373-396.

Krupnick, A.J., M.A. Walls and C.T. Collins. (1993). "Global Warming and Urban Smog: Cost-Effectiveness of CAFE Standards and Alternative Fuels." *The Energy Journal* 14/4, 75-98.

Lawson, D.R., P.J. Groblicki, D.H. Stedmand, G.A. Bishop and P.L. Guenther. (1990). "Emissions from In-use Motor Vehicles in Los Angeles: A Pilot Study of Remote Sensing and the Inspection and Maintenance Program." *Journal of the Air and Waste Management Association* 40/8, 1096-1105.

McConnell, V. and W. Harrington. (1992). "Cost-Effectiveness of Enhanced Motor Vehicle Inspection and Maintenance Programs." *Discussion Paper* QE92-18-REV. Washington, DC: Resources for the Future.

McConnell, V. and W. Harrington. (1993). "Cost Effectiveness of Remote Sensing of Vehicle Emissions." In R. Kosobod (ed.), *Cost Effective Control of Urban Smog*. Chicago: Federal Reserve Board.

OECD. (1975). *The Polluter Pays Principle.* Paris.

*Ostro, B. (1994). "Estimating the Health Effects of Air Pollution: A Method with an Application to Jakarta." Policy Research Department *Working Paper* 1301. Washington, DC: World Bank.

Pigou, A.C. (1920). *The Economics of Welfare.* London: Macmillan & Co.

Pitt, M. (1985). "Estimating Industrial Energy Demand with Firm-Level Data: The Case of Indonesia." *The Energy Journal* 6/2, 25-39.

Roberts, M.J. and M. Spence. (1976). "Effluent Charges and Licenses Under Uncertainty." *Journal of Public Economics* 5, 193-208.

Sandmo, A. (1976). "Direct versus Indirect Pigovian Taxation." *European Economic Review* 7, 337-349.

*Swait, J. and G.S. Eskeland. (1994). "Travel Mode Substitution in São Paulo: Estimates and Implications for Air Pollution Control." Policy Research Department *Working Paper* 1437. Washington, DC: World Bank.

*Ten Kate, A. (1993). "Industrial Development and the Environment in Mexico." Policy Research Department *Working Paper* 1125. Washington, DC: World Bank.

Tietenberg, T. (1992). *Environmental and Natural Resource Economics.* New York: HarperCollins.

*Tybout, J.R. and D. Moss. (1994). "The Scope for Fuel Substitution in Manufacturing Industries: A Case Study of Chile and Colombia." *The World Bank Economic Review* 8/1, 49-79.

United States Environmental Protection Agency. (1986). (AP-42) "Compilation of Air Pollutant Emission Factors." Supplement A, October, Vol. 1, Stationary Point and Area Sources. Research Triangle Park, NC.

Wijkander, H. (1985). "Correcting Externalities Through Taxes on/Subsidies to Related Goods." *Journal of Public Economics* 28, 111-125.

World Bank. (1994). "Chile. Managing Environmental Problems: Economic Analysis of Selected Issues." Latin American Country Development *Report* 13061-CH. Washington, DC.

Chapter 5

GLOBAL CLIMATE CHANGE, ENERGY SUBSIDIES AND NATIONAL CARBON TAXES

Bjørn Larsen (World Bank)

and

Anwar Shah (World Bank)*

INTRODUCTION

We are still no closer to answers on whether the buildup of "greenhouse" gases in the atmosphere could trigger a significant warming of the earth's surface. Nor do we know with any precision what the social and economic consequences of global warming might be. Yet countries all over the world are increasingly showing a willingness to act now, rather than wait for further scientific evidence:

- At the June 1992 "Earth Summit" in Rio de Janeiro, more than 150 countries signed a treaty aimed at stabilizing greenhouse gases -- mainly carbon dioxide (CO_2), which is by far the largest contributor to the greenhouse effect. The UN Framework Convention on climate change specifies a series of obligations for participating countries to respond to this problem. Those obligations took effect when the Convention entered into force on 21 March 1994. Nineteen OECD countries and the European Union (EU) have ratified the treaty as of 1 May 1994 and face a later deadline to submit their climate change action plans to the Convention Secretariat.
- Most industrial countries have adopted national targets of stabilizing CO_2 emissions at 1990 levels by the year 2000.

---

* The findings, interpretations and conclusions expressed in this paper do not necessarily represent the views of the World Bank, its affiliates or its member countries. The authors are grateful for comments by two anonymous referees on a previous draft of the paper.

- Between January 1990 and January 1991, Finland, the Netherlands, and Sweden imposed "carbon taxes" -- that is, taxes on the carbon content of fossil fuels.
- The European Union is considering levying a Union-wide tax on both carbon and energy in general.
- Many developing countries are beginning to weigh the various options, because although industrial counties account for around 43 percent of total CO_2 emissions from fossil fuel consumption, the developing world is fast contributing an increasing proportion.

This chapter reviews some of the policy instruments available to restrict CO_2 emissions in OECD and non-OECD countries that can yield global and local environmental benefits and economic efficiency gains. The following section discusses subsidies to energy and energy complements, presenting recent estimates of subsidies in non-OECD and OECD countries and reviewing CO_2 emission reductions and efficiency effects from energy subsidy removal. Subsequently, a national carbon tax is discussed in terms of its revenue potential, redistributive implications, efficiency impacts, and local environmental benefits. The last section presents a summary and policy implications.

SUBSIDIES TO ENERGY AND ENERGY COMPLEMENTS

Energy Subsidies in Non-OECD Countries

Economists have long argued that subsidizing energy distorts consumption and production decisions and has adverse consequences for government revenues and the protection of local and global environments. Yet such subsidies, here measured as the difference between energy consumption at domestic prices and world prices or long-run marginal costs, have been quite pervasive in emerging market economies and some developing nations.

Kosmo (1989) presents estimates of the level of energy subsidies for a sample of developing countries primarily for petroleum products and electricity in the early 1980s, and concludes that energy is subsidized heavier in energy exporting countries. Sterner (1989) presents a time series of domestic petroleum product prices relative to world prices for Latin American countries. Burgess (1990) evaluates potential CO_2 emissions reductions from efficient electricity pricing in a sample of OECD and non-OECD countries. However, none of these chapters presents estimates indicative of CO_2 reductions that could arise from removal of worldwide energy subsidies. Burniaux et al (1992) presents estimates from the OECD GREEN model of CO_2 reductions from worldwide energy subsidy removal, based on 1985 energy price data.

Estimates presented in this chapter are based on energy price data from 1991 in a sample of twenty-eight countries responsible for 85% of global carbon emissions, and reveal that total energy subsidies in non-OECD countries were on the order of US $270-330 billion in 1991 (presented in Table 1). The difference in the lower and upper bound is primarily due to differing assumptions regarding conversion of domestic prices to world prices in the former Soviet Union. Estimates of electricity subsidies are based on a sample of the 16 largest non-OECD electricity consumers (80-90% of non-OECD electricity consumption) by the Environment Department of the World Bank, and the authors. The estimates of fossil fuel subsidies for 1991 are on the same order of magnitude as presented in Burniaux et al (1994) using 1985 data, indicating no overall improvement between 1985 to 1991, although country-specific variations exist.

Table 1 Energy Subsidies in non-OECD Countries (1990-1991)

	BillionUS $
Coal	37 - 51
Petroleum products	105 - 125
Natural gas	48 - 69
Electricity	80 - 85
Total	270 - 330
Fossil fuels	
Former Soviet Union	145 - 175
Latin America	8 - 9
Middle East and North Africa	15 - 19
Asia	19 - 20
South Africa	1.5 - 2
Eastern Europe	12 - 15
Total	200 - 240

Sources: Larsen (1994); Larsen and Shah (1992).

In absolute dollar values, subsidies to petroleum products accounted for about 40% of total world subsidies, coal for about 15%, natural gas for about 20%, and electricity for about 25% of these subsidies. Among petroleum products, fuel oils receive the largest subsidies in dollar value. Kerosene is usually subsidized, whereas gasoline is taxed in most developing countries. A large number of developing countries priced electricity well below its long-run marginal cost of production. In some cases, when electricity is fossil-fuel generated, this represents implicit fossil fuel subsidies. Oil producers/exporters tend to subsidize petroleum products more often than importers, and coal and natural gas subsidies are almost exclusively found in coal and/or natural gas producing countries.

Energy Subsidies in OECD Countries

Estimated producer subsidies to coal, presented in Table 2, amounted to about US $11 billion in 1991, of which almost 70% were in Germany (DRI/McGraw-Hill, 1994a). These subsidies take the form of direct grants and indirect subsidies to high-cost producers that cannot compete with imported coal. Even with the subsidies, producer prices of coal are substantially higher than import prices, but coal consumers (primarily the electricity sector) are required to purchase domestically produced coal. Dale W. Jorgenson Associates (1994) presents US energy subsidies on the order of US $18 billion in 1992. These subsidies include US $1.7 billion in direct expenditures, US $2.6 billion in tax subsidies, US $4.9 billion to public power, US $2.5 billion for R&D, and US $3.0 billion in federal assumption of legal risk.

Table 2 Subsidies in OECD Countries (1991-1992)

	Billion US $
Coal (OECD)	11
Energy (United States)	18
Energy complements (United States)	50-90

Sources: DRI/McGraw-Hill (1994a,b); Dale W. Jorgenson Associates (1994); and MacKenzie et al (1992).

Although energy subsidies in OECD countries were substantially less than those in non-OECD countries, subsidies or failure to allocate user costs to energy complements can be quite substantial. A study by DRI/McGraw-Hill (1994b) estimated that road transport in the United States is subsidized by about US $50 billion annually, defined as the difference between costs of roadways and services less user fees. MacKenzie et al (1992) estimated road transport subsidies in the United States at about US $90 billion. If the value of free employee parking is included, the figure would be about US $175 billion. These amounts do not include social costs of congestion, estimated at US $100 billion. The difference between the two studies is primarily the type of services included.

Carbon Emissions Impacts of Energy Subsidy Removal in Non-OECD Countries

Estimates of CO_2 reductions from energy subsidy removal in non OECD countries are presented in Larsen (1994) and Larsen and Shah (1992b). The estimates in the two studies are quite similar. Larsen (1994) applies an energy cost function approach to the sixteen largest energy subsidizers, with the resultant equations:

(1)	$S_i\epsilon_{ij} = S_j\epsilon_{ji}$	for all i, j
(2)	$\Sigma_j \epsilon_{ij} = 0$	for all i
(3)	$\epsilon^*_{ij} = \epsilon_{ij} + \epsilon_{EE}S_j$	for all i, j

where $S_{i,j}$ are the fuel-cost shares in total energy cost, ϵ_{EE} is the own-price elasticity of aggregate energy, and ϵ_{ij} are the price elasticities of demand for fuel i and fuel price j. In order to estimate the impact of fossil fuel subsidy removal, values for the own-price elasticity of aggregate energy and the own-price elasticity for each fuel holding aggregate energy constant are assumed. Three fossil fuels (coal, gas, and petroleum products) are considered,[1] and (3) is thereby a three-by-three elasticity matrix of which the three own-price elasticities are assumed. That leaves six unknowns, which are determined by the system of six equations in six unknowns given by (1) and (2). Thus the off-diagonal entries in the elasticity matrix, i.e. the cross-price elasticities, are uniquely determined from the own-price elasticities. The own price elasticities used are in the range of -0.15 to -0.25 in the cases of high rates of subsidies (higher than 50 percent) and up to -0.6 when the rates are lower.

The energy demand function in the relevant region is approximated by a Cobb-Douglas function, i.e. the price elasticity of demand for fuel i, ϵ_{ij}, is constant in the relevant region. Changes in energy demand from subsidy removal are estimated by movement along the demand curve. The change in energy demanded from subsidy removal is thus

$$(4) \qquad \Delta q_i = -\sum_j q_i\left(1-\left(\frac{p_j}{p_j^w}\right)^{-\epsilon_{ij}}\right) \qquad \forall i$$

where q_i is domestic consumption of fossil fuel i, p_j is the domestic price of fuel j before subsidy removal, p_j^w is the world price of fuel j. The impact on carbon

[1] The implicit fossil fuel subsidies of electricity subsidies are not taken into account, which could likely understate the CO_2 reductions from energy subsidy removal. In the case electricity is predominantly non-fossil fuel generated, substitution effects could imply that the estimated CO_2 reductions presented are overstated.

emissions is calculated from (4) by multiplying the change in fuel i by the respective carbon emission coefficients.

The estimates are presented in Table 3. The reductions are quite substantial in many of the large carbon emitting countries. The removal of energy subsidies is estimated to reduce annual carbon emissions relative to the baseline by nearly 25-30% in the former Soviet Union. Even with this change, carbon emissions per dollar of GDP in the former Soviet Union would still be twice as high as in the United States and more than three times higher than in most Western European countries. Negative taxation of energy may have contributed to this relatively poor carbon efficiency in the former Soviet Union. Carbon reductions from subsidy removal in most other subsidizing countries are estimated to fall in the range of 10-20% -- primarily because subsidies are less pervasive than in the former Soviet Union.

Table 3 CO_2 Emissions Reductions relative to Baseline 1990-2010

	Percentage Reductions
Former Soviet Union	25-30
China	5-10
Poland	20-25
India	7- 9
South Africa	13-15
Former Czechoslovakia	16-24
Mexico	4- 7
Brazil	3- 8
South Korea	3- 5
Venezuela	15-22
Indonesia	7-12
Saudi Arabia	7-12
Egypt	11-16
Iran	22-26
Romania	20-24
Bulgaria	10-12

Sources: Larsen (1994); and Larsen and Shah (1992).
Note: The range in emissions reductions reflects a range in subsidies estimated.

Given emissions projections to the year 2010, estimated reductions relative to baseline projections in non-OECD countries collectively are in the order of 10-12%. This relatively low percentage reduction is primarily due to relatively low subsidies, and thus low emissions reductions in China, *and* the fact that emissions are expected to increase at a substantially higher rate in China than in the countries of the former Soviet Union -- so that China's share of non-OECD emissions will be significantly higher in year 2010. Nevertheless, the estimated reductions in non-

OECD countries are equivalent to about 20% of emissions in OECD countries, and it is estimated that the OECD countries would have to levy a carbon tax of US $60-70 per ton in order to achieve the equivalent tons of reductions (see Larsen and Shah, 1992).

The estimated CO_2 emissions reductions from subsidy removal in the non-OECD countries as a group are generally lower than presented in Burniaux et al (1994). The differences, however, can largely be explained by variations in the applied energy-demand elasticities. The relatively low elasticities applied to the countries with a high rate of energy subsidies in this chapter may be considered conservative. However, estimated carbon intensity in, for instance, the former Soviet Union (after subsidy removal) approach that of other middle-income countries with no or minimal energy subsidies and similar primary-energy composition.

Carbon Emissions Impacts of Subsidy Removal in OECD Countries

The CO_2 emissions impacts of energy subsidy removal in the OECD countries are much smaller. Estimated impacts on CO_2 emissions from the removal of coal-producer subsidies and coal-import restrictions are negligible (DRI/McGraw-Hill, 1994a). This is at first sight somewhat surprising, given that removals would lower the price of coal to consumers. However, the main domestic coal-consuming sectors are using coal by requirement at a level substantially higher than is indicated by the high prices paid.

The elimination of energy subsidies in the US is estimated to reduce CO_2 emissions by about 4% below the baseline by recycling the "revenues" through a change in the marginal labor tax and capital tax, while GDP would increase by 0.1-0.2% relative to baseline (Dale W. Jorgenson Associates, 1994). Full cost recovery of road transport in the US is estimated to reduce total CO_2 emissions in the US by 3-3.5% below the baseline, given that road transport contributes "only" about 30% of total emissions (DRI/McGraw-Hill, 1994b).

Based on projected baseline emissions[2] presented in Table 4, world emissions reductions from subsidy removal in non-OECD and OECD countries are

---

[2] Baseline OECD emissions are assumed constant, given that most OECD countries have committed to stabilize emissions at 1990 levels by year 2000, and some to further reduce emissions to 80% of 1990 levels by year 2005. Baseline emissions in the former Soviet Union and eastern Europe are based on Baron (1992), incorporating the impacts of restructuring but not subsidy removal (although subsidy may also induce structural changes). Baseline emissions in all other non-OECD countries are based on GDP projections less 0.5-1.0% annual autonomous energy efficiency improvements that may arise from other factors than energy prices.

estimated at 8-10% below baseline levels in year 2010. However, total world emissions would still be more than 40% higher than in 1990. Moreover, if the reductions in fossil fuel consumption, of which the largest share is in non-OECD countries, induce reductions in world prices of fossil fuels, a substantial share of the reductions in non-OECD countries could be offset by an increase in OECD emissions as a response to lower world prices (see Larsen and Shah, 1992).

Table 4 CO_2 Emissions Reductions in OECD- and non-OECD Countries

	Baseline year 2010 relative to 1990	Percent CO_2 reductions relative to Baseline	Emissions after reductions relative to 1990
World	1.55	8-10%	1.40-1.43
OECD	1.00	3-4%	0.96-0.97
Non-OECD	2.03	10-12%	1.79-1.83

Source: Based on estimates presented in this chapter.

Welfare Effects of Energy Subsidy Removal in Non-OECD Countries

Subsidy removal would have welfare effects, measured as changes in the sum of consumer and producer surpluses, in subsidizing countries as well as non-subsidizing countries. There are two components to the welfare effects, of which the first is improved efficiency of production and consumption decisions from subsidy removal. The second effect stems from a possible fall in world prices if subsidy removal induces a sufficiently large reduction in consumption. The first component is expected to be positive, while the second is negative for fossil-fuel exporters and positive for importers. Thus, accounting for the estimated fall in world fossil fuel prices, welfare effects would be positive for net importers, but may be negative for net fossil fuel exporters if losses in export revenues are larger than the gain stemming from more efficient production and consumption decisions. The following equations have been applied:

(5) $$W_p = \int_0^{q_1} D\delta q - \int_0^{x_1} S\delta x + p_w(x_1 - q_1)$$

where D is the inverse demand function, S is the inverse supply function, q_1 is domestic consumption at subsidized price p, x_1 is domestic production at price p, and p_w is world price; and

(6) $$W_{p_{w'}} = \int_0^{q_{w'}} D\delta q - \int_0^{x_{w'}} S\delta x + p_{w'}(x_{w'} - q_{w'})$$

where $q_{w'}$ is domestic consumption and $x_{w'}$ is production at new world prices $p_{w'}$. Change in welfare from subsidy removal is by linear approximation of D and S,

(7) $$\Delta W = W_{pw'} - W_p$$

Welfare from subsidy removal would increase by US $22 billion in former Soviet Union countries overall when accounting for world price effects, not including the value of improvements in the local and global environments.[3] The net fossil fuel importers Poland, India, Czechoslovakia and Brazil would see their welfare increase by more than US $2 billion. Of net exporters, welfare effects in Saudi Arabia would be a negative US $1.5 billion and in the range of US $150-350 million for Egypt, Venezuela, Indonesia and Mexico. Western Europe, the United States and Japan would see their welfare increase by US $15 billion because reduced fossil fuel prices imply lower import prices. Fossil fuel importing developing countries not in this analysis would for the same reason also experience welfare gains, although insignificant on a global scale. In total, including the subsidizing non-OECD countries and the OECD countries, welfare would increase by US $35 billion, or more than 10% of total subsidies.

NATIONAL CARBON TAXES

The growth in CO_2 emissions in non-OECD countries is projected to increase by 80% from year 1990 to 2010 -- even with the removal of energy subsidies. Combatting the greenhouse effect would thus require additional measures. One such potential measure is a national tax on carbon content of fossil fuels. The idea of imposing a national tax on carbon -- the first tax ever to be aimed specifically at global warming -- has found increasing favor of late, perhaps because governments would welcome a politically popular way of raising revenue. Such a tax would fall most heavily on the fossil fuel with the greatest carbon intensity per unit of energy. This would mean a relatively higher percentage price increase for coal than for alternative fuels (petroleum and natural gas), because coal not only has the highest carbon content per unit of energy, but it also has the lowest price. For example, a $50/ton carbon tax in Western Europe would on average increase end-user prices of coal by 35 percent, natural gas to households by 8 percent, and gasoline by 5

---

[3] The welfare gain is estimated at US $33 billion if the world price effect is negligible.

percent. This compares with price increases in the United States of 90 percent for coal, 13 percent for natural gas, and 12 percent for gasoline.

The carbon-cum-energy tax now being considered by the EU would start at $3/barrel of oil equivalency and increase by $1/barrel each year in real terms until it reaches $10/barrel (roughly equivalent to a carbon tax of $70/ton) in the year 2000. Since the tax is supposed to be "revenue neutral," governments would be expected to lower other taxes by a similar amount. But the chances of the EU adopting this tax anytime soon look poor, as the EU might well wait for others (such as the United States and Japan) to act, worrying about being at a "competitive disadvantage."

Revenue Potential of National Carbon Taxes

The revenue potential of carbon taxes is large. A moderate $10/ton carbon tax, if imposed individually by all nations, could raise $55 billion just in the first year. For some countries (China), such revenues could amount to about two percent of GDP. In countries in which 1987 per capita GDP was less than $900, such a tax would on average yield revenues of more than one percent of GDP and 5.7 percent of government revenues. For the OECD countries, the numbers are lower, but still significant -- 0.21 percent of GDP and one percent of government revenues. Moreover, carbon taxes would likely be easier to administer in developing countries than would personal and corporate taxes and thus would be less prone to tax avoidance and evasion.

Redistributive Implications of National Carbon Taxes

The literature on industrialized countries typically portrays carbon taxes and other economic policy instruments such as GHG emissions taxes, gasoline or btu taxes as regressive, because outlays on fossil fuel consumption as a proportion of current annual income falls with income. But recent studies (see Table 5) using US and European data show that carbon taxes are considerably less regressive relative to lifetime income or annual consumption expenditures than to annual income, at least for industrial countries (see Poterba (1991), Jorgenson, Slesnick and Wilcoxen (1992) and Smith (1992a,b)). Moreover, there is reason to believe that the same holds true for the rest of the world, although for quite a different reason: in developing countries, institutional factors also play an important role (see Larsen and Shah, 1992).

Arguments for progressivity, or at least low regressivity in the developing world, include the following:

- If there is a significant degree of foreign direct investment from countries in which investors are allowed foreign tax credits against domestic income tax liabilities, then a significant tax burden could be passed on to foreign treasuries, producers, and consumers.
- If price controls apply, producers often cannot pass on the tax to consumers in terms of higher prices.
- With binding import quotas or rationed foreign exchange, a tax would reduce the excess profits made by the privileged class.

Arguments for regressivity would be factors such as full market power. In this situation, the producer could increase product prices to fully pass on the carbon tax to consumers. As it turns out, most developing countries exhibit some combination of the above elements, creating a situation in which taxes can be only partially shifted to consumers. This means that a carbon tax would be progressive or much less regressive than most people believe. Further, it is likely to be regressive only for the lowest income groups, which could be protected through direct subsidies or alternative measures. In addition, the overall tax structure could be made even less regressive by using the carbon tax to reduce personal income taxes (the latter are not necessarily progressive because of high tax evasion and urban-rural migration).

Efficiency Impacts of National Carbon Taxes

Various estimates of the welfare effects of carbon taxes are presented in the literature. Most of these studies estimate the welfare effects in terms of GDP impacts. Earlier studies seldom modeled the various options of tax revenue recycling, but assumed lump-sum transfers. Very few studies included the impact on local pollution from a carbon tax, with the exception of Glomsrod et al (1992).

The largest progress in modeling the welfare effects of carbon taxes has been in terms of the recycling of the tax revenues. Shackleton et al (1992) present estimated impacts on GDP in the US for a group of recycling instruments. The models presented are the Jorgenson and Wilcoxen model (Jorgenson and Wilcoxen, 1992), the Goulder model (Goulder, 1991), a DRI model, and the LINK model. The recycling instruments are lump sum, fiscal deficit reduction, marginal labor income tax reduction, corporate income tax reduction, payroll tax reduction, and investment tax credit. The results, in terms of a negative or positive impact on GDP relative to the no-carbon-tax GDP baseline, are presented in Table 6. All four models report that the GDP impacts of a carbon tax recycled by means of lump sum, fiscal deficit reduction, and marginal labor income tax reduction would be negative. A revenue-neutral corporate income tax reduction would yield a positive GDP response according to three of the four models. In half of the models, a revenue-neutral investment tax credit would also have a positive impact. A revenue-neutral employer

Table 5 A Summary of Empirical Evidence on the Redistributive Impact of Economic Instruments

Instrument	Author	Country	Model	Results
Carbon Tax	Bull - Hassett - Metcalf (1993)	United States	computable dynamic general equilibrium model spending behavior may adjust.	tax burden is nearly proportional with respect to lifetime income
Carbon Tax	DeWitt - Dowlatabadi - Kopp (1991)	United States	partial equilibrium model; spending behavior may adjust; expenditure data; no recycling of tax revenues	distributional impact is regressive and varies across regions
Carbon Tax	Jorgenson-Slesnick - Wilcoxen (1992)	United States	computable general equilibrium model; three stages intertemporal optimization for household consumption	carbon tax is either mildly progressive or regressive depending on the welfare function used
Carbon Tax	Poterba (1991)	United States	partial equilibrium model; expenditure and income data;	carbon tax is regressive, but the impact is smaller if expenditure data are used
Carbon Tax	Schillo et al. (1992)	United States	DECO aggregate macroeconomic model; expenditure data; three compensation systems	depending on the compensation system adopted, the carbon tax is regressive to neutral
Carbon Tax	Schillo et al. (1992)	United States	Urban Institute's TRIM2 microsimulation model; two compensation systems	carbon tax is regressive w.r.t. pre-tax income in both scenarios, but it becomes regressive to neutral relative to post-tax income
Carbon Tax	Pearson (1992)	Europe	partial equilibrium model; spending behavior may/may not adjust (two models); Eurostat data	both models indicate that tax on domestic fuels is regressive, while tax on motor fuels is mildly progressive

Table 5 (Continued)

Carbon Tax	Pearson - Smith (1991)	Europe	IFS model of consumer expenditures; compensation system; spending behavior may adjust	Ireland and UK show a regressive impact. For other countries, the burden is weakly related to income.
Carbon Tax	Shah - Larsen (1992)	Pakistan	partial equilibrium model; income and expenditure data; three scenarios	carbon tax incidence is either proportional or progressive in a developing country context
BTU tax	Bull - Hassett - Metcalf (1993)	United States	computable dynamic general equilibrium model spending behavior may adjust.	tax burden is nearly proportional with respect to lifetime income
Gasoline Tax	Greening - Schipper - Jeng (1993)	United States	partial equilibrium model; expenditure data	gasoline tax affects negatively mainly older married couples with dependent children. Income distributional results not reported.
Gasoline Tax	Krupnick - Walls - Hood (1993)	United States	partial equilibrium econometric model; limited adjusting behavior	gasoline taxes are regressive, much more than previous studies since income data are used
Gasoline Tax	Poterba (1990)	United States	partial equilibrium model; expenditure data	gasoline tax is broadly regressive if we ignore the lowest income class. The latter class devotes a smaller share of their budget to gasoline than the lower-middle income class.

Table 5 (Concluded)

Tax on GHG emissions	International Energy Agency (1993)	Europe	partial equilibrium model; expenditure data	regressive effect on households if no compensatory measures with respect to domestic heating. Less clear result for motor fuels
Tax on direct fuel expenditure	Smith (1992a, 1992b)	Europe	partial equilibrium model; expenditure data; spending behavior may/ may not adjust; two compensatory systems	carbon tax is regressive, but if spending behavior adjusts its impact is smaller; only lump-sum transfers make the impact progressive
Tax on industrial energy use	Smith (1992)	United Kingdom	input-output tables + consumer spending simulation program	modest effect of changes in prices on consumers' spending, but negative especially for low-income households

payroll-tax reduction increases GDP, while an employee payroll tax reduces GDP in the two models that distinguish between the employer and employee portion. The strongest positive GDP response is from a revenue-neutral investment tax credit (or least negative in Goulder's model) except for the Jorgenson and Wilcoxen model.

Similar models are rather scarce for developing countries. But can carbon taxes contribute to economic efficiency by replacing other, more distortionary taxes in developing countries? Shah and Larsen (1992) examined four scenarios, using a sample of five countries (India, Indonesia, Japan, Pakistan, and the United States). The framework is partial equilibrium in the absence of CGE models comparable across the five countries.

Case A: Revenue neutral introduction of a $10/ton carbon tax that displaces a personal income tax. Welfare loss (efficiency costs) was measured in terms of additional income required to make the consumer indifferent to a package of policy change, not counting any benefits of reduced global warming or local pollutant reductions. The elasticity of energy demand used was -0.55 to -0.65, while the elasticity of labor supply was 0.1.

Table 6 The Efficiency Value of Carbon Tax Recycling
(GDP Impacts)

Carbon tax recycling instrument	Models			
	Jorgenson and Wilcoxen	Goulder	DRI	LINK
Lump sum	neg	neg	neg	neg
Fiscal deficit reduction	neg	neg	neg	neg
Marginal labor income tax reduction	neg	neg	neg	neg
Corporate income tax reduction	pos	neg	pos	pos
Payroll tax reduction:				
Employee portion		} neg	neg	neg
Employer portion		}	pos	pos
Investment tax credit	0	neg	pos	pos

Source: Shackleton et al (1992).

In this case, the estimated impact is an overall deterioration in welfare, ranging from a low of $1.5 cents per dollar of carbon tax revenues in Indonesia to a high of 17.5 cents in Pakistan (see Table 7). The estimates indicated that a country with a relatively low level of pre-existing energy taxes (Indonesia) would do much better than a country with a high one (Pakistan). Although energy taxes are even higher in Japan than in Pakistan, welfare loss is lower in Japan. This results from the high pre-tax carbon price in Japan, which implies that the percentage increase in energy prices will be low, and thus low carbon reductions, from the additional carbon tax. The loss for India compares well with that for the United States, even though pre-existing taxes in India are much lower, since India has a high consumption of coal and thus its carbon price is only half that of the United States.

Case B. Revenue neutral introduction of a $10/ton carbon tax that partially replaces a corporate income tax. Welfare loss or gain is estimated as under the previous case, with after-tax corporate income being return on savings. Thus, corporate income taxes are viewed as inducing intertemporal substitution in final consumption. The elasticity of savings used was 0.1 with respect to the after-tax rate of return. Under this scenario, with the major exception of the United States, the

Table 7 Costs and Benefits of Carbon Taxes for Selected Countries, 1987

	Pakistan	Indonesia	India	United States	Japan
Fossil fuel consumption (million US $)	3,345	5,330	17,266	246,502	126,100
Carbon (C) emissions (million tons)	13.2	26.6	148.2	1246.1	237.1
Price of carbon ($ per ton)	253	200	117	198	538
Energy Taxes (US$/ton C)	65.13	0.0	10.69	26.64	104.80
Carbon tax (US$/ton)	10	10	10	10	10
Elasticity of energy demand	-0.64	-0.6	-0.651	-0.6	-0.55
Price increase (from carbon tax):					
coal	37.8%	17.5%	26.2%	18.3%	8.7%
petroleum products	3.2%	5.8%	2.3%	3.4%	0.15%
natural gas	2.6%	4.4%	3.0%	4.3%	1.4%
Efficiency costs of a $10/ton carbon tax as percent of carbon tax revenues:					
Case A. Revenue neutral reductions in personal income taxes	-17.5	-1.5	-8.7	-8.4	-11.4
Case B. Revenue neutral reductions in corporate income taxes	+9.0	+8.7	+16.9	-6.2	+9.0
Case C. Lump sum redistribution of carbon taxes	-17.7	-1.5	-8.8	-10.2	-12.3
Case D. Lump sum redistribution of carbon taxes *and* accounting for energy subsidies in Indonesia and India		+0.4	0.0		
Benefit-cost ratio*					
High (SO_2+NOx+PM)	1.8	17.9	9.5	11.2	1.3
Medium (SO_2+NOx+PM)	1.6	12.9	7.5	8.7	1.0
Low (SO_2+NOx+PM)	0.5	2.2	1.9	2.1	0.2

* "High" is based on Glomsrod et al (1990); "Medium" is based on Bernow and Marron (1990); "Low" is based on EPA/Energy and Resource Consultants, Inc. referenced in Repetto (1990).
Source: Shah and Larsen (1992).

countries sampled would be better off -- lending support to a widely held view that corporate income taxes are far more distortionary than are labor income taxes. The estimated net welfare gain varies from a high of 16.9 percent of carbon tax revenues for India to a low of 8.7 percent for Indonesia. The US welfare loss -- about 6.2 percent of carbon tax revenues, or 0.017 percent of GDP -- may result from the country's relatively lower marginal taxation of corporate income. However, three of the four model estimates presented in Table 5 indicate that US GNP may well experience an increase if carbon taxes were introduced in a revenue-neutral fashion as a replacement for taxes on capital.

Case C: Introduction of a $10/ton carbon tax to raise additional revenues, with no change in existing taxes. Here, tax revenues are redistributed lump sum. Here, again, the estimated impact is a deterioration in welfare. But the losses -- ranging from a low of 1.5 cents per dollar for Indonesia (0.005 percent of GDP) -- to a high of 17.7 cents per dollar for Pakistan (0.087 percent of GDP) -- although significant, represent only a small fraction of carbon tax revenues. The welfare losses for India, Indonesia, and Pakistan are only slightly higher than those obtained in Case A, because income tax evasion in these countries is pervasive, meaning that changes in the income tax would have little bearing on the labor supply. The difference in the two cases is larger for the United States and Japan because of the higher pre-existing labor income taxes and higher tax compliance.

Case D: Introduction of a $10/ton carbon tax to raise additional revenues, with no tax change in existing taxes but accounting for fossil fuels subsidies. In this scenario, the outcome depends on the level of existing subsidies, underscoring the point that efficiency costs of carbon taxes are overstated if subsidies are ignored. By incorporating existing energy subsidies into the analysis, the tax would not hurt aggregate welfare in India and Indonesia, where fossil fuel price subsidies were high.

In sum, these four scenarios show that a case for carbon taxes on efficiency grounds alone rests upon introducing them in a revenue-neutral manner, either as a replacement for corporate income taxes or when fossil fuel subsidies exist. Such taxes do not fare so well as a substitute for personal income taxes, at least in countries with pre-existing energy taxes and no subsidies.

Local environmental benefits

Economic benefits aside, however, can carbon taxes be justified on local environmental grounds, particularly through the reduced emissions of pollutants such as nitrogen oxides, particulates, and sulphur dioxides? Shah and Larsen (1992) compared the efficiency costs of carbon taxes to the potential health benefits of local pollutant reduction in the same five countries. The benefits per ton of pollution reduction applied were estimates from three studies, adjusted by purchasing power parity indexes to afford a rough measure of comparability across countries.

The calculations suggest that Indonesia, India and the United States could benefit substantially from a carbon tax on local environmental grounds alone, as benefits exceed costs under all three health-benefit estimates (see Table 7). For Pakistan and Japan, however -- in view of high pre-existing energy taxes and thus high welfare costs of carbon taxes -- the benefit-cost ratios are significantly lower, but one or larger than one for two of the three benefit scenarios.

The efficiency costs applied are the estimates from Case C, i.e., lump sum redistribution of the carbon tax revenues. Thus the benefit-cost ratios would likely be higher if the carbon tax is recycled (as, for instance, in Case B).

SUMMARY AND POLICY IMPLICATIONS

The chapter concludes that (1) the removal of the energy subsidy in OECD countries on the order of US \$30 billion (primarily in the United States and Germany) and subsidies to complements on the order of US \$50-90 (United States) is likely to have little impact on global CO_2 emissions; (2) energy subsidy removal of US \$270-330 billion in non-OECD countries can substantially curb the growth of CO_2 in some large non-OECD countries, equivalent to the impact of a carbon tax on the order of US \$60-70 in the OECD countries; (3) the growth in CO_2 emissions in non-OECD countries is projected to increase by 80% from year 1990 to 2010 even with the removal of energy subsidies; (4) a revenue-neutral national carbon tax, in addition to energy subsidy removal, can improve efficiency *and* yield significant health benefits from local pollution reduction; and (5) carbon taxes are considerably less regressive relative to lifetime income or annual consumption expenditures than to annual income.

Faced with the uncertainty of global warming and the economic costs that concrete actions to stem global change could impose on this and future generations, policymakers must first find tools that can be justified purely on economic and local environmental considerations. For any nation with energy subsidies, the first priority must be the elimination of these subsidies. This would slow the growth of CO_2 emissions, improve the allocation of economic resources, reduce local environmental pollution and release scarce public funds for development projects.

For countries that have eliminated energy subsidies but have little or no taxation of energy, or for countries that cannot overcome political obstacles to removing the subsidies, small carbon taxes would be in order. Such taxes would discourage fossil fuel consumption and reduce local pollution, while raising large amounts of revenues at, likely, lower administrative costs than the prevalent taxes entail. They can also impose lower economic costs for the society as a whole than do corporate income taxes, without unduly hurting the poor.

However, although there are compelling arguments for undertaking these policies, their implementation will not be sufficient to, for instance, stabilize global carbon emissions. In order to achieve substantial global reductions relative to the "business as usual" projections over the longer term, higher national carbon taxes than considered in this chapter and/or an international agreement beyond the participation of OECD countries is viewed as necessary.

References

Bernow, S. and Marron, D. (1990). "Valuation of Environmental Externalities for Energy Planning and Operations," *1990 Update*. Cambridge, MA: Tellus Institute.

Burgess, J.C. (1990). "The Contribution of Efficient Energy Pricing to Reducing Carbon Dioxide Emissions." *Energy Policy* 18/5, 449-455.

Burniaux, J-M., J. Martin and J. Oliveira-Martins. (1994) "The Effects of Existing Distortions in Energy Markets on the Cost of Policies to Reduce CO_2 Emissions: Evidence from GREEN." *Economic Studies*. Paris: OECD.

DRI/McGraw-Hill. (1994a). "The Energy, Environment and Economic Effects of Phasing-out Coal Subsidies in OECD Countries." *Draft*. Prepared for OECD Environment Directorate. Paris.

DRI/McGraw-Hill. (1994b). "Transportation Sector Subsidies -- US Case Study." *Draft*. Prepared for OECD Environment Directorate. Paris.

Glomsrod, S., H. Vennemo and T. Johnsen. (1992). "Stabilization of Emissions of CO_2: A Computable General Equilibrium Assessment." *Scandinavian Journal of Economics* 94/1, 53-69.

Goulder, L. H. (1991). "Effects of Carbon Taxes in an Economy with Prior Tax Distortions: An Intertemporal General Equilibrium Analysis for the US." *Draft*. Stanford University and NBER.

Jorgenson, D.W. Associates. (1994). "The Elimination of Federal Energy Subsidies: Environmental Gains, Tax Considerations and Economic Consequences." *Draft*. Prepared for OECD Environment Directorate. Paris.

Jorgenson, D., D. Slesnick and P. Wilcoxen. (1992). "Carbon Taxes and Economic Welfare." *Brookings Papers on Economic Activity*, 393-431.

Jorgenson, D. and P. Wilcoxen. (1994). "The Economic Effects of a Carbon Tax." In A. Amano (ed.), *Climate Change: Policy Instruments and Their Implications*. Tokyo: IPCC.

Kosmo, M. (1989). "Commercial Energy Subsidies in Developing Countries." *Energy Policy* (June), 244-253.

Larsen, B. (1994). "World Fossil Fuel Subsidies and Global Carbon Emissions in a Model with Interfuel Substitution." Policy Research Department *Working Paper* 1256. Washington, DC: World Bank.

Larsen, B. and A. Shah. (1992a). "Combatting the Greenhouse Effect." *Finance and Development* (December), 20-23.

Larsen, B. and A. Shah. (1992b). "World Fossil Fuel Subsidies and Global Carbon Emissions." Policy Research Department *Working Paper* 1002. Washington, DC: World Bank.

MacKenzie, J., R. Dower and D. Chen. (1992). "The Going Rate: What it Really Costs to Drive." Washington DC: World Resources Institute.

Poterba, J.M. (1991). "Designing a Carbon Tax." In R. Dornbusch and J. Poterba (eds), *Economic Policy Responses to Global Warming*. Cambridge, MA: MIT Press.

Repetto, R. (1990). "Environmental Productivity and Why It Is So Important." *Challenge* (September-October).

Schillo, B., L. Giannarelli, D. Kelly, S. Swanson and P. Wilcoxen. (1992). "The Distributional Impacts of a Carbon Tax." *Draft*. US EPA.

Shackleton, R. et al. (1992). "The Efficiency Value of Carbon Tax Revenues." Washington, DC: US EPA.

Shah, A. and B. Larsen. (1992). "Carbon Taxes, The Greenhouse Effect and Developing Countries." Policy Research Department *Working Paper* 957. Washington, DC: World Bank.

Smith, S. (1992a). "The Distributional Consequences of Taxes on Energy and the Carbon Content of Fuels." In *European Economy: The Economics of Limited CO_2 Emission*. Commission of the European Communities.

Smith, S. (1992b). "Distributional Effects of a European Carbon Tax". *Nota di Lavoro 22.92*. Milan: Fondazione ENI Erico Mattei.

Sterner, T. (1989). "Oil Products in Latin America: The Politics of Energy Pricing." *The Energy Journal* 10/2, 25-45.

PART III

ASSESSMENT PROBLEMS

Chapter 6

HOW A FEE PER-UNIT GARBAGE AFFECTS AGGREGATE RECYCLING IN A MODEL WITH HETEROGENEOUS HOUSEHOLDS

Thomas C. Kinnaman (Bucknell University)

and

Don Fullerton (University of Texas)*

INTRODUCTION

Nearly 2000 communities in the US have implemented user fees to finance garbage collection over the last five years. These user fees require households to pay for each bag of garbage presented at the curb for collection (Skumatz 1993). The revenue raised from these user fees has supplanted the use of general tax revenue to finance garbage collection and disposal costs. Benefits to the community include the social value of less garbage and more recycling. The costs include the social cost of additional litter and the value of resources used to administer the program. The magnitudes of both the benefits and costs depend on the waste removal choices of individual households within the community.

This chapter develops a model of household choice among waste removal options for various types of waste material. Each household in the model compares the cost of traditional curbside garbage collection, the cost of recycling, and the cost of littering for each type of waste material. We assume the marginal cost of traditional garbage collection is the same for all materials, and is zero in the absence of a user fee. The marginal cost of recycling is assumed to differ across materials. Households pay a positive price to recycle some materials,

* We are grateful for helpful suggestions from Lans Bovenberg, Hilary Sigman, and other Congress participants. We are also grateful for financial support from the National Science Foundation Grant SBR-9413334. This paper is part of NBER's research program in Public Economics. Any opinions expressed are those of the authors and do not necessarily represent those of the National Science Foundation or the National Bureau of Economic Research.

several others can be recycled for free if the community offers curbside collection, while still others can be recycled at a negative price if the material has market value. Finally, we assume the marginal cost of littering is zero; however, a fixed cost must be paid if the household chooses to engage in any littering.

A user fee for garbage collection is shown to affect the choice of each household over whether to discard, recycle, or litter each type of material. The user fee's effect on the disposal choices of the household is shown to depend on the household's degree of environmental awareness, on income, and on the availability of curbside recycling. In particular, our model explains (1) why some households participate in a curbside recycling program even in the absence of a user fee, (2) why other households still do not participate in recycling programs even in the presence of a user fee for garbage collection, and (3) why some households choose to litter when others do not.

The effect of a user fee on the disposal decisions of each household determines the communities aggregate changes of garbage, recycling, and litter quantities that would result from the implementation of a user-fee program. These aggregate changes comprise important costs and benefits to a community attributable to the implementation of a user-fee program. The model suggests that aggregate garbage will decrease with the value of the user fee. Aggregate litter increases with the user fee, but, perhaps surprisingly, aggregate recycling may decrease. The magnitude of each of these changes can vary across communities.

The consequence for empirical work, especially the use of a cross-section of communities with different user fees to estimate the demand for garbage collection, is that each local government's choice of user fee is not exogenous. The choice may, in fact, depend on the behavioral responses of households within the community. For example, a local government may be more apt to implement a user fee if the changes in garbage and recycling are expected to be large and the change in illegal dumping is expected to be small. Most empirical studies estimating the effect of a user fee on aggregate garbage quantities have treated the value of the user fee as exogenous.[1]

The next section of this chapter will introduce the model of household choice for methods of waste removal without the option to litter. The model is then expanded to include the littering option. A brief conclusion follows.

---

[1] For example Repetto et al. (1992) and Jenkins (1991).

A MODEL OF HOUSEHOLD DISPOSAL CHOICE

Most models of household solid waste ignore the littering option.[2] In an important study for the World Resources Institute, Repetto et al (1992) claim that any potential increases in dumping can be alleviated by simple measures.[3] To characterize these existing studies, our initial model in this section ignores litter. Our analysis will suggest that a model without the littering option is inadequate to explain how local governments set user fees.

Individual Behavior

Assume that a community is comprised of several households. Each household may engage in consumption of different goods, each comprised of a single type of material, indexed by j = 1,...,J. Let c_j denote the consumption of good j by the household. Consumption generates waste material that must be disposed of. The quantity of waste material j (m_j) is assumed to be proportional to the consumption of good j:

$$m_j = a_j c_j. \tag{1}$$

All waste material must be either collected as garbage (with amount g_j) or recycled (with amount r_j). We add litter to the model in the next section. Since all matter must be removed from the household in one form or another, we have:[4]

$$m_j = g_j + r_j. \tag{2}$$

The household's total garbage and recycling quantities are $g = \Sigma_j g_j$ and $r = \Sigma_j r_j$, respectively.

---

2 For example, Jenkins (1991), Copeland (1991), Sigman (1991), Dinan (1993), Morris and Holthausen (1994), Wertz (1976).

3 They suggest locking dumpsters, vigorously publicizing and enforcing disposal rules in the initial months of the program, reporting households that put out no garbage, and requiring a one-bag minimum.

4 Households could conceivably compost garbage or reduce the quantity of their garbage by demanding less packaging at stores. Since these amounts are relatively small, however, we concentrate here on the major components of the waste stream: garbage and recycling. Later we introduce burning or dumping b_j.

Assume that the household's utility function is defined over its J consumption goods, and the total amount of garbage it contributes to the waste stream (g):

$$u = u(c_1, c_2, \ldots, c_J, g) \tag{3}$$

Using a subscript to denote a first derivative, we assume that households receive positive marginal utility from consumption ($u_{c_j} > 0$), but they dislike their own contribution of garbage to the waste stream ($u_g < 0$). Households may differ with respect to their distaste for contributing garbage: a household with a large value of u_g (in absolute value) dislikes contributing garbage more than a household with a low value.

Each household is endowed with k units of time (or other resources) that can be supplied to a labor market to earn a wage p^k (which can also vary across households), or can be used to separate and store recyclable material. Let k^c denote the amount of household time supplied to the labor market, and let k^r denote the time devoted by the household to the separation of recyclable materials. Therefore, we have:

$$k = k^c + k^r. \tag{4}$$

The amount of time devoted to separating and storing recyclable materials is assumed to be proportional to the amount of recycling conducted by the household:

$$k^r = \sigma r. \tag{5}$$

We assume that σ is the same for all households and all materials (but the value of that time varies with the wage rate, p^k).[5]

With no city program to collect recyclable material, a household that wants to recycle must pay a recycling firm directly for the transfer of the material. Denote this payment by the household as p^r_j per unit of material of type j that is recycled. We assume that p^r_j is particular to the type of material being recycled and can take on a positive or negative value. For example, the cost to the household of paying a firm to recycle toxic waste could be very high, while the household may be able to *sell* precious metals to a recycling firm (p^r_j could be

---

5 We might expect the effort expended to separate and store recyclable material would differ over households and material types, but making this more realistic assumption only serves to cloud the model and adds little to the main conclusions.

negative). For used items sold on the second-hand market, we interpret p^r_j to include advertising costs.

Some materials may be collected from the curb as part of a curbside recycling program. The market price paid by households to recycle these materials is zero. A curbside recycling program represents an implicit subsidy to the household of s^r_j per unit of recycled material, which offsets p^r_j such that the overall cost $p^r_j - s^r_j = 0$. The value of s^r_j is zero for all materials not collected at the curb. The value of s^r_j will be negative for all collected materials that have positive market value. Therefore, s^r_j is a tax on households if the government mandates that households produce these materials at the curb.[6] Aluminum could be an example. A household that is not required to recycle a given material compares the cost of free curbside recycling pickup with the cost of sending used materials directly to secondary markets (bypassing curbside pickup). Thus our model is general enough to encompass second-hand sales.

The local government could also levy a user fee t^g per unit of garbage that is the same for every type of material. This fee is the household's only out-of-pocket marginal cost of curbside garbage disposal. The household may pay property taxes or a monthly fee, but, in the absence of a user fee for garbage collection, the price paid at the margin per bag of garbage is zero.

Household income ($p^k k - p^k k^r$) can be used to purchase consumption goods at market prices p^c_j, to recycle material j at unit cost p^r_j, and to pay any user fee that might be levied on garbage. Therefore, the budget constraint facing the household is:

$$\Sigma_j p^c_j c_j + t^g g + \Sigma_j (p^r_j - s^r_j) r_j = p^k k - p^k(\sigma r) \tag{6}$$

where $s^r_j = p^r_j$ if the recyclable material j is collected for free, and s^r_j is zero for all other materials not collected for recycling.

Each household chooses consumption, and how to dispose of the resulting material, by maximizing utility (3) subject to (1), (2), and the budget constraint (6). The relevant first-order conditions from this maximization are:

$$dL/dg_j:\ u_{c_j} \leq \lambda[p^c_j + a_j t^g] - a_j u_g \qquad \text{for all j} \tag{7a}$$

$$dL/dr_j:\ u_{c_j} \leq \lambda[p^c_j + a_j(p^r_j - s^r_j) + a_j \sigma p^k] \qquad \text{for all j} \tag{7b}$$

---

[6] The model does not include such a law, so households are not expected to recycle valuable materials at the curb.

where λ is the marginal utility of income to household i. Define λ^* and u_g^* to be the values of λ and u_g at the solution to (7). The household's choice between waste removal options for each material is governed by:

$$(8a) \qquad m_j = g_j \qquad \text{iff} \qquad t^g - u_g^*/\lambda^* < (p^r_j - s^r_j) + \sigma p^k$$

$$(8b) \qquad m_j = r_j \qquad \text{iff} \qquad t^g - u_g^*/\lambda^* > (p^r_j - s^r_j) + \sigma p^k$$

The household compares the marginal cost of recycling material j (p^r_j - s^r_j + σp^k) to the overall marginal "cost" of contributing garbage of material j -- which is the money payment (t_g) plus the value of marginal utility lost from contributing a unit of garbage ($-u_g^*/\lambda^*$). These conditions could imply a set of corner solutions. If the marginal cost of recycling a newspaper is less than the overall marginal cost of discarding it, then a household would be expected to recycle *all* of its newspaper, not just some portion of it.[7]

The household could choose an interior solution for a material j in which it discards some portion (g_j) and recycles some portion (r_j) of the material, if t_g - $u_g^*/\lambda^* = (p^r_j - s^r_j) + \sigma p^k$. This equality will hold only for very few materials at the knife's edge, where the cost of garbage collection exactly matches the marginal cost of recycling. However, the household will never choose interior solutions for two materials with different marginal costs. To see this, first remember that garbage and recycling are perfect substitutes in disposal ($m_j = g_j + r_j$) and that different materials provide equal disutility if discarded ($g = \Sigma_j g_j$). Now suppose the household *were* using both garbage and recycling for two kinds of waste materials with different marginal costs of recycling. This situation cannot be an optimal solution because the household could switch one unit of the higher-recycling-cost waste from recycling to garbage, and switch one unit of the lower-recycling-cost waste from garbage to recycling, while consuming the same amount of each material, generating the same total garbage and recycling, and saving the difference in marginal cost. Such switches would continue until the household is discarding all materials that meet condition (8a) and recycling all that meet (8b).[8]

---

[7] This strong theoretical result is supported by casual observation. Households who recycle often devote space for virtually all quantities of certain materials, e.g. one bin for newspaper and another for aluminum cans.

[8] The same maximization provides first-order conditions for consumption, not shown, in which the "effective" price of consuming good j depends on its purchase price, the relative waste generated ($m_j = a_j c_j$), and the cost of the chosen disposal method for that good. We thus capture the possibility that a garbage fee can discourage consumption of waste-intensive commodities.

With a few simplifying assumptions, we can use Figure 1 to illustrate the disposal choices of the household. First, assume that the income effect of an increase in the user fee is zero. In other words, the reduction of income attributable to an increase in the user fee does not change total consumption. Second, we assume that all types of materials can be ranked along the horizontal axis in a descending order according to the cost of recycling (p^r_j). For example, toxic waste would be ranked first and precious metals last. Third, we assume the household must dispose of one unit of each type of waste material on the spectrum. Given these assumptions, the overall schedule of recycling costs can be represented by P^r in Figure 1. Material types are ranked from highest cost to lowest cost from left to right. Discrete materials (j = 1,...,J) would generate a step function, but we show a continuous P^r schedule for ease of exposition. This schedule does not have to be linear or even differentiable, it only has to be non-increasing.

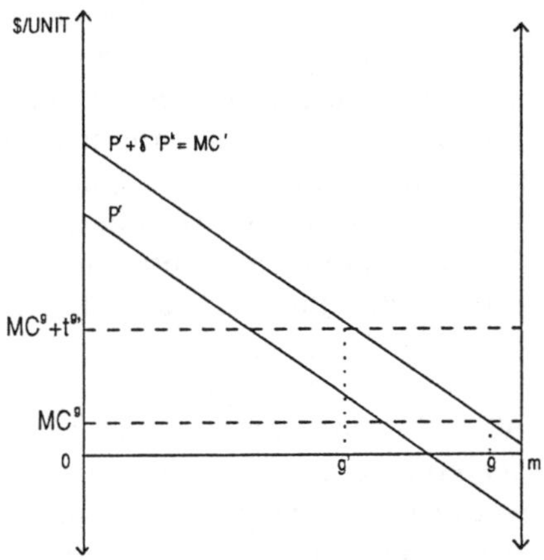

Figure 1 Household disposal choice

The first-order conditions indicate the marginal cost to the household of recycling is equal not only to the market transfer price (P^r), but to the value of resources devoted to separating and storing the material (σp^k) as well. The addition of this cost introduces another curve in Figure 1 higher than P^r but parallel, labeled MC^r. The MC^r curve indicates the overall marginal cost to the household of recycling each material type across the spectrum. The overall marginal cost of placing garbage at the curb in the absence of a user fee is the value of the marginal utility lost by household, $-u_g^*/\lambda^*$, which we assume is constant across materials. This cost is represented by the horizontal line labeled MC^g in Figure 1. The height of MC^g is unique to each household because each household has a different value of u_g^*/λ^*. These marginal cost schedules allow us to illustrate the amounts of garbage and recycling chosen by the household.

According to (8a) and (8b), the household can be expected to remove material j using the lowest cost alternative. This choice is illustrated by finding the lower envelope of marginal waste removal costs across all materials. In the absence of a user fee for garbage collection, the quantity of garbage is determined in Figure 1 by the horizontal distance between the origin and point g (where the cost to discard garbage is less than the cost to recycle). The quantity of recycling is indicated by the horizontal distance between point m and point g. The household transfers these materials directly to secondary markets in the absence of a curbside recycling program. Notice that all materials along the spectrum are either recycled or discarded as garbage. The amounts of garbage and recycling by each household will depend on where that household's unique MC^g curve intersects the MC^r curve. Households with a high opportunity cost of recycling (p^k) and those having a low disutility of throwing out garbage ($-u_g^*/\lambda^*$) will recycle less and throw out more in the garbage.

Figure 1 can also be used to indicate the amount of garbage and recycling produced by a household when a user fee for garbage collection is implemented. A user fee of t^{g}' per bag increases the overall marginal cost to the household of contributing garbage to $MC^g + t^{g}$'. The user fee causes the amount of garbage presented by the household to decrease from g to g' and the amount of recycling to increase from (m-g) to (m-g'). The magnitudes of these changes depend on the slope of the MC^r curve through point g. The flatter MC^r is, the cheaper it is to recycle additional materials, and the greater the change in recycling.

The implementation of a curbside recycling program also affects the waste disposal choices of households. Assume that the local government has agreed to collect, transport, and recycle certain materials at no cost to the household (an implicit subsidy or tax of $s^r_j = p^r_j$). Therefore, the marginal cost of recycling materials that are collected from the curb is only the value of extra time (σp^k) devoted to separating and storing the material. In the absence of a user fee for garbage collection ($t^g = 0$), (8a) and (8b) suggest only households that have a strong dislike for contributing garbage ($-u_g^*/\lambda^* > \sigma p^k$) will participate in the curbside recycling program. All other households will not participate. Of course, all households might still recycle the more valuable materials directly to secondary markets if they receive a price such that the net cost MC^r is less than the cost of any alternative (either MC^g for garbage, or 0 for curbside recycling).

Figure 2 illustrates the disposal choices of households when a curbside recycling program has been implemented in the community. Again, let MC^r represent the schedule of marginal recycling costs and let MC^g represent the value of disutility from contributing garbage. Assume that all materials ranked to the right of point a are collected by the government at no cost to the household.[9] Households must still devote time to separate and store these materials (σp^k). Since this cost is assumed to be constant across all materials, the MC^r curve is flat over the range of materials that are collected for recycling by the government.

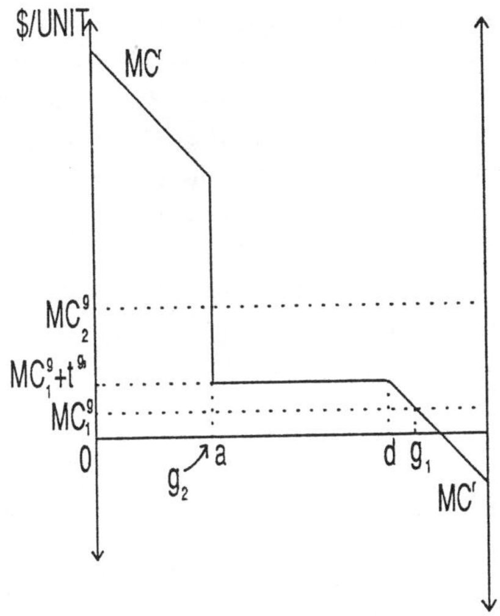

Figure 2 A curbside recycling program

The garbage cost function of two households in the same community, but with different preferences about contributing garbage, are labeled MC^g_1 and MC^g_2 in Figure 2. The household with MC^g_2 has strong distaste for contributing garbage ($MC^g_2 > \sigma p^k$) and will participate in the curbside recycling program in the absence of a user fee for garbage collection. The marginal cost of recycling the materials ranked between a and d is less than the overall marginal cost of throwing the material away. To see this, look again for the lower envelope of marginal costs in Figure 2. Household 2 discards all materials ranked to the left of g_2 (point a), recycles all materials ranked between g_2 and d at the curb, and sells all materials ranked to the right of d in secondary markets (where $p^r_j < 0$ so they get paid for those materials).

The household facing MC^g_1 has little distaste for contributing garbage ($MC^g_1 < \sigma p^k$) and will not recycle at the curb. This household discards all materials ranked to the left of g_1 (where MC^g_1 is less than MC_r) and sells material

[9] The government is not willing to collect materials ranked to the left of point a, perhaps because it costs too much to recycle these materials. We assume that the government's decision over which materials to collect is exogenous to the model.

ranked to the right of g_1 directly to secondary markets. In order to encourage these households to recycle at the curb, the government must implement a user fee.

If $t^{g'} > \sigma p^k - MC^g_1$ is levied on each bag of garbage collected, the household with MC^g_2 will not change its disposal behavior. It will still recycle materials ranked between points a and d at the curb. However, the household represented by MC^g_1 is given the incentive to participate in the curbside recycling program. It discards all materials ranked to the left of point a, presents material ranked between a and d at the curb for recycling (since $MC^r_1 < MC^g_1 + t^{g'}$ for those materials), and sells all material ranked to the right of d in secondary markets. As indicated by equation (8b), t^g must be greater than $\sigma p^k - MC^g_1$ to induce these households to choose to recycle these materials ranked between a and d at the curb. Larger values of the user fee would be needed to encourage households with a lower distaste for contributing garbage to recycle at the curb.[10]

Aggregate Behavior

One point of the theory is to explain how a user fee for garbage collection affects the aggregate amounts of garbage and recycling produced by a community. These aggregate amounts are obtained by summing across the garbage and recycling quantities produced by each household in the community, for any possible value of a user fee. These aggregate quantities generated by two different communities for various values of a user fee are illustrated in Figure 3.

Assume, for comparison, that both communities have the same amount of aggregate material to discard and neither has implemented a curbside recycling program. The total amount of recycling in community 1 is measured by the horizontal distance between point M (the right vertical axis) and the curve labeled D^R_1. Aggregate garbage is the rest of M, and is measured by the horizontal distance between the left vertical axis and D^R_1. In the absence of a user fee for garbage collection, households in community 1 discard G_1 in a landfill and recycle $M-G_1$. A user fee of $t^{g'}$ will encourage this community to reduce its garbage to G_1' and increase its recycling to $(M-G_1')$. These changes arise because

[10] A mandatory recycling law provides incentives similar to those of a user fee. For those materials that households are required to recycle, the expected cost of discarding increases as the expected fine for noncompliance increases or as the probability of getting caught increases. Households with very low values of MC^g would require a high penalty or greater enforcement to convince them to recycle.

households in the community choose to recycle more materials once a user fee is implemented (as in Figure 1).

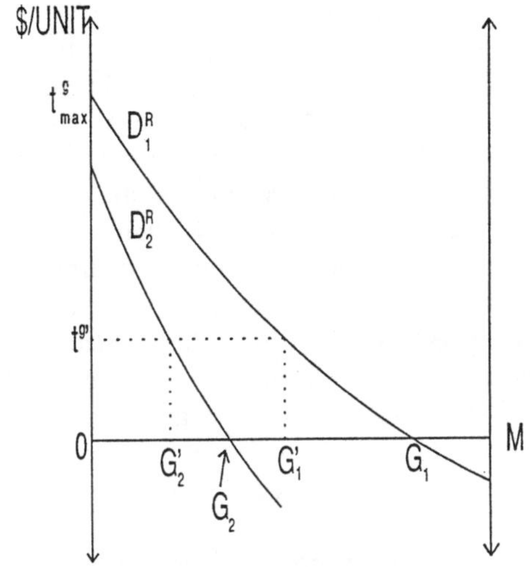

Figure 3 Aggregate behaviour

The D^R_2 function determines the aggregate quantity of garbage and recycling produced by community 2. This community recycles a greater quantity of material than the first community for all possible values of a user fee. For example, if a user fee with value $t^{g'}$ is levied, community 2 discards only G_2' and recycles ($M-G_2'$). Community 2 could be comprised of more households that have strong distaste for contributing garbage or have a low opportunity cost of time.

Empirical observation indicates that some local governments have chosen to implement user fees for garbage collection while most others have not. We assume that a local government will implement a user fee if the benefits of doing so outweigh the costs. Important benefits could be the reduction in garbage and increase in recycling that result from the user fee. Since households in this model engage in no dumping, the only cost is the value of resources used to administer the user-fee program.

As drawn in Figure 3, the change in garbage and recycling quantities resulting from the implementation of a user fee is greater for community one than for community two. The town of those with a strong distaste for contributing garbage (community two) does *not* necessarily experience a large *increase* in aggregate recycling due to a user fee. Households in community two already recycle in large quantities before the implementation of a user fee, leaving little room for additional recycling. Therefore the benefits of implementing a user fee may be greater for community one than for community two. *Ceteris paribus*, we expect communities with flatter D^R schedules (like D^R_1) to be more likely to implement a user fee, since the benefits are greater.

Why is community one more conducive to the implementation of a user fee for garbage collection than is community two? Our model suggests that community one must be comprised of many households that increase their recycling amounts following the implementation of a user fee. These large increases would result if the MC^r schedule in Figure 1 is relatively flat near the horizontal axis, and if many households in the community do not participate in the curbside recycling program before the user fee. A flat MC^r schedule indicates that more materials can be recycled at a low additional cost to the household. In the presence of a curbside recycling program, for example, the marginal cost of recycling is *flat* across those materials collected at the curb. This serves to flatten the aggregate recycling schedule. Therefore, communities with curbside recycling may be more likely to implement a user fee.[11]

Suppose the value of the user fee does not affect administrative costs such as advertising the program, printing the stickers, and enforcing the law. Since the benefits of implementing the user-fee program increase with the value of the user fee, and the costs do not, this simple model without a litter option suggests that governments would be likely to levy a very high user fee.[12] For example, the government in community one could levy a user fee with a value of t^g_{max} and then collect no garbage. All materials would be recycled. The model suggests that all governments would either (1) not implement a user fee, if the administrative costs are greater than the benefit of having no garbage, or (2) charge at least t^g_{max} for each bag of garbage, if the benefits of no garbage exceed the administrative cost of the user-fee program.

Casual observations in the US reject this theoretical implication. Only 2000 communities across the US have implemented user fees for garbage collection. Most of these fees are moderate, like $0.80 per bag. One explanation for the failure of this model to explain government behavior could be that governments *are* concerned with the amount of illegal dumping that would arise following the implementation of a user fee. A more appropriate model would

---

[11] Of a sample of 200 communities with user fees (sample collected by the authors), all have curbside recycling programs.

[12] A user fee may change the relative price of a consumption good that is more waste intensive than others. Therefore, another cost or benefit of implementing a user fee is this effect on consumption.

include illegal dumping as a third removal option for households. We now proceed to develop such a model.

THE MODEL WITH A LITTERING OPTION

Several recent models of household solid waste behavior have considered the littering option. Kennedy and Laplante (1994) solve for the optimal user fee given the option to litter, but do not consider materials with different recycling costs. Fullerton and Kinnaman (forthcoming) argue that the possibility of illegal dumping implies that the optimal user fee for garbage collection could be zero. Sullivan (1987) and Dobbs (1991) also allow for litter in their models. We improve on these models by allowing heterogeneous households to choose among three removal methods for diverse types of goods.

Individual Behavior

Assume that households have the option to burn, dump, or litter garbage:

$$m_j = g_j + r_j + b_j \qquad (9)$$

where b_j is the amount of burning of material j done by the household. Also, assume that all households dislike the act of burning or littering garbage:

$$u = u(c_1,\ldots,c_J,\ g,\ b) \qquad (10)$$

where $b = \Sigma_j b_j$, $u_b<0$, and all other variables are defined as above.

No market or price exists for the littering or burning of garbage, but we assume that the household must pay a fixed cost f if it engages in such practice. This fixed cost could include, for example, the cost of finding a suitable dump site, the fixed portion of the cost of traveling to the dumpsite, the psychic cost of breaking a local ordinance, and the risk of a fine. In other words, dumping two bags is not twice as costly as dumping one bag, in terms of transport cost, the risk of getting caught, and feeling bad about this antisocial behavior. Later we address the possibility that fixed costs arise in recycling.

The new budget constraint facing the household is:

$$\Sigma_j p^c_j c_j + t^g g + \Sigma_j(p^r_j - s^r_j) r_j = p^k(k-\sigma r) \qquad \text{if } b=0 \qquad (11a)$$

$$\Sigma_j p^c_j c_j + t^g g + \Sigma_j(p^r_j - s^r_j) r_j + f = p^k(k-\sigma r) \qquad \text{if } b>0. \qquad (11b)$$

The household maximizes utility (10) subject to (1), (9), and the budget constraint (11) by choosing the amounts of consumption, garbage, recycling, and litter. First-order conditions for disposal are:

(12a)	dL/dg_j: $u_{cj} + a_j u_g \leq \lambda[p^c_j + a_j t^g]$	for all j
(12b)	dL/dr_j: $u_{cj} \leq \lambda[p^c_j + a_j(p^r_j - s^r_j) + a_j \sigma p^k]$	for all j
(12c)	dL/db_j: $u_{cj} + a_j u_b \leq \lambda[p^c_j]$	for all j

where λ is still the marginal utility of income to the household. Again, these results imply a set of corner solutions if the values of the right-hand sides of (13) are not equal to each other. If the household engages in no littering (b=0), then household behavior is described by (8a) and (8b). With some illegal dumping (b>0), choice on the margin over methods of removing the waste material of type j is determined by:

(13a)	$m_j = g_j$ iff	$t^g - u_g^*/\lambda^* < MIN[(p^r_j - s^r_j) + \sigma p^k, -u_b^*/\lambda^*]$
(13b)	$m_j = r_j$ iff	$p^r_j - s^r_j + \sigma p^k < MIN[t^g - u_g^*/\lambda^*, -u_b^*/\lambda^*]$
(13c)	$m_j = b_j$ iff	$-u_b^*/\lambda^* < MIN[t^g - u_g^*/\lambda^*, p^r_j - s^r_j + \sigma p^k]$

where λ^* and u_g^* are the values of λ and u_g evaluated at the maximum.

The household will again choose to remove material j using the lowest cost alternative (where the "cost" of contributing garbage includes effects on utility and the user fee). For most materials, households are again predicted to use only one method of removal.

With no curbside recycling program

Under what conditions will a household pay the fixed cost associated with littering garbage? In the absence of a curbside recycling program, the answer depends on the value of the user fee and the schedule of recycling costs. With a few simplifying assumptions, we can use Figure 4a to illustrate the conditions that create the incentive for a household to pay the fixed costs (f) and litter garbage. First, assume that the value of disutility associated with contributing garbage is equal to the value of disutility of burning or dumping garbage illegally ($u_g = u_b$). This assumption simplifies the graphical exposition while changing few of the conclusions of the model (see footnote 13 for an exception). Second, assume that the income effect of an increase in the user fee is zero. In other words, the reduction of income attributable to an increase in the user fee does not change total consumption. Furthermore, assume that we can once again order materials according to the marginal cost of recycling (MC^r), which includes the cost of household time spent separating and storing material and the market price for the transfer of the material. The marginal cost of functions associated with recycling (MC^r), and with contributing or illegally dumping garbage ($MC^g = MC^b$)

are illustrated in Figure 4a. The fixed costs associated with dumping introduces an average cost curve, which is also included in Figure 4a, and labeled AC^b.

According to (13a), (13b), and (13c), and as illustrated in Figure 4a, the household will discard all materials ranked to the left of g_o, will litter nothing, and will supply the remaining materials ($m-g_o$) directly to secondary markets in the absence of a user fee for garbage collection. These results are again determined by looking for the lower envelope of disposal costs. Notice again that the removal of all materials is accounted for.

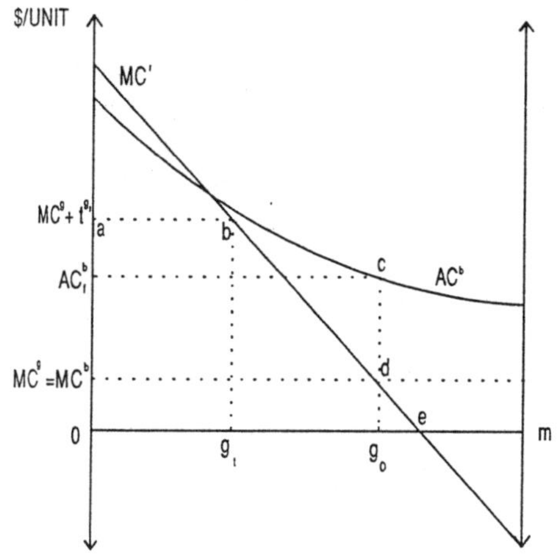

Figure 4a Household disposal choice with littering and no curbside recycling

A user fee for garbage collection will change the households disposal choices because it increases the marginal cost of contributing garbage to the curb. Small increases in the user fee (not shown in Figure 4a) will increase the amount of material that is recycled and will decrease the amount of material that is placed at the curb as garbage. These small increases do not provide the incentive to pay the fixed costs associated with illegally dumping garbage. However, a user fee of $t^{g'}$ increases the marginal cost of contributing garbage to $t^{g'} + MC^g$, and makes the household indifferent between paying the user fee for each unit of garbage and paying the fixed costs f to dump waste illegally. More specifically, a user fee of $t^{g'}$ makes the household indifferent between (1) putting into garbage all materials ranked to the left of g_t and supplying all other materials to secondary markets, and (2) dumping all materials ranked to the left of g_o and supplying the remaining materials to secondary markets. The threshold value of the user fee ($t^{g'}$) is found where the total cost of option (1), denoted by the trapezoidal area a,b,e,0 in Figure 4a (the cost of discarding g_t plus the cost of recycling), is equal to the total cost of option (2), which is AC^b_f,c,d,e,0 (the fixed cost of burning or dumping g_o plus recycling costs). The household would choose the second option for any user fee with value greater than $t^{g'}$.

The information contained in Figure 4a can be captured in another manner using Figures 4b, 4c, and 4d. Figure 4b maps the demand curve for garbage as a function of the user fee. As stated above, the household decreases its quantity demanded for garbage collection (and increases recycling) as the price of garbage increases to $t^{g'}$. Once the price of garbage exceeds this threshold value, the quantity demanded for garbage collection falls to zero. Figure 4c maps the cross-price relationship between the price of garbage and the quantity of recycling. Household recycling increases with larger values of the user fee over the interval between zero and $t^{g'}$. At this point, recycling *decreases* as the household begins to litter garbage. Household recycling efforts are unresponsive to all values of the user fee that are greater than $t^{g'}$. Figure 4d illustrates the effect of the user fee on the quantity of burning or dumping litter. No littering is conducted below the threshold value of the user fee, but the household litters g_o for all values of the user fee that are greater than $t^{g'}$. Once the fixed costs of littering are paid, the marginal cost of littering items ranked from g_o leftward is less than the marginal cost of recycling them.

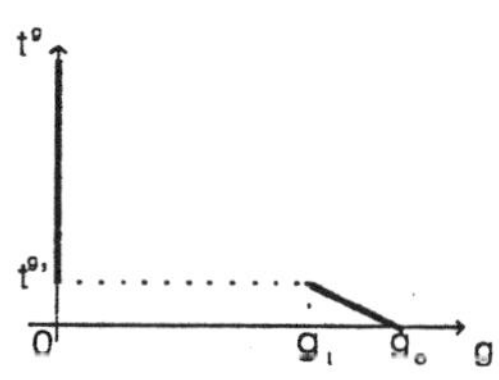

Figure 4b

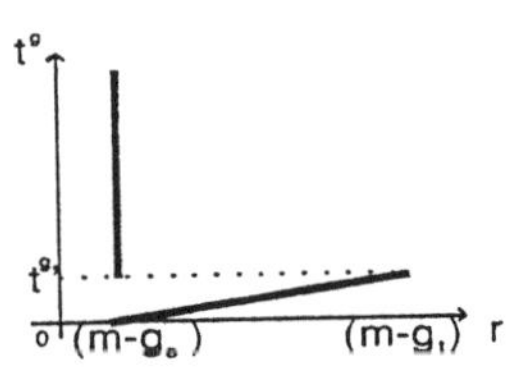

Figure 4c

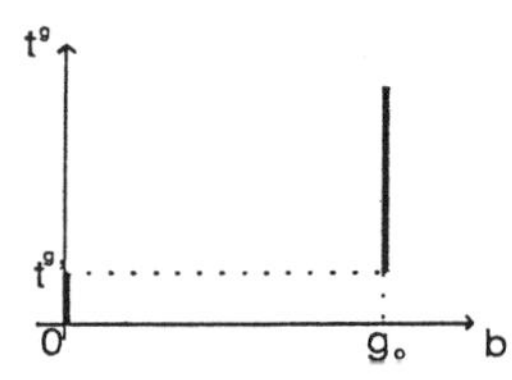

Figure 4d

Several points arise from this simple model. First, the household would never simultaneously engage in positive quantities of littering b and discarding garbage g. The household uses only garbage and recycling at low values of the user fee, and only uses dumping and recycling at high values. Second, the household will never recycle more than $(m-g_t)$ for any value of the user fee. Therefore, a very high user fee may *not* be optimal. In fact, free garbage collection may be optimal if (1) the administrative cost of operating a user fee program is high, and (2) $t^{g'}$ is relatively low.

Recall that $-u_g^*/\lambda^*$ is the marginal "cost" of contributing garbage. The values of g_o, g_t, and $t^{g'}$ in Figure 4a are determined, in part, by the value of $-u_g^*/\lambda^*$ and are, therefore, unique to each household. The values of g_o and g_t decrease as the value of $-u_g^*/\lambda^*$ increases. Households with a strong distaste for contributing garbage will recycle more, discard less, and litter less than households with less of a distaste. The threshold value of the user fee ($t^{g\prime}$) increases as $-u_g^*/\lambda^*$ increases. Therefore, households that have a strong distaste for contributing garbage (and littering) wait for a higher value of the user fee before they are induced to litter their garbage. An increase in $-u_g^*/\lambda^*$ will also serve to shift the curves in Figure 4b and 4d to the left and shift the curve in Figure 4c to the right, while again, raising the threshold value of the user fee. A decrease in income produces similar effects on these variables.

With a free curbside recycling program

How will the implementation of a curbside recycling program influence household response to a user fee for garbage collection? A household will respond to a user fee in one of three ways, according to its value of $-u_g^*/\lambda^*$ relative to σp^k. A first category of households will never recycle at the curb, regardless of the value of the user fee. A second category of households will recycle at the curb for a certain range of the user fee, but will not recycle for values outside this range. A third category of households will still recycle at the curb even when it engages in littering. The households in higher numbered categories hold a greater distaste for contributing garbage ($-u_g^*/\lambda^*$) relative to the opportunity cost of recycling (σp^k).

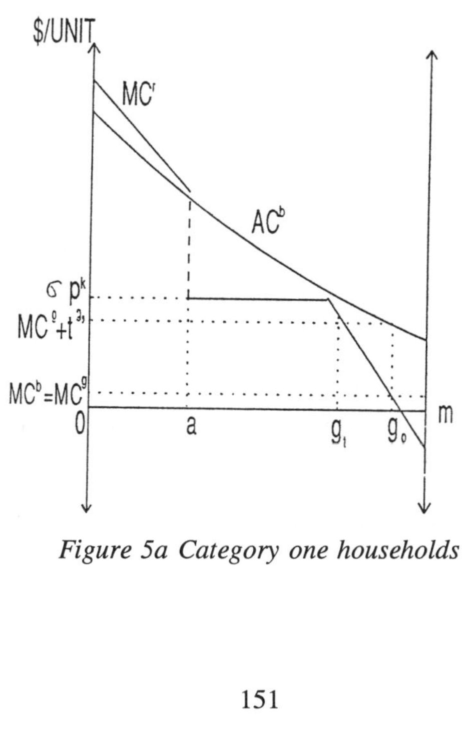

Figure 5a Category one households

The first category of households has the lowest distaste for contributing garbage; its choice among disposal methods is illustrated in Figure 5a. The overall marginal cost of presenting garbage, denoted by MC^g in Figure 5a, is very low relative to σp^k. These households either do not care too much about their garbage contributions or find the opportunity cost of recycling to be too high. Assume that the city is willing to collect materials ranked between point a and point m in Figure 5a at no monetary cost to the household. In the absence of a user fee, the household discards g_o at no monetary cost and sends $(m\text{-}g_o)$ to secondary markets. This household will sell more material in secondary markets as the value of the user fee increases from zero, but will still not participate in a curbside recycling program. To try to induce these households to participate in the curbside recycling program, a user fee would have to be implemented with a value greater than $\sigma p^k\text{-}MC^g$. However, any value of the user fee that is greater than $t^{g'}$ would cause this household to pay the fixed costs and litter all materials ranked to the left of g_o (as explained above). Since $\sigma p^k > t^{g'} + MC^g$, this household will never recycle at the curb. Most communities have households that fit in this category. No community has experienced 100% participation rates in its curbside recycling program, even those with substantial user fees.

Figures 5b, 5c, and 5d summarize the relationships between the value of the user fee and the quantities of garbage (g), recycling (r), and burning (b). A small user fee will induce this kind of household to sell more materials in secondary markets, but any user fee greater than $t^{g'}$ will induce this household to litter.

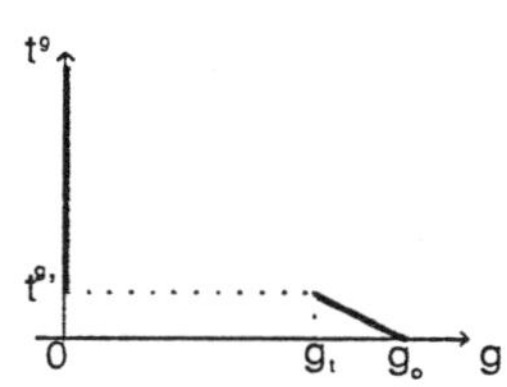

Figure 5b

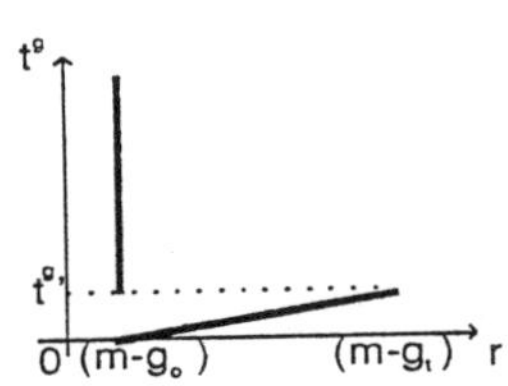

Figure 5c

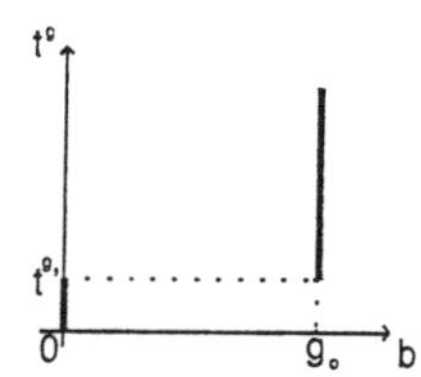

Figure 5d

The second category of households have a greater distaste for contributing garbage than the first category, however, the marginal cost of recycling material (σp^k) is still greater than the overall marginal "cost" of contributing garbage to the curb. This household will recycle at the curb, but only in response to very specific values of the user fee. The household will discard these materials if the user fee is too low or litter them if the user fee is too high.

Figure 6a illustrates the behavior of this category of households. In the absence of a user fee, this category will put into garbage all materials ranked to the left of g_o and will sell the remaining materials in secondary markets. As the value of the user fee increases from zero, the household will gradually discard less material and recycle more. It will recycle at the curb for values of the user fee slightly greater than σp^k. If a user fee of $t^{g'}$ is implemented, the household is indifferent between (1) discarding all materials ranked to the left of g_t and recycling the rest (including materials g_o-g_t collected at the curb), and (2) littering all materials ranked to the left of g_o and sending (m-g_o) to secondary markets. The total cost of method one is the area (f,a,h,c,e,0), while that of method two is ($AC^b{}_f$,b,d,e,0). The threshold value of the user fee ($t^{g'}$) is found where these two areas are equal. The household chooses to litter g_o for all values of $t^{g'}$ greater than the threshold value.

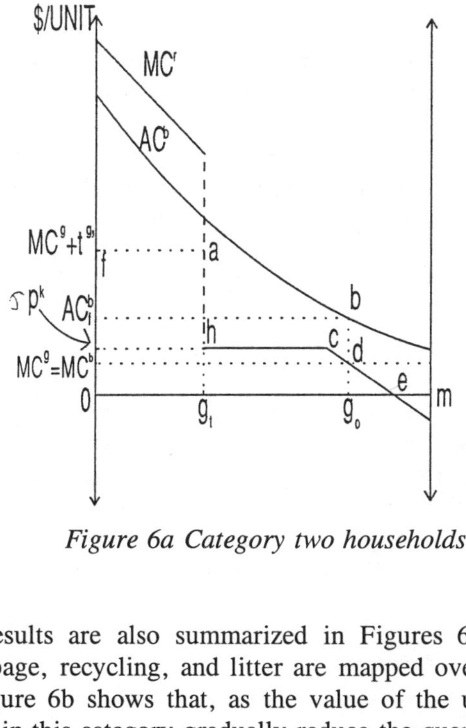

Figure 6a Category two households

These results are also summarized in Figures 6b, 6c, and 6d, where quantities of garbage, recycling, and litter are mapped over all positive values of the user fee. Figure 6b shows that, as the value of the user fee increases from zero, households in this category gradually reduce the quantity of garbage. As the user fee increases above σp^k-MC^g, the household sharply reduces its garbage and begins to recycle at the curb. Once the user fee rises above $t^{g'}$, the household stops discarding garbage, and, as Figure 6d illustrates, begins to litter. This response of recycling to a user fee is somewhat surprising. The household participates in curbside recycling for values of the user fee slightly greater than σp^k-MC^g, but it abandons its curbside recycling efforts once the user fee exceeds

$t^{g'}$. At this high fee, the household has the incentive to pay the fixed cost f associated with dumping. Since the marginal cost of dumping materials (MC^b) is lower than the overall marginal "cost" of recycling them at the curb (σp^k), the household will litter these recyclable materials as well.

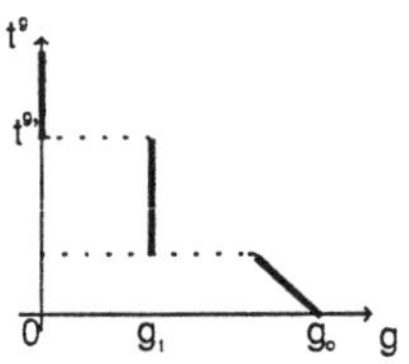

Figure 6b

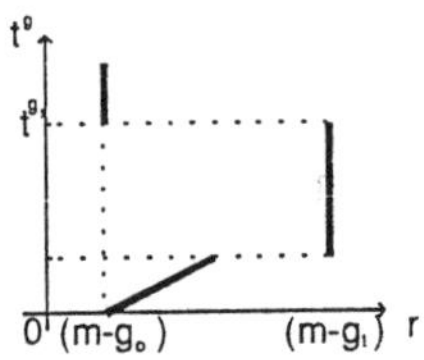

Figure 6c

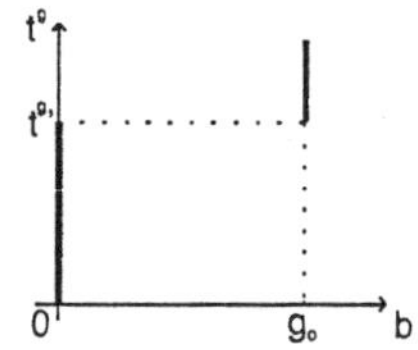

Figure 6d

Households in category three will recycle at the curb, even in the absence of a user fee. The marginal cost to recycle materials collected at the curb (σp^k) is less than the marginal "cost" of discarding garbage in the absence of a user fee MC^g. The logic can be seen in Figure 7a. The household discards all materials to the left of g_o and participates in curbside recycling in the absence of a user fee. Therefore, the implementation of a user fee for garbage collection will have no impact on this household's curbside recycling level. However, if the value of the user fee exceeds $t^{g'}$, the household begins to litter all material ranked to the left of g_o (at marginal cost MC^b) and recycle the rest.[13]

Figures 7b, 7c, and 7d summarize these results. Notice in Figure 7c that the amount of recycling conducted by these households is fairly unresponsive to price. In fact, the only major change in waste removal methods attributed to a change in the user fee is a switch from discarding garbage (g) to burning or littering (b). This switch occurs for values of the user fee greater than $t^{g'}$. This result suggests that garbage fees would have little success in increasing the recycling levels of households that have rather strong preferences for recycling.

---

[13] A possible fourth category of households could involve $MC^b > \sigma p^k > MC^g$. These households would not recycle at the curb in the absence of a user fee. High user fees would still create the incentive for this household to litter, but it will still place those materials that are collected by the city out for recycling.

These households already participate in the curbside recycling program in the absence of a user fee, leaving little room to increase their recycling quantities following the implementation of the user fee.[14]

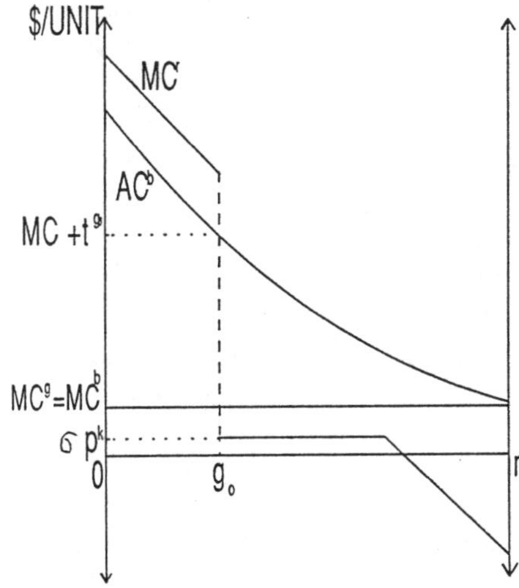

Figure 7a Category three households

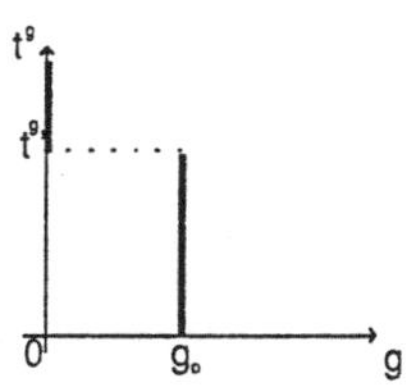

Figure 7b

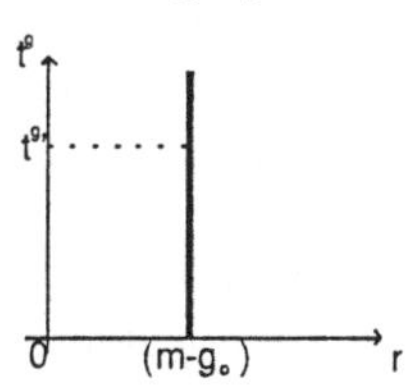

Figure 7c

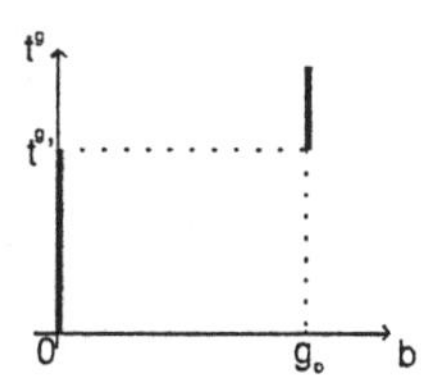

Figure 7d

Aggregate Behavior

The community's demand schedule for garbage collection can be derived by adding the amounts of garbage thrown out by all households in the community, at each value of the user fee. Individual garbage quantities with curbside recycling are depicted in Figures 5b through 7b, where all types of households either reduce or leave unchanged their garbage quantities with increases in the value of

---

[14] A possible fifth category of households will never litter garbage. The overall marginal "cost" of recycling even the most expensive items (the whole MC^r schedule) is less than the cost of burning or littering (the AC^b schedule).

the user fee. A higher user fee never increases the quantity of garbage. Therefore, aggregate garbage never increases with the user fee.

Similarly, Figures (5d) through (7d) illustrate that increasing values of the user fee either increase or leave unchanged amounts of litter for all households. Therefore, the aggregation of household litter must either increase or remain unchanged with higher values of the user fee. Aggregate litter will never decrease.

The same story cannot be told for aggregate recycling levels. One might expect that aggregate recycling would increase with higher values of the user fee, and this result certainly holds for some households over some ranges of the user fee. However, Figures (5c) through (7c) indicate that households can *reduce* recycling over *some* range of the user fee. Different households will decrease recycling by different amounts, and over different ranges of t^g. Therefore, the horizontal aggregation of recycling across all households could either rise or *fall* with increasing values of the user fee.

The logic is a bit different for each category of household, but increases in $t^{g\prime}$ will eventually induce all households to pay the fixed cost and begin to litter waste material. Once these fixed costs are paid, the household may find the marginal cost of dumping (MC^b) to be less than the marginal cost of recycling (MC^r). At this point, a higher value of $t^{g\prime}$ induces less recycling.

The relationships between the user fee and these aggregate amounts are illustrated in Figure 8 for two specific communities. The aggregate recycling function for community 1 (call it $D^R{}_1$) is drawn so that aggregate recycling (measured from the point M leftward) rises with the user fee. The litter function for community 1 (call it $D^B{}_1$) reflects the fact that some households, perhaps in category 1, litter garbage even at very low values of the user fee. As the user fee increases from zero, more households are predicted to pay the fixed cost associated with littering and contribute to aggregate litter.

If no user fee is charged for garbage, the community with recycling function $D^R{}_1$ and littering function $D^B{}_1$ litters an amount $B_1 = 0$, recycles $M\text{-}G_1$, and discards G_1. As long as the marginal cost of littering is positive and the marginal cost to discard garbage is zero, then all communities experience no litter in the absence of a user fee. Once a user fee of $t^{g\prime}$ is levied, the amount of litter increases to B_1', the amount of recycling increases to $M\text{-}G_1'$, and the amount of garbage decreases to $G_1'\text{-}B_1'$.

These functions for a second community are denoted by $D^R{}_2$ and $D^B{}_2$. With user fee $t^{g\prime}$, community two increases litter from B_2 to B_2', increases recycling from $M\text{-}G_2$ to $M\text{-}G_2'$ and reduces the level of garbage from $G_2\text{-}B_2$ to

G_2'-B_2'. Notice that the increase in litter is smaller than experienced by community one. As drawn, the change in recycling for community two is larger than for community one. Therefore, community two may realize greater benefits and lower costs from the implementation of a user fee.[15]

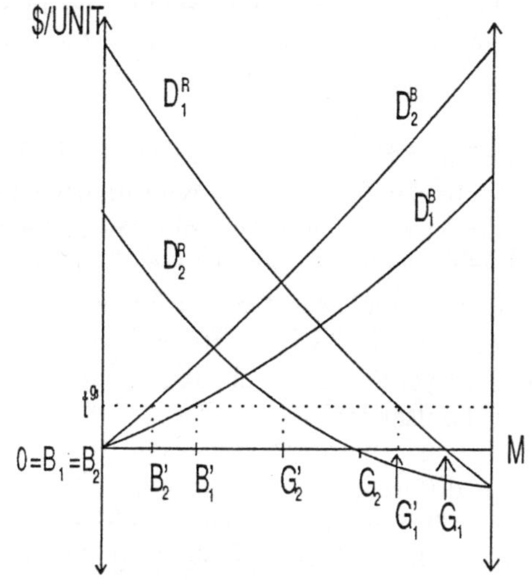

Figure 8 Aggregate behavior with littering

IMPLICATIONS AND CONCLUSION

This chapter has contributed to the literature on user fees for garbage collection in several ways. We introduced heterogeneity into a model of household choice over garbage removal methods. This model explained (1) why some households participate in curbside recycling programs in the absence of a user fee, (2) why other households do not recycle at the curb even in the presence of a user fee, and (3) why some households choose to litter garbage when others do not. We aggregated diverse households in each community to provide a prediction that a user fee will never increase garbage, and will never decrease illicit burning or dumping. At some point, however, the garbage fee becomes high enough to induce some people to pay the fixed cost of dumping and thus to switch away from recycling. As a consequence, the garbage fee might *decrease* aggregate recycling.

---

[15] The model does not suggest that the community with less littering will also experience a greater change in recycling (as drawn in Figure 9). This community could experience a smaller change in recycling.

The model is general enough to encompass fixed costs for recycling instead of for dumping. Individuals and recycling firms may incur search- and transport costs that do not vary with the amount recycled. If so, the same model could be employed to show that a rising garbage fee would generate monotonic decreases in garbage and monotonic increases in recycling. At some point, however, the garbage fee would become high enough to induce some people to pay the fixed cost of recycling and thus to switch away from dumping. As a consequence, the garbage fee might *decrease* aggregate dumping.

We found that the observed low or moderate garbage fees cannot be explained in a model where households cannot burn, dump, or litter. Therefore, we added this option to the model. We described three categories of households, and we aggregated individual behavior to find community demands. We found that individual behavior, and thus aggregate demands, depend upon (1) the community's distribution of household preferences for contributing garbage, (2) the fixed costs associated with dumping, (3) the price of recyclable materials, and (4) the presence of curbside recycling. Therefore, empirical work will require measures of those variables.

The theory also makes some specific predictions that can be tested empirically. In general, the own-price elasticity for garbage should be non-positive. The cross-price effect of the garbage fee on recycling quantities could be negative or positive at high values of the garbage fee (where it could induce people to pay the fixed cost of dumping and thus to switch out of recycling), but it should be non-negative at low values of the user fee.

We thought about how these aggregate outcomes affect the costs and benefits to a community from the implementation of a user-fee program. In particular, the heterogeneity of households implies that some communities will respond more than others. Thus a user fee is more likely to be adopted in communities with low administrative costs, large increases in recycling, and small increases in dumping. The important implication for empirical work is that the town's choice of user fee is not exogenous.

References

Copeland, B.R. (1991). "International Trade in Waste Products in the Presence of Illegal Disposal." *Journal of Environmental Economics and Management* 20, 143-162.

Dinan, T.M. (1993). "Economic Efficiency Effects of Alternative Policies for Reducing Waste Disposal." *Journal of Environmental Economics and Management* 25, 242-256.

Dobbs, I.M. (1991). "Litter and Waste Management: Disposal Taxes versus User Charges." *Canadian Journal of Economics* 24, 221-227.

Fullerton, D. and T.C. Kinnaman. (Forthcoming). "Garbage, Recycling, and Illicit Burning or Dumping." *Journal of Environmental Economics and Management*.

Jenkins, R. (1991). "Municipal Demand for Solid Waste Disposal Services: The Impact of User Fees." *Mimeo*. University of Maryland.

Kennedy, P.W., and B. Laplante. (1994). "Municipal Solid Waste Management: The Optimal Pricing of Garbage and Recyclables Collection." *Mimeo*. World Bank.

Morris, G.E., and D.M. Holthausen. (1994). "The Economics of Household Solid Waste Generation and Disposal." *Journal of Environmental Economics and Management* 26, 215-234.

Repetto, et al. (1992). *Green Fees: How a Tax Shift Can Work for the Environment and the Economy*. Washington DC: World Resources Institute.

Skumatz, L.A. (1993). "Variable Rates for Municipal Solid Waste: Implementation Experience, Economics, and Legislation." Policy Study 160. Reason Foundation.

Sigman, H. (1991). "A Comparison of Public Policies for Lead Recycling." *Mimeo*. Department of Economics UCLA.

Sullivan, A.M. (1987). "Policy Options for Toxic Disposal: Laissez-faire, Subsidization, and Enforcement." *Journal of Environmental Economics and Management* 14, 58-71.

Wertz, K.L. (1976). "Economic Factors Influencing Households' Production of Refuse." *Journal of Environmental Economics and Management* 2, 263-272.

Chapter 7

ENVIRONMENTAL TAXES ON INTERMEDIATE AND FINAL GOODS WHEN BOTH CAN BE IMPORTED

James M. Poterba (Massachusetts Institute of Technology)

and

Julio J. Rotemberg (Massachusetts Institute of Technology)*

INTRODUCTION

One of the central administrative problems that arises in implementing environmental excise taxes is the treatment of imports. While similar issues arise with respect to all indirect taxes, the problems associated with environmental excises are particularly complex because many environmental taxes are levied on intermediate inputs, such as fossil fuels, rather than on final goods. Many nations have border tax adjustments to complement their domestic taxes on intermediate goods, taxing imported intermediate goods at a rate equal to that on domestically produced goods. Davie (1995) describes the border tax adjustments associated with current environmental excise taxes in the United States.

While border tax adjustments for imported intermediate goods are usually straightforward, more difficult issues arise with respect to imports of final goods that are produced using the taxed intermediate good as an input (see Barthold, 1994). Not taxing such imports would place domestic producers of final goods at a cost disadvantage, and would encourage offshore production of these final goods.

---

* We are grateful to the Center for Advanced Study in Behavioral Sciences, the Institute for Policy Reform, the MIT Center for Energy and Environmental Policy, and the National Science Foundation for research support, and to Thomas Barthold, Lans Bovenberg, Bruce Davie, Gunnar Thorland Jepsen, Ralph Landau, Tyrone Montague, Roger Noll, Michael Rauscher, Nancy Rose, Michael Whinston, and two anonymous referees for helpful comments.

Whether residents of the country imposing intermediate-good taxes will regard this relocation of production abroad as desirable depends on a number of considerations. If the intermediate good creates pollution only where it is used, then the country that imposes a tax on domestic use of intermediate goods will experience a reduction in pollution, as use of the intermediate good migrates offshore. This may be viewed favorably by domestic residents. However, if use of the intermediate good generates pollution that flows across borders and consequently affects residents of the taxing nation, as with the global externalities associated with chlorofluorocarbon use, then such a shift in production may not be viewed as favorable. Similar conclusions may emerge if domestic residents derive disutility from pollution both at home and abroad, or if the industry using the intermediate good generates externalities or captures oligopolistic rents, as in some of the "new trade" models considered by Helpman and Krugman (1989). These and related issues are explored in the context of a unilateral carbon tax by Hoel (1994). He shows that if imported final goods cannot be taxed, then it may be optimal to tax the use of the same carbon fuels in different industries at different rates, and in particular to levy lower rates on industries that face international competition.

Concern about the taxation of imported final goods is especially likely to arise in discussions of broad-based fossil fuel taxes, such as the BTU tax proposed and debated in the United States in early 1993. The only border taxes in President Clinton's BTU tax proposal were taxes on imports of fossil fuels. When the House of Representatives considered the BTU tax, however, its legislation included an "imputed BTU tax on imported products identified by the Treasury Department as having direct energy inputs that would have been taxed if produced in the United States in excess of 2% of value of the final product" (Davie, 1995). The proposed legislation, which never became law, did not indicate how "imputed" BTU tax rates would be set. But the proposal for such taxes, and the requirement for similar taxes on imported final goods under a number of current environmental taxes, suggests the importance of border tax adjustments in designing environmental tax policies.

This chapter analyzes the design of environmental taxes on imported final goods when policymakers seek to affect equal changes in the marginal costs of domestic and foreign producers. Identifying such taxes is important, since preventing environmental taxes from being tilted against foreign producers is paramount if one is to ensure that environmental protection does not become an excuse for mercantilist protectionism. We abstract from whether border taxes are feasible under present international trade agreements, a topic addressed in detail by Hufbauer (1993).

We show that when it is possible to measure the amount of the taxed intermediate good that is used to produce an imported final good, an import tax equal to the tax on the intermediate good times the amount of the intermediate good used in *domestic* production of the final good will raise the marginal cost of domestic and foreign producers by the same amount. When final goods are produced jointly, however, we obtain a negative result. Because the amount of the intermediate good that is used in producing any particular final good is not well defined, it is not possible to prescribe appropriate border taxes. Moreover, reliance on arbitrary measures of the amount of intermediate good that is used in producing one of several joint final goods can lead to substantial differences in the tax-induced change in marginal cost for foreign and domestic producers.

This paper is divided into four sections. The first briefly describes the current treatment of imported products under the US environmental excise tax system. The second section considers the taxation of finished goods in a competitive industry that exhaustively uses a taxed intermediate good, and derives the tax on imported final goods that raises marginal costs for foreign and domestic producers by the same amount. The next section generalizes the analysis to the case of a multiproduct firm that uses the intermediate good in the joint production of several final goods. We illustrate our central result for the simplest possible joint production technology: a Leontief technology for two goods. A brief conclusion describes the industries for which our results seem relevant.

CURRENT TREATMENT OF IMPORTS UNDER US ENVIRONMENTAL EXCISE TAXES

The US government currently levies four environmental excise taxes: a Superfund tax on petroleum, a tax on ozone-depleting chemicals (ODCs), a tax on toxic chemicals other than ODCs, and a tax on motor fuels to pay for environmental damages associated with leaking underground storage tanks (the LUST tax). The LUST tax is collected as part of the motor fuels excise tax, so only the first three environmental excises raise issues of border tax adjustments.

The tax on petroleum and petroleum products is levied at a rate of 9.7 cents per barrel of crude oil or petroleum product, or less than one percent of the current world price of crude oil. The related Oil Spill Liability tax, discussed in Davie (1995), collects an additional five cents per barrel. The tax on ODCs is levied on a set of chemical compounds, principally chlorofluorocarbons (CFCs). The 1995 tax rate is $5.35 per pound. Finally, the tax on toxic chemicals specifies a list of forty-two chemicals, primarily hydrocarbons and metal compounds, with associated tax rates. The rate for most hydrocarbons is $4.87 per ton in 1995.

Each of these environmental taxes includes a provision for treatment of imported products. Under the Superfund tax, imports of petroleum products are taxed at the same *per barrel* rate as crude oil received at US refineries. This treatment of imports can place different burdens on US and foreign refiners for two reasons. First, not all refinery inputs are transformed into outputs: some crude oil input is used to supply energy to the refining process. Second, the number of barrels of input may differ from the number of barrels of output because refinery outputs are typically lighter than inputs.

The ODC tax legislation includes explicit provisions for imports of products containing ODCs. The tax on "an imported taxable product ... is computed by reference to the weight of the ODCs used as materials in the manufacture of the product." (US *Code of Federal Regulations*, 26 CFR 52.4682-.3). Table 1 presents examples of items that are listed in the table of imported ODC-using products.

Table 1 US Imputation of Ozone-Depleting Chemical Content of Imports

Product	Imputed ODC Weight	1993 Tax
Household Freezers	2.40 lbs.	$8.04
Computer Keyboards	0.07	0.20
Telephones with Value > $11	0.10	0.27
VCRs	0.06	0.16
Foam Chairs	0.30	1.01
Passenger Cars with Air Conditioning	4.00	13.07

Source: US Code of Federal Regulations, 26 CFR 52.4682.3, 24-27.

The tax on toxic chemicals treats imports in a similar fashion. The tax rate on imported products that embody 50% or more of taxable chemicals, measured either by weight or value, is set equal to the tax that would have been collected if the taxable chemicals used in production had been sold in the United States. In the absence of any information regarding the content of non-ODC taxable chemicals, the Secretary of the Treasury may prescribe a tax rate based on the use of taxable substances in the *predominant method of production* for the imported product,[1] or the Secretary may impose a five percent *ad valorem* tax on imported chemical products.

IMPORT NEUTRALITY WITHOUT JOINT PRODUCTION

This section formalizes the problem of taxing final good imports when producing the product in question exhaustively consumes a given set of inputs. We assume that a final good (q) is supplied by both domestic and foreign producers under constant returns to scale using inputs of labor (L) and an intermediate good (E). Foreign and domestic production are denoted by superscripts F and D, respectively. Domestic and foreign wages, w^D and w^F, may differ, but there is an integrated world market for the intermediate good, which has a unit price of e. Initially, both domestic and foreign firms provide this good to the domestic market, so the prices of their goods must be equal ($p^D = p^F$).

Given constant returns, the total cost function for domestic producers is $q^D * c^D(e, w^D)$, where q^D is the quantity of domestic output and c^D is the unit-cost function, and the analogous cost function for foreign producers is $q^F * c^F(e, w^F)$. The prices of domestic and foreign, goods, p^D and p^F, are related to the two unit marginal costs, c^D and c^F, with equality in the case of perfect competition. When the world price of the intermediate good is fixed, a specific tax of θ levied on its domestic use raises its domestic price to $e + \theta$. Holding constant the quantity of domestic output produced, the associated change in the marginal cost of the domestically-produced good (m^D) is therefore

$$dm^D = c_e^D(e, w^D) * \theta = \frac{E}{q^D} * \theta. \tag{1}$$

The second equality holds because the derivative of total cost with respect to e, $q^D * c^D_e(e, w^D)$, equals the input demand for the intermediate good E. This implies $E = q^D * c^D_e(e, w^D)$, so $E/q^D = c^D_e(e, w^D)$.

---

[1] The IRS *Cumulative Bulletin* (1989-1, page 718) provides an example to help chemical firms compute the tax on imports of non-ODC taxed chemicals. This example uses the Friedel-Crafts alkylation process, which is the predominant means of producing ethylbenzene, to show that since 0.75 pounds of benzene are reacted with 0.28 pounds of ethylene to produce one pound of ethylbenzene, the tax rate on ethylbenzene is $0.75 * \tau_{benzene} + 0.28 * \tau_{ethylene}$.

To raise the marginal cost for foreign producers by as much as the intermediate good tax raises the marginal cost for domestic producers, imports must bear a tax τ such that $\tau = dm^D = (E/q^D)*\theta$. This analysis rationalizes the use of average input requirements as a basis for levying taxes on imported final goods. This result holds for any constant-returns-to-scale production function, not just for the special case of a Leontief production function that we will use below to study joint production.

To raise the marginal cost of foreign and domestic producers by the same amount, the tax on imported final goods must be based on the importance of the intermediate good in *domestic* production. Regulations that allow importers to pay taxes equal to the actual amount of intermediate good used in production, subject to documentation, may therefore not raise marginal costs of foreign and domestic producers by the same amount. Importers will presumably take advantage of their option to provide specific documentation only when their inputs of the taxed intermediate goods are below the inputs for domestic producers.

IMPORT NEUTRALITY WITH JOINT PRODUCTION

When several final goods are produced *jointly* from a given intermediate good, it becomes difficult to quantify the amount of intermediate good that is used for the production of any single final good. Joint production processes are extremely common in some of the industries that produce taxed goods, such as petroleum refining (see Leffler (1979)) and petrochemicals (see Burdick and Leffler, 1990).

To illustrate the difficulties of taxing jointly-produced goods, we begin by considering a general joint production function, and then specialize our analysis to a simple example. We begin by supposing that two goods, q_1 and q_2, are jointly produced so that the total cost of producing them is not additively separable and is given by $C(q_1, q_2, e, w)$. Provided $C(.)$ is differentiable, the marginal cost of good one is the derivative of this cost function with respect to q_1. The derivative of this marginal cost with respect to a tax on the intermediate good is $C_{q1e}(q_1, q_2, e, w)$, and the associated change in the marginal cost of good one from a specific tax of θ on E is $dm_1 = C_{q1e}*\theta$. Since the demand for the intermediate good satisfies $E = C_e(q_1, q_2, e, w)$, the change in the marginal cost of good one with respect to θ can also be written as $dm_1 = (\partial E/\partial q_1)*\theta$. This does *not* necessarily equal $(E/q_1)*\theta$, as in the case of non-joint production above. Without imposing more structure on the joint production function, we are unable to describe the effect of changes in θ on the marginal cost of good one.

For that reason, we now consider the simplest possible joint production technology: the fixed-coefficients or Leontief case. In particular, we assume the following:

(2) $$q_1 = \min[\frac{L_1}{\delta_1}, \frac{E}{h_1}]$$

(3) $$q_2 = \min[\frac{L_2}{\delta_2}, \frac{E}{h_2}].$$

Labor input must be dedicated to the production of one good or the other, but the intermediate good input (E) is "public" in the sense that over some ranges of output, production of one good can be increased without raising intermediate good input. The production technology described in (2) and (3) exhibits constant marginal costs of producing goods 1 and 2, with a total cost function

$$c(q_1, q_2, e, w) = (\delta_1 * q_1 + \delta_2 * q_2) * w + \max[h_1 * q_1, h_2 * q_2] * e.$$

The supply behavior of a price-taking firm facing the production function given by (2) and (3) can be described as follows. If the price of either good one or good two is below its marginal labor cost, $\delta_1 w$ and $\delta_2 w$, respectively, the firm will not produce this good. Even when p_1 and p_2 both exceed their marginal labor costs, however, the firm may not supply any output, since it must also cover the cost of intermediate good inputs. The break-even condition for the firm to produce *both* goods is:

(4) $$\frac{p_1}{h_1} + \frac{p_2}{h_2} = e + [\frac{\delta_1}{h_1} + \frac{\delta_2}{h_2}] * w.$$

This expression equates the marginal cost of jointly producing $1/h_1$ units of good one and $1/h_2$ units of good two to the marginal revenue from selling these goods. The relative quantities of goods one and two in this expression are set by their relative intermediate good input requirements, which dictate that a firm producing both goods will set $q_2 = (h_1/h_2) * q_1$.

We now consider the properties of plausibly defined measures of the direct intermediate good inputs used to produce goods one and two. We refer to such measures as α_1 and α_2. Without joint production, the direct intermediate good input to good one can be defined as $\alpha_1 = E/q_1$, as in the last section. For the linear production technology we are now considering, $E/q_1 = h_1$ so the associated border tax adjustment would be $\tau = h_1 * \theta$. With joint production, however, one unit of E leads to the production of $1/h_1$ units of good one and $1/h_2$ units of good two. Since the *bundle* of $1/h_1$ units of good one and $1/h_2$ units of good two requires one unit of direct intermediate good input, we know that

$$\frac{\alpha_1}{h_1} + \frac{\alpha_2}{h_2} = 1. \qquad (5)$$

This condition ensures that $\alpha_1 < h_1$, so that the direct intermediate good input to good one is strictly less than the ratio of intermediate good input to output of good one. Given the logic of the previous section, this would appear to suggest that the appropriate border tax adjustment for good one is *less than* $h_1 * \theta$. However, (5) is clearly insufficient by itself to uniquely determine α_1, the intermediate good input to producing good one. This is the central problem of assigning intermediate good inputs to joint products.

Designing border tax adjustments for joint final goods is more complex than even the foregoing discussion, which suggests that it is impossible to measure α_1, suggests. Even though we know that $\alpha_1 < h_1$ in the case we have considered, we will now illustrate circumstances under which it is necessary to impose a border tax of *more* than $h_1 * \theta$ to maintain parity between the domestic and foreign marginal costs of good one. We do this by considering how the marginal cost of producing good one changes in response to a tax on the intermediate input.

It is easiest to define the marginal cost of good one if good two has an infinitely elastic demand, and therefore a fixed price (p_2). One can then think of the marginal cost of producing one more unit of good one as the cost incurred as the firm purchases more inputs, minus the revenues that it receives from additional sales of good two. Assuming also that labor is elastically supplied at wage w, the marginal cost of good one is:

$$m_1 = h_1 e + [\delta_1 + \frac{h_1 \delta_2}{h_2}] * w - \frac{h_1 p_2}{h_2}. \qquad (6)$$

If we further assume that good one is traded in international markets, while good two is a non-tradable production by-product, foreign and domestic firms will face the same price for good one, but they may face different prices for good two. We assume that their respective prices for good two are fixed in their domestic markets. Under these assumptions, a tax on the intermediate good will be fully reflected in the domestic marginal cost of good one.[2] Holding w and p_2 fixed, we differentiate (6) with respect to e to find the effect of an excise tax that raises the price of E to $e + \theta$. Provided e is fixed, we find $dm_1 = h_1{}^*\theta$. This implies that the tax on imports of good one that raises marginal costs by the same amount for domestic and foreign producers is $\tau = h_1{}^*\theta$. But we know from above that "natural" measures of α_1 are less than h_1. Thus, a border tax adjustment based on an estimate of the direct intermediate-good input to producing good one will raise marginal costs for foreign producers by less than the cost increase for domestic producers. This finding suggests that intermediate-good intensity can be an unreliable standard for choosing border tax adjustments associated with domestic environmental taxes.[3]

APPLICATIONS

Two industries that may be described by our stylized analysis of joint production with constant returns are petroleum refining and petrochemical production. Leffler (1979) explains that crude oil is a complex mixture of many hydrocarbons. The refining process separates these different components, and in some cases also initiates chemical reactions that transform some component hydrocarbons into others. Refining is a textbook example of a joint production process: it is essentially impossible to produce only a single product from crude oil input. Even better examples of joint production, in which some outputs are tradable and others are not, can be found in the petrochemical industry. This industry is a downstream segment of the refining industry, which produces a range of synthetic hydrocarbons used in plastics, resins, and a wide range of other manufactured products.

The negative character of our results is unfortunate. We show that standard prescriptions for determining the direct intermediate good input to final goods are unlikely to work in the case of joint production. Moreover, we argue that the very notion of the intermediate good input to a joint product is poorly defined, even though this is the concept that typically underlies actual attempts to

---

2 This is a standard tax-incidence problem: how will a tax on an intermediate good be reflected in the prices of the two final goods and the wage? Our assumptions of a fixed wage and fixed price for good two require that all of the tax is shifted forward to the price of good one. Relaxing these assumptions would lead to some adjustment in the price of good two, and the wage, as well.

3 Braeutigam (1980) explains that in regulatory contexts, a number of arbitrary rules, such as the relative physical weights of the two outputs, or their relative market values, or the relative variable costs that can be attributed to each of these products, have been used to solve the joint cost attribution problem.

design border tax policies. Our simple example suggests that the border tax rate that raises marginal costs for foreign and domestic producers by the same amount will depend on conditions in the markets for each of the joint products, as well as conditions in the markets for other factors that are used to produce the joint products. There are no simple, general rules for border tax adjustment in these cases.

References

Barthold, T.A. (1994). "Issues in the Design of Environmental Excise Taxes." *Journal of Economic Perspectives* 8, 133-152.

Braeutigam, R.R. (1980). "An Analysis of Fully Distributed Cost Pricing in Regulated Industries." *Bell Journal of Economics* 11, 182-196.

Burdick, D.L., and W.L. Leffler. (1990). *Petrochemicals in Non-Technical Language*. Tulsa: PennWell Press.

Davie, B.F. (1995). "Border Tax Adjustments for Environmental Excise Taxes: The US Experience." *Mimeo*. Washington DC: US Treasury Department.

Helpman, E. and P.R. Krugman. (1989). *Trade Policy and Market Structure*. Cambridge: MIT Press.

Hoel, M. (1994). "The Role of a Carbon Tax in Environmental and Fiscal Policy." *Mimeo*. University of Oslo.

Hufbauer, G.C. (1993). "The Evolution of Border Tax Adjustments." Washington, DC: Institute for International Economics.

Leffler, W.L. (1979). *Petroleum Refining for the Non-technical Person*. Tulsa: Petroleum Publishing Company.

PART IV

INSTITUTIONAL ASPECTS

Chapter 8

ARE INCENTIVE INSTRUMENTS AS GOOD AS ECONOMISTS BELIEVE? SOME NEW CONSIDERATIONS

Hannelore Weck-Hannemann (University of Innsbruck)

and

Bruno S. Frey (University of Zürich)

THE ISSUE

The Theory of Environmental Economics strongly suggests that the use of incentive instruments, which exploit the relative price effect, constitutes the most efficient and desirable policy. A recent survey on environmental policy (Hahn 1989, p. 95) notes that "the two tools which have received widespread support from the economics community (are) marketable permits and emission charges." Environmental incentive instruments count among the most successful applications of economic theory (Faulhaber and Baumol, 1988; Cropper and Oates 1992). During the last few years, emission charges and even tradeable effluent licenses have moved from being an academic proposal to being part of the policy programme of many parties across the whole political spectrum. Nevertheless, incentive instruments are *little* used all over the world (Hahn, 1989); and an OECD study (1989) also sees little prospect that they will be used on an extensive scale in the future.

This paper completely accepts the economic arguments brought forward in favour of incentive-based environmental instruments. Using the relative price effect in order to improve our natural environment is strongly supported by the authors. The paper argues, however, that the favourable outcomes attributed to environmental incentive instruments *depend on particular conditions*. The domain of favourable outcomes is restricted because the introduction of incentive instruments under identifiable conditions leads to a countervailing effect. Under some conditions, therefore, the application of emission charges or tradeable licenses leads to a *perverse effect,* i.e. that the environmental conditions worsen.

Our argument is not based on the (well-known) transaction costs, i.e. that the costs of administrating and monitoring incentive instruments may be so high that they outweigh the benefits achieved by their application. Rather, our argument is based on a behavioral effect originally identified in experimental social psychology. This relationship has been dubbed *'crowding-out effect'* and states that there exist conditions under which incentive instruments undermine environmental ethics, tending to lead to less environmentally conscious behaviour.

The pertinent question analysed in this paper is: which of the two effects -- the (economic) relative price effect or the (psychological) crowding-out effect -- prevails under what conditions? The following section analyses the crowding-out effect. Then, the relative price effect and the crowding-out effects are compared, and we identify under what conditions the crowding-out effect becomes relevant and may dominate the relative price effect. The last section discusses the consequences for environmental policy.

THE CROWDING-OUT EFFECT

Incentive Instruments and Intrinsic Motivation

Following the rational choice model (Becker, 1976; Frey, 1992a), individual behaviour is determined by preferences and constraints. Changes in constraints are due to changes in relative prices. While standard neoclassical theory analyses human behaviour by taking basic preferences to be constant (Stigler and Becker, 1977), it may alternatively be assumed that preferences are *systematically* affected by pricing or monetary incentive instruments. Introducing or more intensively using prices to influence individual behaviour under identifiable conditions will have a detrimental effect on preferences or, equivalently, on intrinsic motivation. A particularly important intrinsic motivation is morale, in particular environmental ethics. According to Deci (1971, p.105), "one is said to be intrinsically motivated to perform an activity when one receives no apparent reward except the activity itself." Intrinsic motivation is used here in a general way; what matters is the contrast to the relative price effect of microeconomic theory.

Experimental and field research has established (see Deci and Ryan, 1985) that three aspects are of crucial importance for the crowding-out effect to occur:

- *Recognition.* A monetary reward, which makes people feel that the monetary incentive is applied only to bring about more desired behaviour according to other persons' wishes, reduces intrinsic motivation.

- *Fairness*. The application of incentive instruments, which is considered to be unfair, decreases intrinsic motivation.
- Incentive instruments, which are experienced as an increase in the extent to which one's behaviour is determined from the outside, decrease self-determination and thereby reduce intrinsic motivation.

In experimental psychology, the negative effect of monetary (external) rewards on intrinsic motivation has been extensively discussed under the heading of 'hidden cost of reward' (Lepper and Greene, 1978). The reason why external rewards decrease intrinsic motivation and corresponding activity has been attributed to a shift in the locus of control to an external source, which then guides behaviour. One may also speak of 'oversufficient justification' -- of being motivated at the same time, and unnecessarily, by two motivations for the same task. Many experiments support the crowding-out effect (McGraw, 1978, pp. 55-56): "the evidence for a detrimental effect comes from a wide variety of works in which a large number of subjects and methodological parameters have been varied." This holds, in particular, for work motivation (see Frey, 1993b, 1993c; and Barkema, 1995, for an econometric analysis).

Conditions for the Crowding-Out Effect

The previous subsection discussed the psychological basis for crowding-out effects and has thus mainly been concerned with the *possibility* for the effect to occur. This subsection now deals with the *specific conditions* that favour or restrain the crowding-out effect, and therefore make it possible to deduce when and how monetary incentive instruments should be applied to influence environmental behaviour.

Six conditions affecting the size of the crowding-out effect can be distinguished (Frey, 1993a):

a. The more *personal* the relationship between the actors involved, the more a monetary incentive tends to shift the locus of control from intrinsic to extrinsic motivation, and a crowding-out effect is more likely to occur. There is a 'psychological contract' (Lee (1987)) between principle and agent. When it is violated, the workers respond by reducing work effort, an effect that is supported by substantial empirical evidence (see, e.g., Doeringer, 1991).

b. The more *interesting* a task is, the higher is the intrinsic motivation of the actors. Increased monetary incentives in that case tend to reduce self-determination and self-valued competence. This effect has been successfully tested in laboratory experiments (see, e.g., Deci and Ryan, 1985).

c. When a person can *self-organize* his or her tasks to a degree, an application of monetary incentives shifts the locus of control, damaging intrinsic motivation. This particular crowding-out effect has been empirically supported for firms (Kochan et al., 1986). It has also been shown that the more developed the institutions for direct-democratic participation in government are, the higher is tax morale, which is a specific type of intrinsic motivation (Weck-Hannemann and Pommerehne, 1989).

d. Incentive instruments which *fail to distinguish* between persons of high and low intrinsic motivation are considered as unfair and tend to undermine intrinsic motivation. For that reason, the design of fair compensation systems plays a central role in industrial relations theory (e.g. Lawler, 1990).

e. The more incentive instruments are *contingent on performance* desired by the principal, the more strongly the locus of control shifts from intrinsic to extrinsic motivation, and the stronger is the crowding-out effect. This proposition has been empirically supported by laboratory experiments (see Deci and Ryan, 1985).

f. The less incentive instruments *acknowledge* a person's instrinsic motivation, the more strongly it is crowded out because the affected person's self-confidence is reduced.

RELATIVE PRICE AND CROWDING-OUT EFFECTS IN THE ENVIRONMENT

Environmental Ethics as Intrinsic Motivation

There exists an extensive literature on environmental ethics.[1] It is, however, almost exclusively normatively oriented, and there are only few serious explanatory studies about the extent and occurrence of environmental ethics in the population.[2]

A positive approach to environmental ethics starts from the net cost (i.e. the cost after deducting possible benefits) of pursuing ethical concerns in individual behaviour. Following North (1990, pp. 43-44), who suggested a downward sloping demand curve for ideology in general, we hypothesise that the extent of environmental ethics applied is larger, the lower the personal net cost is

---

1 E.g. Attfield (1983), Miller (1990), Des Jardins (1993). There is also a specialised professional journal of *Environmental Ethics*.

2 A major shortcoming of the few studies devoted to a positive analysis is that they do not clearly distinguish environmentally conscious behaviour induced by benefit-cost considerations, and that behaviour induced by 'pure' ethical concepts; see e.g. Rodman (1983) and Des Jardins (1993).

of doing so *(figure 1)*. It is thus suggested that individuals are prepared to apply environmental ethics in their behaviour when it costs them little (see Kliemt, 1986; and Kirchgässner, 1992, on 'low cost decisions'). The more costly it gets, the lower the weight of ethical concerns. When the cost is very high, only ecological extremists still follow the principles of environmental ethics, while ordinary people find a lot of reasons why they should desist from doing so.

It is crucial to carefully identify what 'cost' means in this context. A firm under perfect competitive conditions has a high cost (it faces bankruptcy) if it follows environmental norms and the other firms do not. This statement must, however, be qualified in various respects. First of all, to act in an environmentally conscious way may pay for a firm if it can thereby increase its sales or circumvent a boycott or other trouble from environmental action groups. Such behaviour may also be warranted if it preempts impending government environmental policy and helps the firm to be on good terms with the government and its administration, on which the firm depends, e.g. on future orders. Finally, few firms are subject to perfectly competitive conditions; rather, most of them have sufficient discretionary room to deviate to some extent from profit-maximizing behaviour.

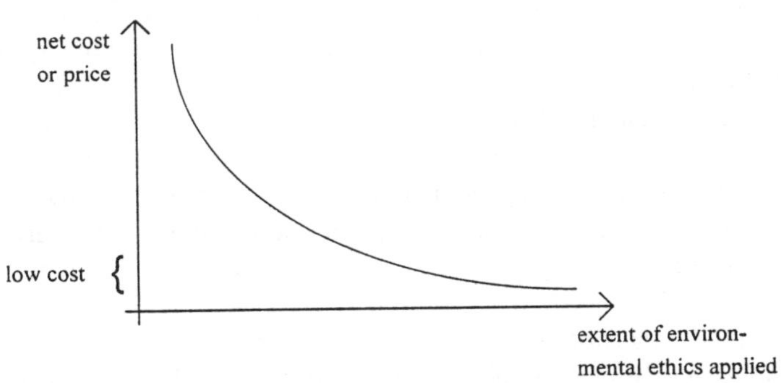

Figure 1 The Demand Curve for Environmental Ethics

The survival argument just discussed for firms is relevant for individuals in wartime or in poor developing countries where people die from hunger if they do not fully exploit all opportunities. Under these conditions, the cost of environmental ethics is exceedingly high. It cannot therefore be expected, for example, that such persons refuse to burn wood or plunder the tree stock -- even if they are aware that there is an unfortunate outcome in the long run (i.e. the erosion of the soil).

For most people in industrial countries, however, the crude survival argument is not relevant. For (professional) environmental activists, such as employees of Greenpeace, the pursuit of environmental ethics is even connected with positive net benefits in terms of prestige, attention, and promotion prospects. The same holds for persons living in groups and contexts in which ecological concerns are taken as a kind of religion (Spretnak, 1986). Even for individuals not living under such conditions, environmental ethics may be connected with low or even zero cost insofar as small actions are concerned. Examples are collecting household refuse according to ecological principles (e.g. separating out glass or batteries), abstaining from throwing trash on the road, or in the case of catastrophies (Baumol and Oates, 1979, pp. 296-99; for post-catastrophy behaviour, see DeAlessi, 1968, 1975). Another important case in which the cost of pursuing environmental ethics is low is voting in elections and popular referenda. Each voter only has a minuscule impact on the total vote outcome and casting a vote in favour of environmental projects is therefore nearly costless.

There are thus important circumstances in which individuals and firms pursue environmental ethics, and this behaviour is at least partially determined by intrinsic motivation, so that the crowding-out effect becomes potentially relevant.

When are Incentive Instruments Effective?

The effectiveness of environmental incentive instruments depends in our context on two factors: (a) the importance of environmental ethics for behaviour, which is determined by the associated cost; (b) the intensity in which the incentive instruments are used, i.e. the extent of external intervention affecting individuals. For simplicity, the interaction of these two determinants is analysed by distinguishing in each case only a low level and a high level. Table 1 illustrates the outcome expected when both, the cost of environmental ethics and external intervention, are either low or high.

Table 1 Environmental Outcomes

		cost of applying environmental ethics	
		low cost	high cost
extent of external intervention (application of incentive instruments)	low intervention	(1) environment preserved	(2) environment destroyed
	high intervention	(3) countervailing effects on environment	(4) environment preserved

The four paradigmatic outcomes are now discussed in sequence.

a. *Low cost of environmental ethics, low extent of intervention*

Under these conditions, individuals and firms exhibit a high amount of environmental ethics. As the corresponding behaviour is not influenced by the application of emission charges or tradeable licenses, the environment will be preserved. The area in which these conditions hold may be called *'voluntaristic environmentalism'*. Examples are the separation of household refuse or careful disposal of wastepaper or chewing gum in public spaces. The voluntaristic environmental area has been neglected by economists, but plays a great role in the thinking of laypeople, who often tend to overestimate its importance. Many environmentalists seem to believe that most environmental issues belong to that area.

b. *High cost of environmental ethics, low extent of intervention*

In this area environmental ethics is not active because it burdens individuals and firms too heavily, nor is behaviour influenced by the application of environmental instruments. There is thus neither an internal nor an external motivation to protect the environment.
Examples for this area are:

- Environmental behaviour in centrally planned economies where the cost of behaving ethically was high, as production plans had to be met or even surpassed (at least in formal terms). At the same time, central economic policy did not care about ecological issues. As a result, grave environmental damages occurred -- as we are now, after the breakdown of the communist systems, fully aware.
- The destruction of forests and the consequent soil erosion in developing countries, where individuals cannot afford to behave ethically towards the environment, and where there is no environmental intervention by the government.
- Newly arising market economies without environmental policy, where people enter the consumption phase and subjectively find it costly to behave according to environmental norms (in particular because of the need to keep up with the 'Jones').
- International environmental problems such as the greenhouse effect or the preservation of oceans, where many people find it difficult to establish a strong connection between their own behaviour and the global ecological outcomes, and where environmental ethics is therefore costly. At the same time, the public-good nature of the problems makes it hard, or even impossible, for governments to cooperate and to externally intervene.

c. *Low cost of environmental ethics, high extent of intervention*

When incentive instruments are used to fight environmental pollution, and individuals follow environmental ethics, a *crowding-out effect* is induced. The question is whether the destructive effect on intrinsic environmental motivation is stronger than the behavioral change induced by the relative price effect of the environmental instruments. The outcome depends on the six conditions generally identified above as prerequisites for the crowding-out effect:

In *personalised relationships* the crowding-out effect is strong. This condition may apply for ecological activity groups that rely strongly on the cohesion between particular persons, where the government intervenes to steer environmental behaviour via relative price changes. Such external intervention is likely to destroy such groups' cohesion because their *raison d'être* is fundamentally questioned. If the government fulfils the tasks previously undertaken by such groups, their members may well lose their sense of purpose and direct their intrinsic motivation to other issues. Whether the environment improves or degrades is open.

For *interesting tasks* the crowding-out effect is strong. People who follow the standards of environmental ethics in their behaviour because they find the underlying ecological issues fascinating, experience a severe loss of motivation when the same task is performed by an externally imposed policy instrument.

When environmental problems have been cared for by *self-organization*, external intervention destroys the intrinsic motivation the participants have. As Ostrom (1990) has shown, many common-property resource problems have indeed been successfully addressed by self-organization, but her research also reveals that outside intervention by government often has disastrous consequences on morale.

When the use of incentive instruments does *not differentiate* between persons with high and low environmental ethics, but all are treated as if they had no intrinsic motivation with respect to the preservation of nature, the crowding-out effect is strong. An example are firms that use their discretionary room to voluntarily safeguard the environment but that are nevertheless subjected to the same regime faced by firms that do not care for ecological issues. This happens when an environmental tax for administrative reasons is based on the sector's average emission. In this case environmentally conscious behaviour is not recognised, and it is likely that the managers of those firms react by adjusting their pollution to the sector's average level (see Wicke, 1991, pp. 369-70).

The more strongly *rewards are contingent on behaviour*, the more environmental ethics is driven out. Economic incentive instruments are favourably based on a close relationship to behaviour so that the relative price effect works at its best. At the same time, however, this makes intrinsic motivation unnecessary, crowding it out. An example is household refuse priced according to quantity. While each household has an extrinsic incentive to minimize its quantity, environmental ethics becomes superfluous.

Environmental ethics is crowded out when the environmental policy instruments applied *fail to acknowledge* affected persons' intrinsic motivation to protect the environment. This crowding-out effect particularly applies in the case of tradeable licenses. Licenses allow a specific amount of pollution and do not morally condemn damaging the environment as long as the emitter stays within these limits. In this particular sense, *once* a license has been acquired, there is a 'license to pollute.' All intrinsically motivated environmental concerns by this procedure are implicitly declared to be irrelevant. An environmental policy built on tradeable licenses thus actively undermines environmental ethics (for a similar argument, see Kelman, 1981).

d. *High cost of environmental ethics, high extent of intervention*

In this area, environmental incentive instruments work best. Environmental ethics does not affect behaviour, so that the relative price effect of emission charges and tradeable licenses fully determines behaviour, provided the other conditions necessary for the application of these instruments apply. In particular, (i) the transaction cost of applying incentive instruments must be so low that it pays to employ them; (ii) politicians and administrators must understand the working of incentive instruments to apply them properly; (iii) interest groups that lose from an efficient environmental policy must not have the power to block or undermine incentive instruments; (iv) public officials must be made to effectively apply the incentive instruments. This last point is especially non-trivial -- as bureaucrats often prefer direct regulations, as these give them more discretionary power.

CONSEQUENCES FOR ENVIRONMENTAL POLICY

Our analysis helps to explain why the environmental incentive instruments have been applied so little for such a long time. Even today their use is quite restricted and experts do not expect that they will play a major role in future environmental policy. While there are many politico-economic reasons why such instruments are not applied (see Weck-Hannemann, 1994), our discussion suggests quite another reason (see also Frey, 1992b): people may be aware of the damaging effects incentive instruments may have on environmental ethics. They abhor the moral vacuum with which incentive instruments are connected.

People may also be concerned about the *spillover effect* prompted by the destruction of environmental ethics induced by incentive instruments.That the environment is seen as a whole is supported by the fact that in many countries there are green parties and interest groups based on this one issue (Parkin, 1989). Decisionmakers are thus afraid that the use of incentive instruments has a double effect: the quality of the environment can be improved in those areas in which environmental charges and/or tradeable licenses are applicable, but environmental quality is lowered in all other areas because the guiding environmental ethic is weakened or is completely destroyed. This destruction of environmental ethics hampers individuals' willingness to accept any kind of action to fight pollution, i.e. political support for environmental policies would also be negatively affected. This explains why even those groups that claim to fight for green interests may oppose pricing as an instrument of environmental policy: it would erode the basis of their popular support. Provided the motivation, crowding-out and spillover effects are sizeable enough, a strong counterproductive effect takes place and 'overall' environmental quality falls when incentive instruments are used. Under these conditions their application leads to the opposite of what was intended.

Pollution is viewed as a mainly moral issue, both by many decisionmakers as well as by the general population (see e.g. Blinder, 1987, pp. 147-54). In the United States, where the use of pricing for resource allocation is more accepted than it is in Europe or Japan,[3] motivation crowding-out and spillover are of less importance, leading to the proposition that incentive instruments in environmental policy are used more in the US than in Europe. Another proposition derived from the analysis is that the motivation crowding-out and spillover effects are stronger for consumers than for firms, which cannot pursue environmental ethics if they are in a strongly competitive situation. It is, moreover, to be expected that firms supplying intermediate products are less subject to motivational effects of incentive instruments. Finally, the theoretical analysis suggests that those instruments are preferred which entail a moral condemnation of pollution (regulations and, even more so, subsidies that reward environmentally conscious behaviour) compared to tradeable permits and charges, which imply that no moral wrong is connected with pollution. This leads to the empirical propositions that environmental incentive instruments are little used for a policy directed at consumers but rather at firms, and in particular, at firms supplying intermediate products and which are oligopolistic or monopolistic. These propositions are consistent with the evidence (OECD, 1989; and Hahn,

---

[3] One piece of evidence supporting this statement is that American economists, other factors held constant, have a stronger preference for pricing solutions than do European economists (Frey et al., 1984). That American workers are more prepared to accept the price system than are their Japanese and European counterparts; see also Beer et al. (1984, p. 116) and Aoki (1990).

1989): environmental charges or tradeable permits are rarely if ever applied directly to consumers, but rather to firms, though standard economic theory argues that it works well (Baumol and Oates, 1979). Indeed, the only successful application of tradeable permits in the United States is to oligopolistic petroleum refineries to induce them to reduce the lead content of gasoline (Hahn, 1989). On the other hand, regulations, subsidies, and deposit-refund programs (supporting environmental ethics) are predominantly used in all countries -- compared to those pricing instruments that tend to be interpreted by actors as a direct or indirect 'license to pollute.'

Our analysis also makes it possible to draw conclusions with respect to what environmental policy is *desirable*. It suggests that there are definite *limits* to the application of pricing to fight pollution (and elsewhere). Economists who have rightly stressed the strong behavioral effects of relative price changes have not considered sufficiently, if at all, that there are areas in which incentive instruments are much less effective due to the countervailing effect of crowding out environmental ethics. Our analysis aims to identify those areas for practical environmental policy where the crowding-out effect is likely to be strong, and where it need not to be taken into account.

Moreover, environmental incentive instruments should be modified in order to minimise possible crowding-out effects. Most importantly, economists should change the attitude with which they argue for incentive instruments. Most people concerned about the environment find it cynical to state that once one has paid a price one is completely free to damage the environment. This Friedman type of thinking and arguing[4] (while perfectly acceptable among professional economists) is strongly objected to by practically all non-economists. Indeed, Blinder (1987, p.148) has found that "environmentalists ... want business to reduce pollution because it is the right thing to do, not because it is in their financial interest." Economists should acknowledge that the motivational structure of individuals is more complex than in their model. Once they accept that behaviour is not solely motivated by extrinsic inducements, but to a significant extent also depends on intrinsic motivation, they must become aware that their cynicism has considerable cost by damaging environmental ethics and prejudicing voluntary action.

Incentive instruments should be applied in conjunction with statements stressing that a person's or firm's intrinsic motivation to protect the natural environment is highly valued, and the use of incentive instruments should mainly be suggested for areas in which environmental ethics is weak or non-existent.

---

[4] Summarized by his well known statement: "The only business of business is to make profit."

What we thus propose is a partial rehabilitation of moral appeals in environmental policy -- without giving up incentive instruments.

References

Aoki, M. (1990). "Towards an Economic Model of the Japanese Firm." *Journal of Economic Literature* 28, 1-27.

Attfield, R. (1983). *"The Ethics of Environmental Concern."* New York: Columbia University Press.

Barkema, H.G. (1995). "Do Top Managers Work Harder When They Are Monitored?" *Kyklos* 48/1, 1942.

Baumol, W.J. and W.E. Oates. (1979). *"Economics, Environmental Policy, and the Quality of Life."* Englewood Cliffs: Prentice Hall.

Becker, G.S. (1976). *"The Economic Approach to Human Behavior."* Chicago: Chicago University Press.

Beer, M. et al. (1984). *"Managing Human Assets."* New York: Macmillan.

Blinder, A.S. (1987). *"Hard Heads, Soft Hearts."* Reading: Addison-Wesley

Cropper, M.L. and W.E. Oates. (1992). "Environmental Economics: A Survey." *Journal of Economic Literature* 30/2, 675-740.

DeAlessi, L. (1968). "The Utility of Disasters." *Kyklos* 21, 525-533.

DeAlessi, L. (1975). "Toward an Analysis of Postdisaster Cooperation." *American Economic Review* 65/1, 127-138.

Deci, E.L. (1971). "Effects of Externally Mediated Rewards on Intrinsic Motivation." *Journal of Personality and Social Psychology* 18, 105-115.

Deci, E.L. and R.M. Ryan. (1985). *"Intrinsic Motivation and Self-Determination in Human Behavior."* New York: Plenum.

Des Jardins, J.R. (1993). *"Environmental Ethics."* Belmont: Wadsworth.

Doeringer, P.B. (1991). "The Socio-Economics of Labor Productivity." In R.M. Coughlin (ed.), *Morality, Rationality, and Efficiency: New Perspectives on Socio-Economics*. Armouk: Sharpe.

Faulhaber, G.R. and W.J. Baumol. (1988). "Economists as Innovators." *Journal of Economic Literature* 26/2, 577-600.

Frey, B.S. (1992a). *"Economics as a Science of Human Behaviour. Towards a New Social Science Paradigm."* Boston: Kluwer.

Frey, B.S. (1992b). "Pricing and Regulating Affect Environmental Ethics." *Environment and Resource Economics* 2, 399-414.

Frey, B.S. (1993a). "Motivation as a Limit to Pricing." *Journal of Economic Psychology* 14, 635-664.

Frey, B.S. (1993b). "Does Monitoring Increase Work Effort? The Rivalry with Trust and Loyalty." *Economic Inquiry* 31/4, 663-670.

Frey, B.S. (1993c). "Shirking or Work Morale? The Impact of Regulating." *European Economic Review* 37, 1523-1532.

Frey, B.S. et al. (1984). "Consensus and Dissension among Economists: An Empirical Inquiry." *American Economic Review* 74, 986-994.

Goodin, R.E. (1994). "Selling Environmental Indulgences." *Kyklos* 47/4, 573-596.

Hahn, R.W. (1989). "Economic Prescriptions for Environmental Problems: How the Patients Followed the Doctor's Orders." *Journal of Economic Perspectives* 3, 95-114.

Kelman, S. (1981). *"What Price Incentives? Economists and the Environment."* Boston: Auburn House.

Kirchgässner, G. (1992). "Towards a Theory of Low-Cost Decisions." *European Journal of Political Economy* 8, 305-320.

Kliemt, H. (1986). "The Veil of Insignificance." *European Journal of Political Economy* 2/3, 333-344.

Kochan, T., H. Katz and R.B. McKersie. (1986). *"The Transformation of American Industrial Relations."* New York: Basic Books.

Lawler, E. E. (1990). *"Aligning Organizational Strategies and Pay Systems."* San Francisco: Jossey-Bass.

Lee, C. (1987). "The New Employment Contract." *Training* (Dec.), 45-56.

Lepper, M. and D. Greene (eds.). (1978). *"The Hidden Costs of Reward: New Perspectives on the Psychology of Human Motivation."* New York: Erlbaum.

McGraw, K.O. (1978). "The Detrimental Effects of Reward on Performance: A Literature Review and a Prediction Model." In M. Lepper and D. Greene (eds.), *The Hidden Costs of Reward: New Perspectives on the Psychology of Human Motivation*. New York: Erlbaum, 33-60.

Miller, G.T. (1990). *"Living in the Environment,"* 6th ed. Belmont: Wadsworth.

North, D.C. (1990). *"Institutions, Institutional Change and Economic Performance."* Cambridge, MA: Cambridge University Press.

OECD. (1989). *"Economic Instruments for Environmental Protection."* Paris.

Ostrom, E. (1990). *"Governing the Commons. The Evolution of Institutions for Collective Action."* Cambridge, MA: Cambridge University Press.

Parkin, S. (1989). *"Green Parties. An International Guide."* London: Heretic Books.

Rodman, J. (1983). "Four Forms of Ecological Consciousness Reconsidered." In Donald Scherer and Thomas Attig (eds.), *Ethics and the Environment*. Engelwood Cliffs: Prentice Hall.

Spretnak, C. (1986). *"The Spiritual Dimension of Green Parties."* Santa Fe: Bear &. Co.

Stigler, G. and G.S. Becker. (1977). "De gustibus non est disputandum." *American Economic Review* 67, 76-90.

Weck-Hannemann, H. (1994). "Die politische Ökonomie der Umweltpolitik." In Rainer Bartel and Franz Hackl (eds.), *Einführung in die Umweltpolitik*. München: Vahlen.

Weck-Hannemann, H. and W.W. Pommerehne. (1989). "Einkommensteuerhinterziehung in der Schweiz: Eine empirische Analyse." *Schweizerische Zeitschrift für Volkswirtschaft und Statistik* 125, 515-556.

Wicke, L. (1991). *"Umweltökonomie,"* 3rd ed. München: Vahlen.

Chapter 9

THE POLITICAL ECONOMY OF THE ENVIRONMENT IN DEVELOPING COUNTRIES: MARKET FAILURE AND INSTITUTIONAL RESPONSE

Eduardo Wiesner (Consultant)

INTRODUCTION

At a recent symposium on new institutional economics, Ronald Coase (1993, p. 97) stated that the "invisible hand" was an incomplete theory and that he "wanted to move on from Adam Smith." His remark was not intended to refute classical economics, but to underscore the importance of creating the *conditions* that would enhance the capacity of the market forces to deal with externalities and with the problems that arise when property rights are not well defined and when transaction costs are high or undeterminable.

Following Coase's illustrious directive, this chapter[1] focusses on measures and policies that middle-income developing countries (DCs)[2] could adopt to strengthen the capacity of their markets and increase the effectiveness of governmental intervention to deal more adequately with their environmental problems.

---

[1] For reasons of brevity, this paper leaves out several important issues such as carbon taxes, global warming, and transnational pollution. The paper also does not deal with the role played by multilateral institutions in developing countries.

[2] "Developing countries" is a generic term with limited analytical applicability, as there are a myriad of "developing-country conditions." In this paper, the concept relates largely to Latin American countries and to others experiencing a rapid process of industrialization and development (e.g., Indonesia, Thailand, Korea, and The Philippines).

The chapter approaches the problem of pollution control from the perspective of political economy,[3] recognizing that policy advice must deal with real-world restrictions and that policy reform needs a strategy[4] to circumvent resistance and to effectively initiate a process of change. The premise is that some specific measures directly concerned with the environment, combined with others more veiled, Rawlsian,[5] and distant (dealing principally with the rules under which public and private decisionmaking take place), can have a positive effect on the quality of public and market choices.

THE CHOICE OF POLICY INSTRUMENTS

Governments concerned with the environment have two broad sets of interdependent instruments to implement their policies: first, command-and-control instruments and, second, market-based incentives. Using a mix of these instruments, governments should, in principle, be able to correct for market failures and for negative externalities. The question that arises is whether governmental intervention is indispensable to correct for negative and even for positive externalities.

According to Coase (1960, p. 109), if victims of pollution hold secure "rights" to an unpolluted environment, polluters will want to buy and transact those "rights" to allow some level of pollution. Similarly, if polluters have the right to pollute, victims will try to "bribe" them to pollute less. As long as

---

[3] The underlying theme of political economy is the classical question of under what circumstances are markets or governments the preferred institution for allocating societal resources. The tenet of this paper is that in most cases *both* institutions are needed. The challenge is to find the right mix of instruments and policies to improve the performance of the market *and* of governments. This approach is eclectic and pragmatic and starts out from the premise that to simply tell governments to change is often a futile exercise. To actually achieve change, structural and organizational conditions must be put into place. These "remedies must necessarily be structural, altering in some fundamental way the rules which govern public sector decision-making" (Inman 1987, p. 739).

[4] On the importance of having a strategy for policy reform in DCs, see Harberger (1993).

[5] The concept of the "veil of ignorance" was coined by John Rawls in his widely read *A Theory of Justice* (1971) in which he sets out to develop a set of principles to apply to the basic structure of society. For a more recent discussion of the "veil of ignorance," see his *Political Liberalism* (1993, p. 24), in which the author introduces the distinction between a "thick and a thin veil."

negotiations and transactions are not costly, the resulting amount of pollution would be optimal.

But how realistic is it to assume that the costs of negotiating and enforcing agreements are negligible? When polluters and victims are few, and the number of beneficiaries from an agreement is fixed, negotiations can provide for the internalization of externalities. But in practice, public intervention will, in many instances, be necessary to ensure at least the conduct of efficient negotiations.

Given that intervention is required in most cases, what instruments should be used: command-and-control or market-based instruments?

Price-based Versus Quantity-based Incentives

In a world of perfect knowledge, marketable emission permits would be fully equivalent to unit taxes. As long as no uncertainty exists about abatement costs, price-based incentives, such as taxes, and quantity-based incentives, such as tradable permits, should have the same effect.

However, in a setting of imperfect information about the marginal benefit and cost functions, the results of the two approaches may differ considerably. When producers have information that the regulator does not have, the quantity-based-instruments (which can guarantee an emission level) seem preferable if unexpected emissions are large and particularly costly, as is the case with leaks from nuclear facilities. In contrast, when the marginal damage caused is relatively flat, price-based instruments are better, ensuring that the marginal abatement costs do not exceed the tax rate.

As might be expected, a combination of the two instruments may be the best strategy, i.e., using both price and quantity instruments. The regulator can establish the level of permits that equates expected marginal benefits and costs. He then offers a subsidy for emission reductions in excess of those required by the permits. In addition, the regulator establishes a unit tax to cover the risk should the control costs turn out to be much higher than anticipated. As Oates and Cropper (1992, p. 30) put it:

> "In this way, a combination of price and quantity instruments can, in a setting of imperfect information, provide a larger expected welfare gain than an approach relying on either policy instrument alone."

Complementing the mixed use of market-based and command-and-control instruments is the role of indirect instruments. When environmental damage cannot be tackled directly because of high monitoring costs (or for other reasons), indirect instruments aimed at the outputs and inputs[6] of the polluting activity can be used. Indirect pollution taxes applied to gasoline and coal are examples.

Market Failure and Policy Failure

Although much of the mismanagement of the environment and of natural resources is the result of market failures, i.e., prices that do not reflect the true social costs and benefits of resource use, policy failure can often also lead to serious environmental degradation. Ideally, government intervention -- through taxation, regulation, institutional reform and incentives -- should correct or mitigate market failure.

However, government intervention does not automatically correct market failures. In practice, government intervention and policies often introduce additional distortions. The reasons are many and varied. Government intervention and policies often lead to unintended consequences and have unforeseen side effects. Policies that are seemingly unrelated to the environment may affect it substantially and negatively. Policy failures include both the failure to intervene when necessary and beneficial, and the failure to refrain from intervention when unnecessary and detrimental (Panayotou, 1993).

Many DCs encourage the consumption and production of energy by providing subsidies for coal, gas, and electricity. Oil consumption is subsidized in oil-exporting countries. Larsen and Shah (1992), who have estimated that total world energy subsidies exceeded US$230 billion in 1990, state

> "Getting energy prices right would prima facie represent a first-order priority in any economic policy response designed to curtail greenhouse gas emissions."

Poterba (1993, p. 47) pointed out that fossil fuel subsidies in DCs are an important impediment to efficient global action to reduce greenhouse gas emissions.

---

[6] For a discussion on whether outputs or inputs should be utilized, see Sandmo (1976).

An intriguing example of the relationship between macroeconomic policies and the environment has been found by Shome (1994b). At a time when many Latin American countries are adopting an open and more market-oriented strategy, the authorities have continued to reduce domestic excises on automobiles.

In brief, the choice of instruments is likely to end up being an exercise in how to articulate the proper mix of command-and-control regulations, on the one hand, with economic and policy incentives, on the other. The proper dosage will require a careful judgement of the potential offered by each instrument vis-a-vis the specific monitoring capabilities and institutional conditions of each country and sectors within countries. The dividing line between the two traditional sets of instruments is not always clear and is seldom a static one. At the end of the day, the "optimal" choice is going to involve a combination of policy instruments: some making use of economic incentives and others not (Rudas, 1994).

POLITICAL ECONOMY AND THE ENVIRONMENT IN DEVELOPING COUNTRIES

With regard to the specific situation of developing countries, the political economy approach developed in this chapter recognizes that policy outcomes are largely influenced, if not virtually determined, by who gains and by who loses when policies change, and not by overall social welfare.

The corollary of this tenet is that, generally speaking, there is little logic in collective action (Olson, 1965) and that the social welfare aggregate does not flow automatically either from the "invisible hand" or from groups that would, in principle, be expected to act in support of their own self interests. In other words, it is extremely difficult to get "individuals" to pursue their joint welfare in contrast with their individual welfare (Ostrom, 1993). This is the well-known problem of the free-rider.[7]

Interest Groups vs Social Welfare

But why would some groups act collectively in the pursuit of their self interest, while other groups are unwilling or unable to do likewise? What would explain

[7] Ostrom (1993) holds that the tendency of individuals and institutions to "free-ride" can be mitigated through particular institutional arrangements to counteract perverse incentives. These arrangements often are a mixture of market and state conditions. See Ostrom, Schroeder and Wynne (1993).

this different behavior? A plausible explanation is that the degree of information that each group possesses about how specifically it is going to gain or to lose, as well as its capacity to organize itself and to utilize that information, are the key determinants of whose collective logic and collective action triumphs. The key elements, then, appear to be "information" and "organization" and the links between the two.[8]

The Quest for a Stronger Invisible Hand

In general, says Stiglitz (1991): "when risk markets are incomplete and information is imperfect... the invisible hand does not work." In this case, the information and the perceived risks by the two groups are asymmetrical. The "organized group" sees clearly the risks of losing or the possibility of gaining under a given possible policy outcome. The other "larger" group sees less clearly the risk of losing, or assumes that the mere "possibility" of an efficient market outcome will render an optimal social outcome. In both cases information, its organization and institutions, are the factors that determine how the "gains" and "losses" are finally distributed.

In brief, institutions, secure property rights, transaction costs, institutional structure, organizational arrangements, the rules of the game, and information appear to be the key elements to strengthen the markets.

A STRATEGY FOR MANAGING THE ENVIRONMENT IN DEVELOPING COUNTRIES

As we have seen, information, institutions and economic organization matter. Their interrelationships, along with their interplay with market forces, determine the political economy realm within which a given level of economic efficiency and social welfare is attained. Sometimes these factors interact to produce dynamic change leading to greater efficiency and welfare; other times, these factors fail to cooperate, as often happens in developing countries.

When a political economy realm is less than propitious to reform, what should policymakers in developing countries do?

The answer offered in this chapter is to implement a strategy that *gradually* builds an institutional and organizational framework favoring a stronger

---

8 On this subject, the seminal article appears to be that of Alchian and Demsetz (1972).

role for the "invisible hand." This answer follows Simon (1955) in the sense that the strategy should not aim at "maximizing" efficiency or welfare but at "satisficing" or finding a course of action that is good enough and that offers the potential for opportune correction. The essence of the proposed strategy has three interdependent fronts of action:

a. Environmental federalism.

b. The dissemination of information concerning the results of independent evaluations.

c. A mixed policy of command-and-control instruments together with economic-based incentives.

The first two lines of action fall under what was called in the introduction "Rawlsian measures," in the sense that they are placed "behind a veil" and removed from direct relevance to the specifics of environmental policy. Although they may only be indirectly linked to the environment, yet they are of crucial importance. They respond to Coase's search for *conditions* to strengthen the workings of the market. In essence, they both enhance competition and lead to more learned "bounded rationality."

Environmental Federalism

In many developing countries, particularly in Latin America, a strong fiscal and political decentralization drive is currently taking place. "The reform agenda of these countries is giving special priority to the subnational public sector and to issues of economic efficiency, equity and governance" (Wiesner, 1994). The basic strategy in these countries is to utilize the advantages of decentralized choice to make their public sectors more efficient. Within this reform agenda, environmental federalism fits aptly, as some of its requirements may be easier to satisfy under a local public-choice framework.

While some environmental problems, like acid rain and global warming, transcend local (and even national) boundaries, others do not. "Where they do not, decentralized regulation may reflect local conditions and tastes better than top-down regulation" (Portney, 1990, p. 284).

Within the particular conditions of some developing countries, the set of environmental problems that falls under the boundaries of local jurisdictions (e.g., waste treatment, watershed protection and air pollution) have been largely neglected, precisely because the attempts to deal with them have been designed from a national perspective.[9]

Interjurisdictional environmental problems may be approached through a process resembling a Coasean bargaining process between local public sectors. Although, obviously, the Coasean conditions of, *inter alia*, secure property rights and low transactions costs are not there yet, the question is whether they would emerge faster under a vertical control structure dictated from above.

In other words, the argument for horizontal interjurisdictional management is a relative one. At a time when fiscal federalism is part of the reform agenda of many developing countries, environmental federalism offers an attractive opportunity for improved policy performance. Transfers and grants from the national government could be used to encourage local governments to develop their institutional base to better manage environmental problems.

The importance of property rights in dealing with environmental issues is widely recognized. The strategy of decentralizing some environmental issues seeks to strengthen the notion of property rights at the local level. That is, devolving to the local citizenry the responsibility for making choices about their private and public property. Local citizens would have to make fiscal and political choices about water, parks, the air, noise, emissions, subsidies, etc.

The jurisdictions that feel that a clean environment is preferable may be able to attract taxpayers that would (Tiebout, 1956) "vote with their feet" by moving to those locations, thus inducing interjurisdictional competition.[10] These localities may levy taxes either for Pigovian reasons or for revenue reasons. Although a proper mix of these two purposes is difficult (Oates, 1991), a strong case can be made for placing the authority for pollution taxes largely in the hands of local authorities. From the perspective of the assignment of expenditure, the national government should address externalities that affect the economy as a whole (Shome, 1994a).

Markets are particularly inefficient in developing countries (Oates and Portney, 1992, p. 94). Under those conditions market-preserving federalism (Weingast, 1989) can make a significant contribution to their functioning and to

---

[9] For a discussion of environmental regulation from an intergovernmental-level perspective, see Estache and Zheng (1993).

[10] Whether local government will establish taxes and emission standards at inappropriate levels to attract investment is debatable. For a rigorous examination of this issue, see Oates and Schwab (1988).

competition. And competition matters.[11] Without competition, markets fail and, if unwarranted intervention follows, policy failure may also ensue, particularly in developing countries. Competition between different jurisdictions will compel "each polity to establish efficient property rights" (Eggertsson, 1993).

Box No. 1
Local Public Choices for the Environment *vis-à-vis* National Public Choices

A public choice can be defined as the antonym of a market choice. The first refers to a choice made by a policymaker, a politician, or a bureaucrat in the public sector. The second refers to a choice made by an individual or a firm in the private sector. How can it be argued that a decentralized choice would lead to better environmental management when, after all, a local public choice is still a public choice?

Enhanced interjurisdictional competition is the answer.

Even though a local public choice is still a "non-market choice," and thus prone to reflect largely the interests of the local politicians or bureaucrats, it is, nevertheless, more subject to competition than is a "national public choice." Interjurisdictional competition will force local public choices to reflect better the environmental preferences of the local constituency.

Regional Corporations charged with environmental responsibilities seem to be an attractive option to deal with some specific environmental problems such as water and air pollution control and the management of renewable resources. Colombia's new legislation contemplates a regional system of corporations through which the country's environmental federalism is being implemented.

The vast literature on rationality, altruism, paradoxes, institutions, information, property rights and transaction costs point out the importance of getting all these factors "right." And, indeed, they need to be "right." But, how can developing countries supply these inputs, given their restrictions and political economy conditions?

---

[11] For a discussion of regulatory federalism and interjurisdictional competition, see Oates (1991).

Box No. 2
Colombia's Environmental Federalism

A good example of the application of the environmental federalism approach is the case of Colombia. After a Constitutional reform in 1991, which established the broad mandate for environmental policy, Colombia's new Law 99 of 1993 created the Ministry of the Environment and the National Ambient System (SINA). The central axis of SINA is a set of 34 "Autonomous Regional Corporations," through which environmental management will be effectively further decentralized.

Colombia is not improvising a federalist approach. Already 18 Regional Corporations are in place and have been dealing with environmental responsibilities for many years. The new strategy builds on that experience and brings local authorities closer to the management of the Corporations.

A special feature of the new system is the allocation of special attributes and resources to Corporations responsible for areas in which cities with more that one millon inhabitants are located. It is anticipated that these large cities will create their own specialized environmental administrative units within their local public sectors.

The government's basic policy goal is to increase citizen participation and public discussion of local environmental issues.

This is largely what Prud'homme calls the problems of "supply efficiency.[12] For reasons of brevity, I cannot discuss here the issues of production efficiency (Campbell et al., 1991). However, it appears that, in principle, the local public choice realm is more conducive to getting those conditions "right" than would be a national framework. "Public choice may no longer be as exciting as it used to be" (Frey, 1993, p. 95), but this may be so in more developed countries. The presumption offered in this chapter is that in developing countries the local public choice model may have considerable potential for developing the "right" conditions for better environmental

[12] For an excellent synthesis on the risks of exaggerating the advantages of decentralization, see Prud'homme (1994).

management. In brief, "local governments have a key role to play." (Prud'homme, 1991).

The Dissemination of Information and Independent Evaluations

The acceptance of "bounded rationality," in the sense that "human beings are capable only of very approximate and bounded rationality" (Simon, 1993, p. 156), still leaves unanswered the question of why some individuals are apparently able to be more rational or less bounded than others are in protecting their self interests.

The answer is that some individuals and groups are better informed than others and are thus more motivated to act. They belong to associations, to labor unions, cartels, lobbies, collusion groups, etc, which have evolved from and around failed government interventions. They constitute what the literature calls rent-seekers.[13] They are the ones who gain from "bad" policies. The losers do not always know how much they are losing, while the winners incur costs to hide information from them (Birdsall and James, 1990). In this context information becomes policy. New information opens up the possibility of new and better policies.

But for us to conclude that because information is important there should be more information is a tautological recommendation. Where would the new information come from and why would it become available in the first place? These are two valid and interdependent questions. The new information should come from independent external evaluations commissioned by the national or by the local authorities. In hundreds of cities and regions in the world there are conspicuous examples of environmental damage that are regarded by local communities as a "public bad" but little is done to correct these situations.

At the heart of the problem is the scant information that the public has about the specifics of the problems (Pearce, 1991). Case studies of contamination of rivers, lakes, the air, and the soil would be the primary source of new information.[14] The dissemination of the results of independent and very concrete evaluations would contribute to the development of a local political constituency

---

[13] For a useful discussion of the costs of rent-seekers to society, see Tullock (1990).

[14] For a review of the studies commissioned by the Colombian government as part of its process of reform, see Rodríguez (1994), and Uribe and Yamiro (1994).

for the environment.[15] This is the ground on which to build political support for taxes and for the application of the Polluter-Pays-Principle.

Political leadership will be needed for the case studies of pollution to be commissioned. Granted that political leadership is an imponderable factor, the argument here is that under environmental federalism it may be a more plausible development than it would be under a national vertical structure.

The Mix of Instruments in DCs

It is virtually impossible to determine *ex ante* the precise combination of instruments that developing countries should adopt.[16] What this chapter suggests is that the best strategy is to give a significant role to local jurisdictions in determining the "right mix" of instruments. A local public choice positions citizens closer to the most pressing environmental problems in developing countries. Local residents are the first to suffer or to gain from contamination or environmental protection. The combination of "information," as discussed above, with the proximity to the problems and to the potential solutions "supports the case for environmental federalism" (Oates and Cropper, 1992). Under a decentralized strategy, information requirements appear less difficult to satisfy. Taxes and the application of the Polluter-Pays-Principle, along with other market-based incentives, have a better chance (Panayotou, 1991).

Some of the more refined instruments are difficult to apply in developing countries. Auction systems, for instance, tend not to be used (Boadway and Flatters, 1993). Given their particular conditions, it appears that a reasonable policy line would be the reduction of subsidies and the increase of taxes. The application of the Polluter-Pays-Principle (PPP) could be utilized more, even though the optimal[17] conditions for its implementation may not be found. According to Uribe and Ernesto (1994), taxes and fees will play a key role in Colombia's new environmental policy.

---

[15] For the importance of external and independent evaluations to reform public sectors, see Wiesner (1993).

[16] According to Pearce and Turner (1990) "The main form of regulatory instrument used throughout the world is the *standard*."

[17] For the discussion of the Polluter-Pays-Principle, PPP, its possibilities and restrictions, see OECD (1989).

SUMMARY AND POLICY CONCLUSIONS

When developing countries find that the markets fail in their attempts to deal with their environmental problems, should they follow their proclivity for government intervention? The answer suggested in this chapter is that a better overall strategy would be to strengthen the *conditions* under which the markets *and* government intervention are less likely to fail. Instead of reacting to market failure with direct government intervention and presence, DCs should gradually build an institutional and macroeconomic organizational structure that favors a stronger role for the "invisible hand" *both* within the markets and within the government itself.

The proposed strategy has three interdependent components: (a) environmental federalism, (b) the dissemination of the results of independent evaluations of specific local or regional pollution problems, and (c) the reduction of subsidies and a more intensive use of pollution charges.

Environmental federalism relies on decentralized choice and on interjurisdictional competition to induce better local environmental public choices. Independent evaluations would provide new information with which to ameliorate the political economy restrictions prevailing in many middle-income DCs. With regard to the mix of instruments, a more intensive use of economic-based instruments is considered to be the preferable policy.

References

Alchian, A. and H. Demsetz (1972). "Production, Information Costs, and Economic Organization." *American Economic Review* 62, 777-795.

Birdsall, N. and E. James. (1990). "Efficiency and Equity in Social Spending: How and Why Governments Misbehave." *Working Paper* 274. Washington, DC: World Bank.

Boadway, R. and F. Flatters. (1993). "The Taxation of Natural Resources." *Working Paper* 1210. Washington DC: World Bank.

Campbell, T. *et al.* (1991). "Decentralization to Local Government in LAC: National Strategies and Local Response in Planning, Spending and Management." World Bank, Regional Studies Program, *Reports* 5, 6. Washington DC: World Bank.

Coase, R. (1960). "The Problem of Social Cost." *Journal of Law and Economics* (October).

Coase, R. (1993). "Coase on Posner on Coase," *Journal of Institutional and Theoretical Economics* 149/1, 96-98.

Eggertsson, T. (1993). "Mental Models and Social Values: North's Institutions and Credible Commitment." *Journal of Institutional and Theoretical Economics* 149/1, 24-28.

Estache, A. and K. Zheng. (1993). "Pollution Control in a Decentralized Economy: Which Level of Government should Subsidize what in Brazil?" *Working Paper* 1066. Washington, DC: World Bank.

Frey, B. (1993). "From Economic Imperialism to Social Science Inspiration." *Public Choice*, 95-105.

Harberger, A. (1993). "The Search for Relevance in Economics." *American Economic Review* 83/2, 1-16.

Inman, R. (1987). "Markets, Governments and the New Political Economy." In A.A. Auerbach and M.F. Feldstein, *Handbook of Public Economics* II. Amsterdam: North Holland.

Larsen, B. and A. Shah. (1992). "World Energy Subsidies and Global Carbon Emissions." Background Paper 25. *World Development Report* 1992. Washington DC: World Bank.

Oates, W. (1991). "Pollution Charges as a Source of Revenue." Egon Sohmen Conference on Economic Evaluation and Environmental Concerns. Linz, Austria.

Oates, W. and M. Cropper. (1992). "Environmental Economics: A Survey." *Journal of Political Literature* 30/2, 675-740.

Oates, W. and P. Portney. (1992). "Economic Incentives and the Containment of Global Warning." *Eastern Economic Journal* 18/1, 85-98.

Oates, W. and R. Schwab. (1988). "Economic Competition Among Jurisdictions: Efficiency Enhancing or Distorting Inducing?" *Journal of Public Economics* 35, 333-354.

Olson, M. (1965). *The Logic of Collective Action*. Cambridge, MA: Harvard University Press.

Ostrom, E. (1993). *Governing the Commons: the Evaluation of Institutions for Collective Action*. Cambridge University Press.

Ostrom, E., L. Schroeder and S. Wynne. (1993). *Institutional Incentives and Sustainable Development*. Westview Press.

OECD. (1989). *Economic Instruments for Environmental Protection*. Paris.

Panayotou, T. (1991). "Economic Incentives in Environmental Management and Their Relevance to Developing Countries." *Environmental Management in Developing Countries*. Paris: OECD.

Panayotou, T. (1993). "Policy Failures and Environmental Degradation." In *Green Markets: The Economics of Sustainable Development*. Cambridge: Harvard Institute for International Development.

Pearce, D. (1991). "New Environmental Policies: The Recent Experience of OECD Countries and its Relevance to the Developing World." *Environmental Management in Developing Countries*. Paris: OECD.

Pearce, D. and K. Turner. (1990). *Economics of Natural Resources and the Environment*. Baltimore: Johns Hopkins University Press.

Poterba, J.M. (1993). "Global Warming Policy: A Public Finance Perspective." *Journal of Economic Perspectives* 7/4, 47-63.

Portney, P. (1990). *Public Policies for Environmental Protection*, Washington DC: Resources for the Future.

Prud'homme, R. (1991). "Can Local Governments Stop Environmental Degradation in Central Europe?" Paris: OECD Conference on "National, Regional and Local Development in Multi-Party Democracies: Will Decentralization Succeed?" Jointly organized with the Kennedy School of Government, Harvard University.

Prud'homme, R. (1994). "On the Dangers of Decentralization." Policy Research Department *Working Paper* 1252. Washington DC: World Bank.

Rawls, J. (1971). *A Theory of Justice*. Cambridge, MA, Harvard University Press.

Rawls, J. (1993). *Political Liberalism* New York: Columbia University Press.

Rodríguez, M. (1994). "Medio Ambiente y Desarrollo: La Agenda de Rio." *Revista de Planeación y Desarrollo* 25. Bogotá: Departamento Nacional de Planeación.

Rudas, L.G. (1994). "Instrumentos Económicos para la Protección Ambiental: una Alternativa ante los Errores de Política y las Fallas de Mercado." *Revista Planeación y Desarrollo*, 25/1. Bogotá: Departamento Nacional de Planeación.

Sandmo, A. (1976). "Direct versus Indirect Pigouvian Taxation." *European Economic Review* 7, 337-49.

Shah, A. and B. Larsen. (1992). "Carbon Taxes, the Greenhouse Effect, and Developing Countries." *Working Paper* 957. Washington, DC: World Bank.

Shome, P. (1994a). "Fiscal Federalism: Revenue, Expenditure, and Macro Management." Paper Presented at CEPAL's VI Regional Seminar on Fiscal Policy, January 24. Santiago, Chile.

Shome, P. (1994b). "Recent Tax Policy Trends and Issues in Latin America." Paper presented at a Seminar Co-sponsored by the Catholic University of Río (PUC), The Central Bank of Brazil, and the International Monetary Fund, Mangaratiba, March 16-19. Río de Janeiro, Brazil.

Simon, H. (1993). "Altruism and Economics." *American Economic Review* 83/2, 156-161.

Stiglitz, J. (1991). "The Invisible Hand and Modern Welfare Economics." *Working Paper* 3641. Cambridge, MA: NBER.

Tiebout, C. (1956). "A Pure Theory of Local Expenditures." *Journal of Political Economy* 64.

Tullock, G. (1990). "The Costs of Special Privilege." In J.E. Alt and K.E. Shepsle (eds.), *Perspectives on Positive Political Economy*. Cambridge, MA: Cambridge University Press.

Uribe, E. and M. Yamiro. (1994). "La Contaminación Industrial en Colombia," *Revista de Planeación y Desarrollo* 25. Bogotá: Departamento Nacional de Planeación.

Uribe, B.E. and S. Ernesto. (1994). "Instrumentos para el Control de la Contaminación Industrial" *Revista de Planeación y Desarrollo*. Bogotá: Departamento Nacional de Planeación.

Weingast, B.R. (1989). "Constitutions as Governance Structures: the Political Foundations for Secure Markets." *Journal of Institutional and Theoretical Economics* 149/1, 286-311.

Wiesner, E. (1993). "From Macroeconomic Correction to Public Sector Reform." *Discussion Paper* 214. Washington DC: World Bank.

Wiesner, E. (1994). "Fiscal Decentralization and Social Spending in Latin America: the Search for Efficiency and Equity, Economic and Social Progress in Latin America." *Working Paper*. Washington DC: Inter-American Development Bank.

Chapter 10

POLLUTION TAXES AS A SOURCE OF BUDGETARY REVENUES IN ECONOMIES IN TRANSITION

Tomasz Żylicz (Warsaw University)

POLLUTION TAXES IN POLAND

Like other countries in Central and Eastern Europe, Poland has a mixed system of standards and environmental and resource taxes to address environmental problems. Through a series of rate increases, Poland's pollution taxes are among the highest in the world and generate substantial revenue that is used to finance environmental improvements. A number of tax measures are also employed to encourage environmental investments, recycling, and consumption of "green" products.

Environmental and resource taxes have existed in Poland since the 1970s. Early on they were introduced to motivate polluters to adjust their behaviour with regard to environmental protection requirements, and many environmental analysts in Poland insisted that such taxes were not only superior to alternative instruments, but also a prerequisite for a rational environmental management system. The analysts seemed to overlook the fact that financial instruments were doomed to fail in an economy where all essential inputs were allocated administratively and plant managers had little incentive to pay attention to price stimuli. Instead of addressing the real issues, environmental policymakers complained that the taxes were too low, and poor environmental performance was linked to the inadequate level of these taxes.

Initially, Poland's environmental taxes were indeed rather low. However, they were raised repeatedly, reaching their peak pre-reform level in 1987. It should have been obvious, though, that even these relatively high taxes would have no effect in the context of the old system of central planning.

In the long run, one would expect that even low taxes might have worked in the right direction, yet their actual incentive effect throughout the 1980s was negligible. The revenues were earmarked for environmental investment, but even in this role these taxes largely failed. Under the central planning regime, financial flows were of secondary importance, as resource allocation was predominantly achieved through administrative measures (e.g. rationing). Needless to say that administrators attached low priority to environmental goals. Consequently, resources were allocated to other purposes, irrespective of where the money was.

Faced with increasing inflation, and lacking confidence in "environmental attitudes" of the government, environmentalists managed to introduce legislation in 1989 pegging tax rates to the official inflation index. The purpose of this legislation was to maintain the real level of the taxes and protect the taxes from political manipulation. Unfortunately, proponents of the legislation did not realize that the taxes were pegged to the *ex post* index which, at best, is not available until the middle of the year *after* the taxes are collected. As a result, three-digit inflation in Poland during the second half of 1989 resulted in serious erosion in the real value of tax rates in 1990.

In 1990, the Ministry of Environment (MoE) persuaded the Parliament to decouple tax rates from the inflation index. At the same time, the MoE persuaded the Council of Ministers to significantly increase the rates to make up for historical slippage due to inflation and to restore taxes to their 1987 levels in real terms, as well as to peg rates to the *ex ante* (predicted) rate of inflation. The MoE was successful in arguing for higher taxes and, in fact, taxes were increased *above* their 1987 levels. The new rates became effective in July 1990 and were increased in real terms until 1992. Because of a series of chaotic political developments in 1992, both charging principles and charging rates were somewhat changed throughout the year. This resulted in revenues of environmental funds being 32% lower than anticipated. On April 1, 1993, new rates became effective that for the most part restored rates to their levels in early 1992, although the new taxes were moderately lower in real terms. For instance, the SO_2 tax -- the single most important source of "environmental" revenue in Poland -- decreased from its peak value of $80 in 1991/92 to $72 per ton (the latter at the prevailing exchange rate of 16,600 zł/$ in April 1993). Throughout the year its real value decreased even further (due to double-digit inflation). At the beginning of the second quarter in 1994, despite another increase, the SO_2 rate fell just below $70 (at the exchange rate of 22,000 zł/$).

Poland has become a country with fairly high pollution taxes. A sequence of significant rate increases supplemented with a new system of more frequent collection changed the operation of the system dramatically. Polish taxes are no longer negligible and, in fact, they rank overall among the highest in the world. Taxes are levied on several hundred pollutants. On a per-ton basis, the rates range from less than $0.05 for CO_2 to more than $45,000 for benzo(a)pyrene and other carcinogenic compounds (see Żylicz, 1994b, for a detailed review of this system). However, almost one half of the pollution tax revenues come from the following three pollutants:

- sulphur dioxide (SO_2): 25%
- nitrogen oxides (NOx): 19%
- saline coal-mining water: 2%.

In principle, the rate structure is rooted in toxicity considerations and has been kept stable for many years. Successive increases are based largely on uniform scaling factors. The rates for newly introduced pollutants have been based on rates already in force and characterized by similar health impacts. It is difficult to trace what determined the initial level of rates in the late 1970s. Most likely it was a compromise between some estimates of marginal damages from pollution and the notion of "what industry could tolerate without significant financial repercussions." Additionally, arguments were put forward that revenues should provide a meaningful contribution to financing abatement projects.

Even though the system that gradually emerged from this is far from consistent, it nevertheless does constitute an indispensable component of Poland's environmental policy. Especially after 1989, it became a key factor in boosting the country's environmental investment expenditures. However, its main effect comes about not so much from charging but rather from the earmarking mechanism, which creates a sizable source of money for abatement project financing.

THE ADMINISTRATION OF THE PRESENT POLLUTION TAX SYSTEM

The most distinctive feature of environmental management in Poland is its extensive system of pollution and resource taxes collected and redistributed to firms and municipalities for abatement and conservation purposes through special funds. These funds are:

- The National Fund for Environmental Protection and Water Resource Management (National Fund)
- Forty-nine regional environmental funds operating in all of the state administration units in the country
- In mid-1993 municipal environmental funds were legally established in all of the 2460 municipalities. However, these funds have not yet been set up everywhere.

The allocation of pollution tax revenues between the funds is regulated by the following rules. Ninety percent of the NOx and 100% of the saline water taxes go the National Fund, where they are further earmarked for nitrogen and saline-water abatement projects, respectively. The centralization of these revenues has been motivated by the fact that successful solutions to the pollution these taxes aim at call for a nationwide approach. (Municipalities, however, do receive 10% of the NOx taxes.) All the other pollution taxes are allocated to the

National, regional, and municipal funds in fixed proportions of 36%-54%-10%, respectively. The SO_2 tax is further earmarked; its revenues are to be spent on sulphur abatement projects. Additionally, there exist water withdrawal fees, which act as resource taxes. For the time being their revenues are channelled into the same three categories of funds as the revenues of the other pollution taxes. The government has made plans, however, to change the proportions in favor of the River Basin Agencies that were established recently.

There are also waste disposal fees and tree-cutting taxes (which do not apply to forestry activities). Revenues from waste disposal fees are allocated to municipal and regional funds on a 50%-50% basis. Tree-cutting taxes flow into municipal funds.

In addition to pollution and resource taxes, "geological" fees have been in use. Throughout the 1980s, a fund based on "extraction fees" was operated. The fund sponsored geological exploration and prospecting activities. In 1991, the fund was discontinued because it was considered inefficient, and geological activities were financed largely out of the state budget. In 1992, geological fees were reestablished, but collection methods were changed and rates were increased from their 1990 levels. Geological concession fees are based on the estimated deposit value net of exploration costs. Mining (extraction) fees are based on the price of extracted minerals. According to the Mining Act, 50% of these fees is distributed to municipalities (without earmarking) and the other 50% is channelled to the National Fund and earmarked for the following expenditures:

- 40% for geological exploration and prospecting
- 40% to assist the mining sector in meeting environmental requirements
- 20% for general environmental purposes of the National Fund.

The mining sector in Poland has been notorious for its environmental impact. Likewise, its products (e.g. coal) are often major sources of pollution at further points of production processes. Nevertheless, geological fees were not established as a proxy for pollution taxes -- where it is easier to tax input than effluent. Rather, they are considered to be resource taxes aimed at capturing the rent. They reflect thus the state ownership of the country's main natural resources. Allocating one half of the revenues from geological fees to the national fund should be seen as a decision to permanently earmark a certain portion of the state budget for environmental purposes.

The Polish system includes non-compliance fees as well. While its deterrent role is disputable, its fiscal role is negligible. The collection rate of non-compliance fees is on the order of 20%. Overall, the non-compliance fees contribute merely two percent of environmental funds' revenues. As a result, the funds rely mainly on pollution taxes, whose collection rate is much higher -- on the order of 70% -- and which contribute approximately 80% of tax/fee revenues.

The Polish environmental funds enjoy a fair degree of flexibility in administering their revenues. Apart from a couple of internal earmarking rules, they can spend revenues on whatever environmental or natural-resource saving projects they choose and in whichever form -- subsidy, concessional loan, or joint-venture -- they deem appropriate. Loan repayments to the National Fund accounted for almost 30% of its total revenues in 1993. In other funds, loan repayments are negligible or nonexisting for the time being, since until recently these funds could spend the money only in the form of subsidies.

Finally, there are other entities -- such as the Forestry Fund or Farmland Protection Fund -- which administer fees somewhat resembling resource taxes. They will not be addressed here. The former is a very peculiar fund ($300-400 million annually) serving redistributive purposes in the State Forest Enterprise. The latter is of negligible revenue interest (less than $10 million annually). The funds operate separately from the environmental funds mentioned above.

Most interesting in this chapter are the pollution taxes, which presently raise as much as $400-500 million per year. Revenues vary on account of the varying rates and collectability of these taxes rather than on variation in emissions. (These have been slowly but steadily declining since 1990.) Pollution taxes raise substantial revenues for environmental funds, where they meet with other proceeds and are spent on environmental projects.

OTHER ENVIRONMENT-RELATED FISCAL INSTRUMENTS

To the extent that the pollution taxes in Poland can be deducted from taxable incomes and are applied for environmental policy purposes, they can be viewed as indirect budgetary contributions of the state to environmental investment programs. In principle, the state budget in Poland does not directly contribute to environmental investment expenditures.

There are a number of taxes or duties (other than pollution and resource taxes) that can be regarded as "environmental," since it would be difficult to conceive of a fiscal measure that is fully "neutral" with respect to natural resource scarcity or the environment, i.e. provides neither incentives nor disincentives for better environmental performance. However, these taxes have had only minimal environmental effects, and their introduction was motivated almost entirely by non-environmental considerations. Virtually any tax has an environmental impact. While there were no studies conducted to assess the environmental aspects of general taxation in Poland, several observations can be made on the basis of a cursory examination of the tax rules, rates, and exemptions.

Generous tax allowances for enterprises that invested in environmental projects were discontinued in 1991. Nevertheless, some allowances still exist that marginally improve the profitability of environmental investment, recycling, conservation, etc. (Paradysz, 1992).

The sales tax in Poland (replaced by the Value Added Tax, VAT, in July 1993) differentiated among various product and service categories. As a rule, basic consumption goods were taxed lower (7% VAT) than the others (22% VAT). Excise taxes are levied on several categories of goods and services such as consumer electronics, alcohol, and cigarettes. Also liquid fuels are subject to high excise. Taxes on gasoline are now on the order of 50% of retail prices, much higher than taxes in the United States, but still lower than in the European Union. The MoE has consistently recommended differential gasoline taxes so as to make the price of lead-free brands slightly cheaper than leaded brands. While in principle no one objected to this idea, the inability of tax officials to effectively check for the lead content of imported fuels (declared as lead-free) postponed its implementation until 1992. (The difference is on the order of $0.02 per liter now.)

The MoE has also proposed differentiation of import duties so as to provide incentives for importers to substitute environmentally friendlier products for "dirtier" ones. However, only in the case of passenger cars has such a principle been implemented. There is a complete ban on the importation of cars that are more than ten-years old. For second-hand cars, the amount of duty increases with their age. Other proposals were to lower or, perhaps, even abandon, duties on environmental protection goods and monitoring equipment. Likewise, there was a proposal to levy very low duties on products whose domestic manufacturers were heavy polluters. Few of these proposals were incorporated into the new import duty system established in 1991. The government decided not to overburden the system with excessive exemptions, as it was already one of the most liberal in the world in terms of the number and value of allowances. Administrative efficacy was another argument for limiting the differentiation.

A number of the fiscal measures implemented by the government and described above act as *de facto* environmental taxes/subsidies. While their incentive role is moderate, they positively contribute to the process of fund-raising. Since they can be seen as indirect subsidies, their efficiency requires serious analysis, especially now -- when the state budget is coming under closer scrutiny in Poland. Given the fact that these measures were introduced either on an *ad hoc* basis -- as "green" ornaments to various bills and regulations -- or to expedite compromise between various departments within the government, their efficiency (and probably their equity and effectiveness) is questionable, even though no existing rigorous studies corroborate this assertion.

THE CURRENT ROLE OF POLLUTION TAXES

It has been well documented that pollution tax systems have failed to protect the environment in Poland and elsewhere in Central and Eastern Europe (see e.g. Żylicz, 1994a). Two factors were responsible for the low effectiveness of taxes: (1) low rates; and (2) the lack of incentives to encourage cost minimizing behavior. Both of these deficiencies have been corrected or at least partially addressed, and some evidence now suggests that the reformed pollution tax system in Poland is starting to achieve positive results in terms of pollution reductions. These positive results, however, cannot be entirely attributed to the increased taxes and the emergence of cost-minimizing behavior. Rather, they are the outcome of the overall market reform, more consistent use of direct regulations, as well as higher taxes.

First of all, switching to the logic of a market economy led to the elimination of some of the most outrageous examples of wasteful uses of resources. This, in turn, contributed to lower emissions of many pollutants (Żylicz, 1994a). Also, after 1989, the government attempted to solve the enforcement problem by making more courageous use of administrative instruments, including forced shut-downs. The government also attempted to change the concept of the non-compliance fee so that most violations of environmental requirements would trigger administrative action. This change, however, turned out to be more difficult to implement than expected, and non-compliance fees still are collected (or at least assessed) in some "chronic" cases of violations as well as for accidental spills.

Some taxes in Poland provide relatively strong incentives to abate. The tax on SO_2 emissions was \$80 per ton at one time. Even after some slippage in the real rate to \$70/ton or so, the tax is still higher than in nearly all other countries. A number of mines are reported to have undertaken investment in coal-washing equipment in response to the declining demand for high-sulfur coal. With respect to other pollutants, tax rates are also high in comparison to similar taxes in OECD countries. However, while the tax on SO_2 provides adequate incentives to stimulate switching to lower sulfur coal, it is still lower than the *marginal* costs of achieving a permanent 30% reduction through installation of scrubber technology -- estimated at \$500-\$600 per ton of SO_2 abated.

The cost-effective or "Pigouvian" level of taxes, as understood by policy theorists -- i.e. equal to the *marginal* cost of abatement needed to meet policy objectives -- would need to be much larger (approximately eight times higher for SO_2 and even higher for other pollutants) than current taxes in Poland. Theoretically, an efficient Pigouvian tax equates marginal abatement costs to marginal benefits. However, because the marginal benefit function is difficult to determine, a variant of the Pigouvian rule is that a tax is cost-effective if the tax is set high enough to encourage polluters to meet environmental requirements. In other words, for an emission standard, the tax would be set high enough to equal the marginal cost of achieving the standard.

Although higher than the rates typically charged, *average* abatement costs are only a fraction of marginal costs. Thus, the current level of taxes in Poland generates revenues but, as in other countries where taxes have been introduced, does not create incentives to reduce pollution. The taxes ensure that there are adequate resources to finance the public's share of environmental protection and restoration. Although the financing needs in Poland are rather large because of the backlog of past neglect, the public's share of environmental investment expenditures should be kept within reasonable limits. One would not like to free polluters from contributing to an abatement project with their own money and thus having incentives to identify least-cost solutions.

As a result of policy steps taken in 1990-91, Poland's environmental public and private investment expenditure *sensu stricto* (i.e. excluding water retention and water supply systems) reached $840 million. Many environmentalists would argue that water retention and water supply is more a part of the environmental problem than its solution. The remainder of the analysis is confined to this narrow definition of environmental protection.

Expenditures on environmental investments in 1991 in real terms were almost 50% higher than they were in 1990 (the highest previous level). As a percentage of the country's GDP, environmental expenditures grew from 0.7% of GDP in 1990 to 1.1% of GDP in 1991. For the first time, environmental investment levels in Poland, in terms of percentage of GDP, were commensurate with levels in OECD countries. In 1992-1993, environmental investment expenditures were kept stable at the approximate level of $850 million. In relation to the growing GDP, however, their share decreased slightly (to 1% in 1993).

The increase in environmental expenditures after 1989 was achieved despite eliminating -- as economically inefficient -- major tax allowances (which had existed for a long time) for environmental investors. 1990 was the first year that Polish enterprises faced hard budget constraints. (In centrally planned economies, enterprises faced so-called "soft" budget constraints -- see Kornai, 1985). In other words, they were constrained not by financial but rather by physical flows or administratively allocated resources. 1990 was also the last year of income tax allowances for enterprises making environmental investments. Therefore, 1990 was the *only* year ever for those allowances to work. Interestingly, the elimination of these tax allowances in 1991 was more than offset by increased expenditures of public environmental funds. Environmental improvements referred to at the beginning of this section were achieved despite a far from perfect performance of the funds.

Virtually all incremental expenditures after 1990 can be attributed to the funds' spending. The funds' share in financing environmental investment projects in Poland is now just below 50%. While positive in terms of short-run environmental effectiveness, this high percentage is nevertheless undesirable from the point of view of economic efficiency. Even though the funds in principle apply abatement-effectiveness criteria in selecting projects, strong evidence suggests that the projects actually selected for financing may not be cost-effective in their respective categories (Jędrzejewska and Stańczuk, 1994). For instance, the cost-effect ratios of the cheapest and the most expensive sulphur-abatement projects financed by the National Fund differ by as much as four orders of magnitude. Even if one confines to the projects between the 10th and 90th percentiles (ranked according to their effects), the ratios are on the order of 1:10.

TO EARMARK OR NOT TO EARMARK?

Obvious efficiency arguments can be brought to bear against earmarking pollution taxes. Nevertheless, a recent OECD workshop devoted to environmental taxation concluded that earmarking could be an acceptable transition measure.

> *In transition economies environmental expenditures will probably contribute not more than 1-2 percent of GDP in the next couple of years (i.e. much less than expenditures on other activities like education or health care, requiring budgetary support). Thus, the risk of inefficiency conventionally associated with earmarked funds is constrained by the absolute level of spending, and it is of minor importance when confronted with risks from letting the environment compete with short-term political priorities reflected in budgetary debates.* (Smith and Żylicz, 1994, p.14)

Consequently, little doubt exists about the usefulness of keeping the Polish environmental funds alive. However, even at their current sub-Pigouvian rates, the pollution taxes in Poland provide financing of almost 50% of the country's environmental investment expenditure. Is it, therefore, advisable to increase these rates to Pigouvian levels?

As noted above, the gap between the tax rate and the marginal cost of meeting policy objectives is eight-fold in the case of SO_2 and higher in the case of other important pollutants. At the same time, these policy objectives for the rest of the decade are rather moderate and imply no radical reductions in emission levels. One important effect of this fact is that increasing the rates by the factor of 8-10 would result in revenues increased by the factor of 6-8. If all of these revenues were then earmarked -- even in the absence of private spending -- they would boost Poland's environmental expenditures to at least 3% of the GDP. While environmentalists would certainly welcome such an effect, this is neither realistic nor perhaps even desirable.

One important corollary is that applying the Pigouvian level of taxes precludes earmarking of the total revenues. In other words, if pollution taxes are meant to raise higher revenues, they ought to be decoupled from environmental expenditures or at least split into a part flowing to environmental funds and a general-budget part. A similar design was proposed in the case of a fuel charge advocated by many environmentalists. Such a charge levied on liquid fuels could substitute for the entire National Fund revenues by simply adding an earmarked 4% on top of the existing 50% excise tax. If introduced, the charge would set a precedent in creating a hybrid tax mechanism. For the time being, there are only two major taxes in Poland whose revenues are split between two entities. These are the personal income tax (shared by the state and respective municipal budgets in an 85%-15% proportion), and corporate income tax (shared 95%-5%). The fuel charge would implement a design of sharing between the state budget and a separate agency. The legal feasibility of such an arrangement is thus a prerequisite for increasing pollution taxes without losing environmental funds.

POLLUTION TAXES AND THE STATE BUDGET

As documented elsewhere (Żylicz (1994a)), the pollution tax system in Poland is rather extensive. It covers significantly more pollutants than do systems in most other countries, and it would be impossible -- at least in the short-run -- to increase its revenues by adding new items. Thus, the increase can be accomplished only by raising tax rates. As noted above, assuming that the collection rate would not deteriorate, revenues could be raised on the order of $2.5 billion per annum by increasing pollution taxes to Pigouvian levels. This sum corresponds to 10% of the state budget and it might thus be of potential interest. If such taxes could substitute for the top 10% of distortionary ones, this would be a substantial improvement in public finance.

However, several issues must be addressed before firm policy recommendations can be made. The impact on polluting industries has to be assessed, the future of environmental funds needs to be clarified, and the stability of tax revenues ought to be demonstrated.

First, increasing pollution tax rates by the factor of eight would have very serious financial consequences for polluting industries. While pollution taxes account on average for less than 1% of production costs, in the group of the top

(112) industrial polluters this share is 2.7%. A recent study of that group revealed important social and political constraints (employment being the most important one) to the internalization of their externalities (Broniewicz, Poskrobko and Żylicz, 1994). The constraints can be softened if the increased pollution tax burden was compensated by rebates on other (distortionary) taxes. The design of such a package may prove to be a difficult task -- as the target group consists of large industrial plants that struggle anyway to meet a host of challenges caused by the period of economic transition.

Second, channelling pollution tax revenues to the general budget would deprive environmental funds of their privileged position. It could be arranged that a portion of the revenues continues to be earmarked -- for the environment's sake -- but, for the time being, environmentalists in Poland have failed to introduce a similar mechanism even at a smaller scale. Without a provision saving the environmental funds, at least for the period of economic transition, such a tax reform is not likely to benefit the environment.

Third, in the long run, pollution taxes act so as to erode the tax base itself. It is therefore necessary to gradually increase tax rates in order to maintain the volume of revenues at a constant level. One constraint (perhaps a distant one) on such increases is the level of marginal benefits from abatement. Taxes exceeding this level would become as distortionary as the ones they were supposed to substitute. But before that ever happens, the very idea of constantly increasing rates may meet with resistance. The rates planned for each period can be announced well in advance to let businesses adjust their investment decisions. However, determining components of the budget too early may be found politically infeasible. Alternatively, it may turn out to be ineffective in keeping the revenues stable, anyway.

Sooner or later any pollutant can essentially be eliminated by an appropriate tax. An example of this process can be found in the Polish air pollution tax system. Pollution inspectors were reporting decreasing industrial emissions of whatever volatile organic compounds that were included in the list of pollutants subject to charging. After a while it turned out that the emissions of one compound could be "substituted" with the emissions of a similar one which -- perhaps as a more exotic one -- had not been included yet on the monitoring list. There are no limits to substitution. Thus the only firm tax base is the one that is "substitution-proof" in this sense. The only taxable item that cannot be avoided easily is energy, which, however, generates a substantial share of budgetary revenues in most countries anyway (in Poland almost 10% of the state budget is raised from excise tax on gasoline). Ultimately, the question thus boils down to taxing fossil fuels more heavily and has received much attention already in the past (see e.g. Weizsäcker and Jesinghaus, 1992).

It has been suggested that water is another "substitution-proof" input into the production process (Pittevils, 1994). However, while there is little doubt that the economy cannot function without water, it can (and, indeed, should) recycle water on a large scale if motivated by a sufficiently high tax. Thus the crucial difference between energy and water as goods to be taxed is that, by the laws of thermodynamics, the former cannot be recycled. Consequently, only energy seems to possess the characteristics that make it a stable tax base.

THE FUTURE OF POLLUTION TAXES IN ECONOMIES IN TRANSITION

Unlike most of the traditional distortionary (capital and labor) taxation, pollution taxes can raise budgetary revenues without imposing efficiency losses. This observation has led to the so-called *double dividend* argument which, in a nutshell, states that environmental taxes not only raise revenues but at the same time correct market failures caused by environmental externalities (e.g. Gee, 1994). There is a growing consensus that things would improve if the tax burden was shifted from capital and labor to pollution (and exhaustible resources).

The arguments above apply to any economies. Those in transition from centrally planned to market systems have to cope with the backlog of environmental neglect of the old regimes as well as to reform their public finances. This creates a unique opportunity for capturing *both dividends* that are so badly needed there (Smith and Żylicz, 1994). Even though pollution taxes seem to make a particularly attractive policy instrument in economies in transition, their role is not a more prominent one than the role they play in developed market economies. In other words, pollution taxes everywhere are typically set below their Pigouvian levels. They also raise revenues that are too low to substitute for other taxes.

The task of increasing pollution taxes to a level that is meaningful both environmentally and in terms of revenue turns out to be difficult (if at all possible) in economies in transition. Evidence suggests that developed market economies start to experiment with taxation only when the degree of environmental disruption has decreased significantly below the level now prevailing in Central and Eastern Europe. A question could thus be asked what are the pros and cons for the European economies in transition of experimenting with Pigouvian taxes now.

The answer to this question has important consequences for both environmental and tax policies. A discussion of the first context was offered in an earlier paper (Żylicz, 1994b). The author argued that if -- e.g. for political reasons -- governments are reluctant to raise pollution taxes, environmental policies can nevertheless achieve cost-effectiveness when alternative instruments are applied. This chapter has put the emphasis in another context, i.e. on fiscal aspects of a likely evolution in pollution taxes. Poland's experience is referred to mainly, but the conclusions are relevant for other economies in transition, too.

Poland is not a typical economy in transition in that the pollution taxes it levies are higher than they are elsewhere. At the same time, the state budget claims a rather low portion of GDP (less than 30% in 1993). Hence their (currently) moderate fiscal role is nevertheless much higher than in other economies in transition. As the conclusion of this chapter envisages a very limited potential revenue role of pollution taxes throughout the rest of the decade, it *a fortiori* applies to other countries struggling with the heritage of the centrally planned system.

To sum up, the recent growth in environmental investment expenditures can be attributed to the rapid increase in pollution taxes that are recirculated through the system of funds. It is too early to observe any spectacular effects of the newly implemented policies. The government has not succeeded in restoring the environment yet. Nevertheless, it would be inappropriate to suggest that its new environmental policies have been fruitless even during this transition period (Żylicz, 1994a). Few, if any, signs point to an ecological miracle in Poland. However, there have been few ecological miracles anywhere else in the world. In all regions that succeeded in recovering from severe environmental pollution such as London, Pittsburg or the Ruhr Basin, success was achieved gradually, as a result of investment programs spanning many years. The effects of these programs became visible only after years of steady investment. Seen from this point of view, the pollution tax system in Poland has become an effective mechanism for generating revenues and has provided needed funds for environmental investment and recovery -- despite some doubts as to the system's efficiency.

THE FUTURE OF ENVIRONMENTAL FUNDS IN POLAND

Pressures from local governments resulted in a fairly extensive decentralization of pollution and resource taxes in 1993. The effectiveness of the revised mechanism is debatable. Its economic efficiency will probably decrease. The same amount of money will be dispersed more evenly, which will stretch the average construction period of projects. Moreover, nationwide environmental priorities will be less likely to receive adequate support. Nevertheless, arguments to return pollution taxes to local entities where the pollution actually takes place are difficult to contest in the absence of analysis and a stronger lobby for regional and national priorities.

Another source of the environmental funds' fragmentation is the anticipated reform of water withdrawal (and, perhaps, sewage discharge) taxes. Plans exist to channel the entire revenues from these taxes -- or at least the revenues of the National Fund -- to the seven River Basin Agencies. Thus, seven new funds are likely to be established.

Anticipating that tax decentralization was inevitable, the MoE in 1990 tried to introduce a fuel charge (on top of the existing excise taxes), averaging 4% of respective retail prices and earmarked for environmental protection. The charge on liquid fuels, natural gas, and lignites could raise as much as $300 million per year (hard coal should be added to the list above as soon as its price is fully freed from state subsidies). A certain degree of rate differentiation was also suggested so as to account for different qualities of fuels (e.g. 8% for leaded vs. 2% for unleaded gasoline, and so on). The proposal met with almost unanimous opposition from political groups and was abandoned (Żylicz, 1994a). Nevertheless, in one form or another, the proposed product charge resurfaces on the political agenda almost every year, either as initiatives of groups of "green" members of Parliament or new ministerial drafts.

The resistance to product charges -- like the one described above -- which are viewed as a source of revenues for the National Fund, is inconsistent with the widespread expectation that 100% of emission taxes and non-compliance fees should be retained at the regional or even municipal level. As long as it is also expected that some funding from a national fund is desirable, the only way to reconcile the two propositions -- without totally ignoring the *Polluter Pays Principle* by charging the general taxpayer -- is to accept the idea of *product charges*. These charges (levied on environmentally harmful goods), which cannot be convincingly claimed by any single regional or municipal administration, would provide a natural source of centrally managed revenues. Therefore, the MoE has proposed that pollution taxes be fully decentralized -- but only when product charges are established to make up for the resulting "losses" by the National Fund. However, prospects for introducing such charges are dim.

SUMMARY AND CONCLUSIONS

Taxing environmental bads is an attractive alternative to conventional distortionary (capital and labor) taxes. Poland has developed a system of pollution taxes that -- even at their sub-Pigouvian levels -- raise significant revenues. These financial resources do not substitute for other taxes but rather are earmarked and spent on environmental protection through a network of special funds. In this role they have proved to be extremely useful in providing a significant public share in financing environmental recovery after decades of neglect. At the rates applied currently, such taxes could not meaningfully substitute for other sources of the state budget.

Environmental programs can progress cost-effectively even in the absence of Pigouvian taxation. As demonstrated elsewhere (see e.g. Pezzey, 1992), under certain assumptions, whatever could be accomplished through such taxes can also be accomplished through marketable pollution permits. In some circumstances the latter can even make a preferred policy instrument. This is why Poland's environmental policy does not view Pigouvian taxes as a prerequisite for its success.

Pollution taxes set at the Pigouvian level corresponding to medium-term environmental policy objectives could possibly raise as much as 10% of the state budget in the short run. In order for them to substitute for distortionary taxes, they would have to be decoupled from the environmental funds to which they now contribute.

In the longer run, the problem is still open how to reconcile the need to build budgets on firm and predictable tax bases with the environmental improvement mechanism which makes this particular tax-base self-eroding. While the emissions of most pollutants can be halved in years rather than decades, energy -- which generates externalities in one form or another anyway -- is perhaps the only solid tax base in this respect.

References

Broniewicz E., B. Poskrobko and T. Żylicz. (1994). "Internalizing environmental impacts of industry in Poland: Preliminary empirical evidence." Paper presented at the 5th Annual Conference of EAERE. Dublin.

Gee, D. (1994). "Eco-nomic Tax Reform: Shifting the Tax Burden from Economic Goods to Environmental Bads." *European Environment* 4, 17-22.

Jędrzejewska, A. and I. Stańczuk. (1994). "Analiza efektywności kosztowej inwestycji finansowanych przez Narodowy Fundusz Ochrony Środowiska i Gospodarki Wodnej." ["Cost-effectiveness analysis of projects financed by the National Fund"]. *Mimeo*. Warsaw University.

Kornai, J. (1985). "The Soft Budget Constraint." *Kyklos* 39, 3-30.

Paradysz, J. (1992). "Economic Instruments in Environmental Protection in Poland". Ministry of Environment. *Mimeo*. Warsaw.

Pezzey, J. (1992). "The Symmetry between Controlling Pollution by Price and Controlling it by Quantity." *The Canadian Journal of Economics* 25/4, 983-991.

Pittevils, I. (1994). "Pollution Taxes as a Source of Budgetary Revenues in Economies in Transition: A Discussion." 50th Congress of the International Institute of Public Finance. Cambridge, MA.

Smith, S. and T. Żylicz. (1994). "Workshop on Taxation and Environment in European Economies in Transition: Rapporteurs' Report of the Discussions and Conclusions." In *Taxation and the Environment in European Economies in Transition*. Paris: OECD.

Weizsäcker, E. and J. Jesinghaus. (1992). *Ecological Tax Reform: A Policy Proposal for Sustainable Development*. London: Zed Books.

Żylicz, T. (1994a). "Environmental Policy Reform in Poland." In T. Sterner (ed.), *Economic Policies for Sustainable Development*. Dordrecht: Kluwer.

Żylicz, T. (1994b). "Taxation and Environment in Poland." In *Taxation and the Environment in European Economies in Transition*. Paris: OECD.

PART V

THE NEED FOR COORDINATION

Chapter 11

THE ROLE OF THE EUROPEAN UNION IN ENVIRONMENTAL TAXATION

Stephen Smith (Institute for Fiscal Studies and University College London)[*]

INTRODUCTION

Many European countries are currently discussing -- and, in some cases, implementing -- environmentally-motivated tax reforms. This chapter considers the extent to which such policy initiatives are purely a matter for policymakers in the member states in the European Union (EU),[1] and the circumstances in which environmental tax policies might require coordination or other action at the EU level. Many of the issues are not, of course, unique to the EU, and are encountered in varying degrees in other federal, or multi-tier, political systems.

The division of policy responsibilities between the EU and its member states is imprecisely defined, and has evolved over time, both as a result of European Court judgements and new agreements between the member states on the development of the Union. Within the field of environmental taxation, policies at the national level are constrained both by the need to avoid measures that protect domestic producers against competition from other member states and by the terms of existing EU agreements on the permissible forms of indirect taxation in member states. Furthermore, policies at the EU level have been

---

[*] The author is grateful to two anonymous referees for detailed comments on an earlier draft. The paper draws on results from the IFS research programme on taxation and the environment, funded by the Environment Programme of the European Community, and by ESRC Research Centre funding for IFS. The opinions expressed are those of the author alone, and not of the sponsors, nor of the IFS, which has no corporate views.

[1] In conformity with current general practise, this paper uses the term "European Union" to include all activities and functions of the European Community; for simplicity, the term is used systematically throughout the paper, including references to the European Community prior to the Maastricht Treaty, except in direct quotations referring to the "European Community."

heavily constrained by the concern of some member states, notably the UK, to prevent any extension of the Union's mandate in the field of taxation.[2]

The principal proposal for environmental taxation made in the EU countries in recent years was the carbon/energy tax proposed by the European Commission in 1991 (Commission, 1991; Pearson and Smith, 1991). The proposed tax formed part of a programme of measures to contribute to the objective of stabilising CO_2 emissions at 1990 levels by the year 2000, a target that the EU member states accepted at the Rio Convention in 1992. The tax was to be levied on the carbon content of fossil fuels and the energy content of all non-renewable forms of energy, at an initial level equivalent to $3 per barrel of oil, rising in stages to a level of $10 per barrel ($88 per tonne of carbon) in the year 2000. The Commission's original plan was for the tax to be introduced during 1993; however, discussion of the original proposal was protracted, and the initial proposal for a substantial, uniform, EU-wide tax now seems unlikely to be implemented. Although member states endorsed the principle of fiscal instruments as a means of controlling CO_2 emissions, some opposed the proposal because of concern about a possible adverse impact on industrial competitiveness, whilst the UK was implacably opposed in principle to increasing the EU's powers in taxation. At the time of writing (early 1995), EU discussion has turned to a common framework to govern the structure of a carbon tax in those member states which might choose to introduce one.

Since the signing of the Maastricht Treaty in 1992, the criterion of "subsidiarity" has governed assignment issues in areas not unambiguously assigned by the Treaty to either the Union or its member states,[3] although the practical implications of subsidiarity in particular areas of policy have yet to be fully tested. The criterion of subsidiarity forms the main organising principle for the discussion in this chapter. In the next section the chapter describes the meaning of "subsidiarity" in the EU, in drawing a boundary between the roles of member states and the Union, and the application of subsidiarity to tax policy. The following section considers the application of "subsidiarity" to environmental policy, and, specifically, to the assignment of policy responsibility for environmental taxation between the EU and national governments. Next, the implications for assignment of the need for integration of environmental taxation with other areas of EU and national policy are considered. A concluding section evaluates.

---

2 Unanimity is required in EU decisions concerning taxation.

3 The earlier Treaty revisions under the Single European Act of 1986 had required the subsidiarity test to be applied to the EU's environmental policies.

"SUBSIDIARITY" IN THE EUROPEAN UNION

"Subsidiarity" as a Criterion for Assignment

The principle of subsidiarity is expressed in the Treaty on European Union agreed at Maastricht in the following way:

> "In areas which do not fall within its exclusive competence, the Community shall take action, in accordance with the principle of subsidiarity, only if and insofar as the objectives of the proposed action cannot be sufficiently achieved by the member states and can therefore, by reason of the scale or effects of the proposed action, be better achieved by the Community" (Article 3b of the Treaty on European Union).

The principle of subsidiarity thus embodies a presumption in favour of decentralisation: transfer of government functions to the European level should only take place where there are good reasons for such an assignment (Padoa-Schioppa et al., 1987). The main argument for decentralisation in government is that it may help to ensure that the decisions taken by government adequately reflect local preferences and interests.

In its references to "scale" and "effects", the Treaty definition also echoes the two principal "good reasons" for EU assignment identified in the 1977 report of the MacDougall Committee (MacDougall, 1977), which drew on theories of fiscal federalism (eg. Olson, 1969; Oates, 1972) to assess the EU's future role in government. This report had identified the existence of economies of scale and the existence of cross-country spillover effects from national policy as the two main circumstances in which a higher-tier, EU, involvement in policy would be warranted. Policy areas where economies of scale or policy spillovers across national boundaries are significant will tend to be those where the EU has a justifiable role, and where transfer of responsibility for policy from member states to the EU may be one solution to the inefficiency of national policy.

However, assigning full responsibility for a particular policy function to the EU is not the only possible response to policy spillovers, and may not be the appropriate choice in many cases. Often it may be possible for the interests of other countries in the policy decisions taken by national governments to be reflected in an agreement between the countries concerned, specifying the way in which policy will operate.

Indeed, the EU plays a mixed role -- part government, part forum for international negotiation. It therefore provides scope for a range of possible degrees of constraint on national policymaking, ranging from ad hoc negotiation over cross-country effects of member states' policies, to the comprehensive takeover of member states' responsibilities in a particular field. The current emphasis on subsidiarity and decentralisation implies that the former route is to be preferred wherever possible. When, however, is minimal coordination, of this form, likely to suffice?

Gatsios and Seabright (1989) discuss the circumstances in which agreements on policy coordination or harmonisation will be adequate to deal with the problem of cross-country policy spillovers. They argue that the main gains from actually handing over responsibility for an area of policy with large cross-country policy externalities to the EU arise in the form of greater credibility of central operation. Where it is difficult to monitor compliance with international agreements (for example, where their implementation requires a large amount of judgement, based on more information than other parties to the agreement are likely to possess), signatories to an international agreement to implement national policy taking international externalities into account may not have confidence that the other signatories will comply. Where this is the case, compliance is likely to become an increasingly unattractive strategy, for the usual "prisoners' dilemma" type reasons.

The key issues in deciding the functions that need to be assigned to EU level are thus those of information and monitoring. Those functions that the EU itself needs to exercise are those where there are substantial cross-country policy externalities, and where compliance with agreements between national governments cannot easily be monitored by national governments.

"Subsidiarity" and Taxation

In the case of tax policy, there are some strong arguments for retaining tax policy at the national level, except where the gains from coordinated reform would be significant. One reason for this is that the optimal structure of taxation in an economy will generally vary, depending on the preferences and economic behaviour of individuals and enterprises. Also, the optimal level of taxes in member states of the EU may differ because member states have different requirements for budget revenues from taxation. This is a key difference between the assignment of taxation and other policies; the assignment of taxes affects the availability of financial resources, and hence the ability of different tiers of government to carry out other policy functions. In addition to these arguments for diversity, there are also reasons to avoid unnecessary instability in taxation; asset prices and private-sector decisions will reflect the existing tax arrangements, and adaptation to a new system could be costly.

Arguments for centralisation or EU involvement in taxation based on economies of scale are likely to be of less importance than those relating to spillover effects from uncoordinated national policies. The unit cost savings which arise from a transfer from the national level to the EU level are likely to be relatively modest, since the national units are already large, and differences in language, legal systems and corporate structures are likely to reduce quite sharply the potential for administrative savings from EU-wide tax administration. This is not, of course, to deny that there may be potential for beneficial administrative cooperation between the tax authorities of EU member states -- in ways that fall short of full administrative integration.

Spillover effects from national policies (policy externalities) provide much more substantial reasons for EU involvement in taxation, and may take a number of forms. Inter-country spillovers may arise from national decisions about the rates and structure of taxation, and also from the level and pattern of enforcement activities at national level. The spillover effects may also appear in different ways, as effects on private-sector decisions (inter-country distortion, or non-neutrality), and effects on the allocation of tax revenues between member states.

Where national tax policy is liable to have cross-country effects on economic activity and revenues, there is also concern that unregulated decisions at the national level may result in sub-optimal outcomes for the EU as a whole. In particular, there has been considerable recent concern that these cross-country effects may place downward pressure on "exposed" tax rates in member states, requiring member states to place undue reliance on tax bases less exposed to international competition. Restricting the ability of member states to undercut each other in taxation may be of net mutual benefit; indeed, there may be circumstances (as demonstrated by Kanbur and Keen 1993) where it would benefit all parties concerned.

As with other areas of policy, there is then a further question about the form of EU involvement in taxation. Should this take place through coordination of national policies (for example through harmonisation of tax rates), or should it involve complete assignment of particular taxes to the EU level? One set of considerations in this choice is of course the impact on the financial resources available to each level of government; assigning taxes to the EU would increase the EU's powers in other areas of policy, whilst reducing the financial resources under the control of member states. Other considerations relate to the adequacy of coordination measures, as opposed to assignment, in dealing with cross-country tax policy spillovers.

Thus, for example, it may be comparatively straightforward to design credible agreements to limit the extent of cross-country policy spillover effects to limit the impact of national decisions on tax rates on private-sector activity and the allocation of tax bases between member states, since the compliance of member states with a given set of rules regarding permissible tax rates can be easily observed.

In other areas, however, the information needed to monitor member states' compliance with any agreement may be harder to obtain -- for example, where spillovers arise from member states' tax enforcement activities. Even here, however, it may be possible to maintain a substantial degree of decentralisation, which stops well short of the unattractive option of transferring tax administration responsibilities to the EU. Thus, the institution of a EU mechanism for auditing member states' tax enforcement practises and procedures would help to enhance the credibility of agreements about the conduct of member states' tax enforcement activities, and would thus help to sustain continued assignment of the day-to-day operation of administration and enforcement at the member state level, even where enforcement activities involve significant cross-country externalities.

APPLICATION OF THE PRINCIPLE OF SUBSIDIARITY TO ENVIRONMENTAL TAXES

"Subsidiarity" and Environmental Policy

The practical development of EU environmental policy has occurred through a piecemeal accumulation of functions, partly in areas where other EU policies raised environmental issues. As other discussions of the implications of "subsidiarity" have argued (eg. Begg et al, 1993), some of the EU's most significant environmental policy measures, such as the Directives relating to water quality, are difficult to reconcile with subsidiarity. To a large extent, the EU's involvement in these areas does not reflect a process of considered decisionmaking about the need for centralisation, so much as political competition between levels of government over individual areas of policy. This inter-tier competition will almost certainly continue to influence the division of responsibilities between the EU and member states, although it has now been constrained by the presence of the subsidiarity criterion in the Treaty. In considering how far the EU will develop new policies on environmental taxation, future developments will be -- at least in part -- governed by the strength of the "subsidiarity" case for a greater EU role in environmental policy generally, and also with respect to particular environmental taxes.

As with tax policy, the application of subsidiarity to define the EU's role in environmental policy requires a balance to be drawn between the benefits from retaining discretion over policy at the national level, and the benefits from greater coordination at the EU level. There are two main lines of argument against EU harmonisation of environmental policy, where there are no strong arguments for coordination.

One is, again, that diversity in incomes, preferences and economic conditions between member states will typically call for diversity in policy measures. Assuming environmental quality to be a normal good, demand for environmental quality will tend to rise with higher incomes, and richer member states will then, other things being equal, wish to implement more stringent environmental policies than will poorer member states. There may, in addition, be important differences in the need for particular environmental policy measures, reflecting differences in the assimilative capacity of the environment in different member states. Where the environmental effects of national policy are confined to the country concerned, different standards of environmental protection would, in these circumstances, appear desirable.

The second is that there are also some important reasons to prefer the status quo in environmental policy, over the introduction of a new, harmonised, EU-wide approach. Individuals and firms will already have adjusted their behaviour to the requirements of the various existing systems of pollution control and regulation in member states; in some cases these adjustments will have required costly changes in technology, or the sunk costs of industrial relocation. Changing the form of policy, even without any increase in the stringency of policy, could require substantial re-adaptation to the structure of incentives and requirements of the new policy, imposing adjustment costs without necessarily achieving any reduction in pollution.

These arguments amount to a substantial case for avoiding unnecessary harmonisation of environmental policy. Nevertheless, there are a number of important environmental policy problems and policy measures where national environmental policies have either environmental or economic spillover effects affecting other member states, warranting some form of EU intervention. Whilst some of these issues have implications for the coordination of environmental policies in general, there are also specific considerations relating to the coordination of environmental taxes in particular.

The environmental spillover effects arise where environmental problems cross national boundaries, for example through air flows which transport national sulphur emissions and deposit them as acid rain in other countries, or through the impact of national emissions of carbon dioxide on global environmental problems such as climate change. In these cases, which have been the subject of an extensive literature (eg Barrett, 1991; Mäler, 1991, and many of the 1994 IIPF congress papers), coordination of national policies is central to achieving efficient environmental policies, although it is also, as the literature makes clear, extremely difficult to achieve.

The economic spillover effects arise where the policy measures adopted at national level have economic effects that cross national boundaries, and that may in some cases have negative implications for living standards in other countries, or for the welfare of all countries taken as a group. The significance for policy of these economic spillover effects has been controversial, and two aspects of the economic spillovers are distinguished here: (1) economic spillovers from "indirect protection" arising through the form of policy measures adopted at national level, and (2) economic spillovers arising from the level of protection adopted at national level which give rise to a case for international coordination or control of national policy-making.

In particular policy contexts, the environmental and economic spillovers from national environmental tax policy are likely to vary in relative significance. For certain local environmental problems, the need for EU involvement relates only -- if at all -- to the regulation of economic spillovers, whilst for regional or global environmental problems it is the cross-country dimension of the environmental problem that gives rise to the greatest need for coordination of national policies.

International Environmental Spillovers

International bargaining and negotiation has become a key feature of environmental policy, as countries negotiate over how best to regulate international environmental problems. As far as the EU is concerned, such negotiations have both an external and an internal dimension.

With regard to the external dimension, the EU has become a key player in the process of international bargaining over global environmental problems. How far the EU's involvement in international negotiations leads to more efficient bargaining and better outcomes depends, to a large extent, on the relationship between the EU and member states in the negotiation. It is possible that, collectively, the EU may be able to secure a better deal for its members than if each were to participate in international negotiations on its own behalf. Indeed, EU involvement may increase the chances of an agreement being reached at all, by reducing the number of independent participants in international negotiations. On the other hand, in practise, the involvement of the EU in international bargaining has introduced further complexity into some aspects of the process. In some cases, where the EU and member states have joint responsibility in international environmental negotiations, the involvement of the EU increases the number of actors in international negotiation rather than reduces it. Moreover, even were the EU to have a single seat at the negotiating table, the EU's position in negotiation could not be formulated without reference, at some stage, to the member states; this introduces a second layer into the process, and may reduce the flexibility of the European position and slow down the process of agreement.

Internally, the EU faces a number of environmental problems which cross member state boundaries; as in the case of the global environmental problems, there is a need to coordinate the actions of member states, so as to ensure that the cross-country externality is adequately reflected in national policy measures. Whilst such coordination is an important function of the EU, its difficulties should not be under-played; the incentive and information problems which arise in reaching agreement on efficient global environmental policies apply with equal force to decision-making amongst the member states of the EU.

The form of most such negotiations, both globally and amongst the European countries, has been for countries to negotiate over their contribution to global abatement measures, in terms of the proportionate reduction of emissions from some initial baseline. Environmental targets are thus set by international agreement, whilst the measures for implementation are left to individual countries to decide. This division of functions between target-setting and implementation is reflected, as Haigh (1989) notes, in the use of Directives in EU environmental policy, which generally set targets, whilst leaving member states discretion over how these are to be implemented in national legislation. Where cross-country environmental problems are involved, how far is such a division of responsibilities adequate, and consistent with the requirements of subsidiarity?

In some circumstances it would, in principle, be desirable for international negotiations to go further than simply specifying quantitative targets for emissions reductions in individual countries, and for the EU role thus to extend to agreeing the form of policy, as well as its intended outcome. One is that negotiating over tax rates rather than over quantitative emissions targets may in certain circumstances achieve a more efficient pattern of abatement across countries. For example, where global pollutants are concerned, agreement on a uniform tax rate to apply in all countries may be more efficient than agreeing that all countries implement the same percentage reduction in pollution. Another reason for specifying the form of policy, rather than simply emission reduction targets, may be the credibility of any agreement; agreeing to introduce a tax at a particular rate may make it easier for countries to verify that the bargain was being implemented, than where the agreement was simply to undertake unspecified measures to achieve a quantitative target for emissions reduction at some future date.

In some cases, therefore, the distinction between the EU's role in target-setting and the role of the member states in choosing how to meet the targets, is unsustainable. For a number of reasons, including the "credibility" issues raised above, the EU should have a substantial and legitimate interest in the form of environmental policy (eg in the choice between environmental taxes and regulation) as well as in its targets and objectives.

In addition, although the limited administrative resources available to the EU will frequently dictate a substantial role for national governments in administrative implementation of policies (including both conventional tax policies and environmental policies), even where the EU has primary responsibility for the determination of policy targets, there will be a need for this to be accompanied by facilities for monitoring and audit at the EU level, in order to ensure the credibility of policy.

Indirect Protection

Inter-country spillovers from national environmental policy may have an economic as well as an environmental dimension. One aspect of this, which applies both to cases where national policies have cross-country environmental effects, and to cases where purely national (or local) environmental problems are involved, is the risk of indirect protection. Unrestricted policymaking by individual member states carries with it the risk that measures may be introduced by member states (in the guise of environmental protection) that have their principal rationale in their role as instruments of indirect trade protection. It is possible to think of a range of environmental characteristics of particular goods that could be used to justify the imposition of environmental taxes that would in practise have the effect of protecting the goods supplied by the national producers of member states. "Clean car" incentives for vehicles meeting certain emissions standards have already been a concern of the EU; similarly, heavy taxes on non-recyclable drinks containers would tend to disadvantage importers of drinks relative to domestic suppliers (which have larger market shares, able to bear the fixed costs of systems for returning used bottles, and which would, in any case, have to return the bottles over shorter distances).

Although the sections of the Treaty of Rome that govern the conduct of EU environmental policy explicitly leave room for member states to operate more stringent environmental policies than the EU as a whole, fiscal measures that have the effect of indirect protection conflict with other key objectives of EU policy, especially those concerning the free flow of goods between member states. It is under these provisions of the treaty that the Commission attempted to restrain member states' own measures on fiscal incentives for lower vehicle emissions. There is clearly a difficult balance to be drawn between the requirements of the Single Market, and leaving member states any scope for independent environmental policymaking. Where that balance should be drawn is inevitably unclear, and guidelines for member states' policy are difficult to develop. The least that can be said, however, is that it is surely an issue entirely unsuited to resolution though judicial processes, and work on some form of detailed guidelines is long overdue.

Similar "indirect protection" issues arise in the case of the proposed European carbon tax. The least-cost pattern of abatement would be achieved if all sources of carbon dioxide emissions were to face the same tax. However, there were, in practise, demands that the tax should provide scope for selective exemption of certain categories of industrial energy user which might be adversely affected in international competition with firms outside the EU area, either through exemption of certain energy intensive industries, or through more selective exemption of particular firms facing severe competitive pressures. Whilst the arrangements for sectoral or discretionary exemption were proposed as part of the structure of the Community-wide tax, implementation of the exemptions would, inevitably, have been undertaken by the national authorities, and, especially in the case of selective exemption of firms, might have provided considerable scope for national authorities to ease the burden on national producers. Designing the arrangements for exemption in such a way as to minimise the amount of national discretion in their application is obviously desirable, if they are not, in effect, to become a route for selective subsidy of certain sectors. Detailed EU coordination of the carbon tax is then clearly helpful, not only in reducing the need for such exemptions, but also in constraining the circumstances and form in which they are made.

"Downward competition"

A second possible source of economic spillovers from national environmental policy relates to the possibility that cross-country economic considerations might induce countries to set inefficiently-low standards of environmental protection for environmental problems that are purely national in character. The issue is whether, in the context of international economic competition, an environmental policy analogue exists of the "downward competition" argument in tax policy, in the sense that the economic spillovers from national environmental policies may induce countries, acting alone, to choose inefficiently-low standards of environmental protection? Cumberland (1981) has argued that there is, and that international agreements should set a floor to national levels of environmental protection, to prevent considerations of economic competitiveness inducing a process of downward competition in environmental standards. This would then imply that there may be a role for the EU in setting minimum standards of environmental protection, even for problems which involve no cross-country environmental spillovers.

Oates and Schwab (1988), on the other hand, have set out a formal model in which economic competition between independent jurisdictions does not result in excessively-low standards of environmental protection. The model is one in which there is perfect competition, and countries compete with each other for an internationally-mobile capital stock through the standards that they set for permitted pollution levels. By setting less stringent environmental standards, the country may be able to attract more investment. The advantage to countries of attracting more capital is the impact of a higher capital stock on wage levels in the country; the disadvantage is the lower environmental quality that the country has to accept in order to attract more investment. The country thus faces a tradeoff between income and environmental quality, and in a situation where all residents share the same interests, would choose to set environmental standards at the point on this tradeoff where the marginal gain in income from more investment is just equal to the marginal willingness to pay for greater environmental quality. In this case, international economic competition does not induce countries to set inefficiently-lax environmental standards.

Oates and Schwab note, however, that there are circumstances where environmental competition might lead to downward pressure on standards. One would be where taxes on the internationally-mobile capital were needed to finance public expenditures. Another would be where residents' interests differed, and where policymaking was dominated by those with greater preference for income than for environmental quality; the opposite outcome would arise where policy was dominated by those with less interest in wage levels than in environmental quality.[4] But in an important group of cases, where markets are perfectly competitive, where governments have adequate revenue instruments, and where there are no major differences in the interests of groups in the domestic population, systematic downward pressure on environmental standards would not be the outcome from international economic competition. There is not, therefore, a general case for the EU, or international agreements, to specify minimum standards of environmental protection, or minimum levels of environmental taxes, for problems which do not involve cross-country environmental spillovers.

---

[4] A recent paper (Markusen, Morey and Olewiler, 1995) employs a game theoretic approach to model the income and environment choices that competing governments would make in relation to the location decisions of mobile polluting firms with a technology exhibiting increasing returns to scale. They show that, in contrast to the perfectly competitive model of Oates and Schwab, a wider -- and more complex -- range of circumstances arises in which the uncoordinated decisions of governments would lead to inefficient outcomes, involving excessive pollution in some cases, and excessive restriction of pollution in others.

INTEGRATION WITH OTHER EU POLICIES

Much of the EU's initial role in environmental policy arose through the environmental aspects of other areas of policy -- agriculture, transport, the regulation of the internal market, etc. In particular, policies for the development of the internal market have had considerable influence on the development of environmental taxes within the EU. These have included EU efforts, as discussed above, to limit indirect protection arising through national environmental policies. Also, removal of border controls within the internal market has constrained the administration and enforcement of certain types of environmental taxes at the national level; the scope for member states to set differing levels of environmental excises on goods is limited by the free movement of goods across borders, whilst fiscal deposit-refund arrangements at the national level may require special controls if they are not to be undermined by cross-frontier movements of goods. A third way in which internal market policy has affected the relative involvement of the EU and member states in environmental taxation has been where the EU has tried to regulate tax competition by setting limits on member states' tax rates for certain commodities. EU rules on the taxation of motor fuels, for example, have inevitably expanded the EU's involvement in the environmental dimension of transport taxation.

Over and above these developments, there are important questions about the interdependence of assignment decisions in the future evolution of the EU. Assignment of responsibility for environmental tax policy clearly needs to reflect decisions taken about assignment of other areas of policy. In principle, environmental tax policies need to be coordinated both with other environmental policies, and with other fiscal policies. In an ideal world, the level of environmental taxes would be set with regard to the economic costs and environmental benefits of a particular tax level, and with regard to the marginal excess burden of raising revenue through other fiscal instruments. However, perfect coordination of environmental taxes with both environmental and fiscal policies may not be feasible. Where these are assigned to different levels within the EU (eg. fiscal policy to the national governments of member states, and environmental policy to the Union, or vice versa), it will be inevitable that the assignment of responsibility of environmental taxes will result in better coordination with one of these related decisions, and poorer coordination with the other. A choice then has to be made, either of poor coordination between environmental taxes and other environmental policy, or between environmental taxes and other fiscal policy.

A similar issue of coordination, discussed by Oates (1991), arises in the single-country context, where a choice has to be made as to which government agency should be given responsibility for environmental taxes: the Finance Ministry, so that the taxes could be coordinated efficiently with other elements of tax policy, or an environmental ministry or agency, so that taxes could be coordinated with other environmental policies? Either assignment is liable to have some costs; Oates recommends a split assignment, in which the determination of the structure and rate of environmental taxes should be under the control of environmental policymakers, whilst the revenue should accrue to the general exchequer, rather than to the environmental agency's budget. Oates' argument rests on the observation that the range of instruments available to fiscal policymakers for revenue-raising is wide, so that setting environmental taxes without regard to considerations of efficient revenue-raising is unlikely to significantly increase the aggregate excess burden of taxation. On the other hand, the range of alternative environmental policy instruments is smaller, and setting the level of environmental taxes with principal regard to their revenue rather than environmental effects is likely to lead to appreciably sub-optimal environmental policy outcomes. Nevertheless, whilst it may be more efficient to locate responsibility for setting the rate of environmental taxes with the environmental policy authorities, there would be a risk of inefficiency if they derived budget revenues from the environmental tax; the agency's budgetary needs might encourage it to set an excessive tax rate, from the environmental point of view.

This line of argument might also be used to suggest that priority should be given to assigning policy responsibility for environmental taxes in the EU to the same level as the assignment of environmental policy, so that an efficient balance can be drawn between environmental taxes and other environmental policy instruments; coordination with fiscal policies is less critical. In some cases, of course, where both environmental and fiscal policies are assigned to the national level, the issue does not arise. But in the case of global warming policies, efficient coordination of tax and non-tax instruments demands that the tax rates should not be determined by the national fiscal authorities, but should be determined in conjunction with other environmental policy measures relating to global warming.

As in the EU carbon tax proposal, however, there should be no presumption that the revenue should be assigned to the same level of government responsible for determining the rate of the environmental tax. In general it is probably desirable that the revenue from environmental taxes should accrue to member states rather than the EU, regardless of the assignment of responsibility for the environmental tax rate. Since member states have many other potential revenue sources, they are unlikely to seek to drive environmental tax levels beyond the level which is efficient from the point of view of environmental policy (and general revenue-raising efficiency); the EU, on the other hand, which is highly constrained in terms of budgetary resources, might well be inclined to seek to increase environmental taxes above the efficient level, if the revenues from such taxes accrued to the EU budget.

CONCLUSIONS

This chapter has described the place that the principle of subsidiarity now occupies in determining the respective roles of the European Union (EU) and the national governments of its member states. It has drawn on the theory of "fiscal federalism," in which both policy spillover effects and administrative scale economies provide grounds for some form of higher-tier involvement. Following Gatsios and Seabright (1989), the discussion has placed greater emphasis than do conventional theories of fiscal federalism on the significance of "credibility" and administrative discretion in determining how far policy coordination is an adequate response to spillover effects, and how far higher tier assignment is required.

In applying the theoretical framework to assess the extent to which the EU should become involved in issues of environmental taxation, the chapter draws a distinction between the assignment of policy responsibility and the assignment of administrative implementation. One interpretation of the implications of subsidiarity for environmental policy that has had some currency is that the EU may have a valid role in setting environmental targets, but not in choosing the method or details of implementation. The chapter has argued that this distinction is unsustainable; for a number of reasons, including the "credibility" issue, the EU should have a substantial and legitimate interest in the form of environmental policy (eg. in the choice between environmental taxes and regulation) as well as in its targets and objectives.

The chapter has considered the relationship between environmental tax policy and other policy assignment decisions. It has argued that environmental tax policies need to be coordinated both with other environmental policies, and with other fiscal policies. Where these are assigned to different levels within the EU (eg. fiscal policy to the national governments of member states, and environmental policy to the Union, or vice versa), it will be inevitable that the assignment of responsibility of environmental taxes will result in better coordination with one of these related decisions, and poorer coordination with the other. It is argued that the decision about environmental tax rates should be taken by the level of government responsible for other, related, environmental policies, whilst the revenues from environmental taxes should accrue in general to member states in preference to the EU.

References

Barrett, S. (1991). "The Problem of Global Environmental Protection." In D. Helm (ed.), *Economic Policy towards the Environment*. Oxford: Blackwell.

Begg, D. et al (eds.). (1993). *Making Sense of Subsidiarity: How Much Centralisation for Europe*? London: Centre for Economic Policy Research.

Commission of the European Communities. (1991). A Community Strategy to Limit Carbon Dioxide Emissions and to Improve Energy Efficiency: Communication from the Commission to the Council. SEC(91)1744 final.

Cumberland, J. (1981). "Efficiency and Equity in Interregional Environmental Management." *The Review of Regional Studies* 10/2, 1-9.

Gatsios, K. and P. Seabright. (1989). "Regulation in the European Community." *Oxford Review of Economic Policy* 5/2, 37-60.

Haigh, N. (1989). *EEC Environmental Policy and Britain*. London: Longman.

Kanbur, R. and M.J. Keen. (1993). "Jeux Sans Frontières: Tax Competition and Tax Coordination When Countries Differ in Size." *American Economic Review* 83/4, 877-892.

MacDougall, D. (1977). *The Role of Public Finance in European Economic Integration*, Vols 1 and 2. Commission of the European Communities.

Mäler, K-G. (1991). "International Environmental Problems." In D. Helm (ed.), *Economic Policy Towards the Environment*. Oxford: Blackwell.

Markusen, J.R., E.R. Morey and N. Olewiler. (1995). "Competition in Regional Environmental Policies When Plant Locations are Endogenous." *Journal of Public Economics* 56, 55-77.

Oates, W.E. (1972). *Fiscal Federalism*. London: Harcourt Brace.

Oates, W.E. and R.M. Schwab. (1988). "Economic Competition Among Jurisdictions: Efficiency Enhancing or Distortion Inducing?" *Journal of Public Economics* 35, 333-354.

Oates, W. E. (1991). Pollution Charges as a Source of Public Revenues. *Working Paper* 91-22. University of Maryland.

Olson, M. (1969). The Principle of Fiscal Equivalence: The Division of Functions Between Different Levels of Government." *American Economic Review* 59, 479-487.

Padoa-Schioppa, T. et al. (1987). *Efficiency, Stability and Equity: A Strategy for the Evolution of the Economic System of the European Community*. Oxford University Press.

Pearson, M. and S. Smith. (1991). The European Carbon Tax: An Assessment of the European Commission's Proposals. London: Institute for Fiscal Studies.

Chapter 12

PURE GLOBAL EXTERNALITIES: INTERNATIONAL EFFICIENCY AND EQUITY

Peggy B. Musgrave (University of California, Santa Cruz)*

POLLUTION AS A SOCIAL BAD

The pollution issue raises the familiar problem of social goods and the prisoner's dilemma that follows. There, the supply of the social good, if left to individual choice, will fall short of the optimal level, since each actor will allow only for own internal benefits. In the case of social bads, the damage-generating activity will be over-extended, since the individual actor will allow only for internal damage cost, while disregarding the external cost imposed outside and on others. Hence there arises a similar need for a cooperative solution, with a twofold concern for securing an efficient and equitable outcome.

To focus on the essential features of that problem, this chapter addresses a situation in which the production of a good Z in any one country or region generates an emission of X (here, carbon dioxide) such that there results a bad (global warming) that imposes a globally uniform per capita damage. The exercise thereby sets aside the complications that arise if external damages are allowed to fluctuate by origin or if their impact varies by receiving regions. I then simulate the outcome of various arrangements for setting the level of emissions based on a set of hypothetical countries or regions, designed to reflect the relevant characteristics of typical real-world players on the international scene. The results, especially regarding absolute amounts, lay no strong claims to validity as empirical findings. Nevertheless, the relative patterns that emerge contribute to an understanding of the issues that must be faced in arriving at acceptable international forms of cooperation to deal with a pure global externality. The following section explains the model on which the calculations of costs and benefits are based.

---

* I wish to acknowledge the helpful coments of the two discussants, Bernard Herber and Robert Kohn, as well as those of an anonymous referee.

A SIMPLE MODEL

Figure 1 illustrates the analytical basis for the estimation of benefits and costs to a country (region) resulting from a reduction in emissions (ΔX) by it and the rest of the world.

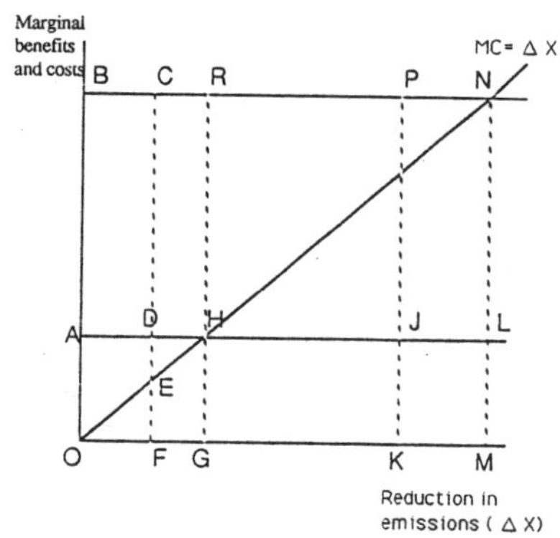

Figure 1 Costs and Benefits of Emissions Reduction

Marginal benefits and costs of reduction of the effluent X are measured along the vertical axis, while the amount of reduction in X from O, the initial position, is measured along the horizontal axis. (The initial position might be thought of as an inefficient Nash equilibrium.)

Marginal cost is shown for the i-th country and has the linear form $MC = \underline{c}\ \Delta X$, in which $\underline{c}$ equals the loss of consumer surplus, as the offending product is cut back, plus the loss of consumer surplus in markets for substitute products, as the price of the latter rises. The value of $\underline{c}$, of course, differs for each country.

Marginal benefits of the reduction in X are assumed to represent the reduction in the death rate, which has the same impact world-wide (this being a case of a pure global externality). Such marginal benefits for each country will thus be a linear function of its population, i.e. $MB = mv\ \Delta X\ n$, where $\underline{m}$ is the marginal change in mortality rate per unit of emissions, $\underline{v}$ is the value assigned to life and n is population. On the diagram, OA represents marginal benefits to the i-th country from i's cutback in X, while AB represents its marginal benefit to the rest of the world.

Let us say that the i-th country chooses to cut back its emissions by an amount OF, while the rest of the world combined cuts back its emissions by an amount FK. Then costs and benefits to that country are as follows:

Total benefits from its own cutback = OADF
Total spill-in benefits from cutbacks in rest of world = FDJK
Total costs of own cutback = OEF
Net benefits to i-th country = OADF + FDJK - OEF
Spill-out benefits to rest of world from i-th country's cutback = ABCD
Benefits to rest of world from own cutbacks = DCPJ
(Costs and net benefits to the rest of world are obtained in similar fashion to those of i-th country). Note that since the marginal value of life is taken to be constant (though it may differ among countries), the country that finds the spill-in damage reduced will have no occasion to change its own abatement level.

However, following the dictate of national efficiency, the i-th country would extend its cutback of emissions to OG (where $MB_i = MC_i$), with an ensuing additional increase of net benefits of EDH for itself and DCRH to the rest of the world.

If the i-th country were unilaterally to apply a standard of international efficiency by internalizing all costs both internal and external, it would reduce emissions to where $MB = MC_i$, i.e. by an amount OM. In this case, its costs and benefits would be as follows:

Benefits from own cutback = OALM
Costs of own cutback = ONM
Internal net benefits from own cutback = OAH - HNL
Spill-out benefits to rest of world = ABNL

DATA BASE AND ASSUMPTIONS

Country Characteristics

I will begin by specifying the most important features that will determine a country's contribution to total emission, as well as the level of benefits and costs that it would meet from various international arrangements. Table 1 offers relevant information for five hypothetical groups of countries covering most of the world and arranged in descending order of per capita income. By way of abbreviation, each of the five groups will be referred to as a "country." Country groups are arranged to include components that are more or less similar regarding the various relevant characteristics. The magnitudes shown for country I reflect Japan, while those for countries II to V, respectively, reflect approximate values for the United States and Canada combined, the European Community, Eastern Europe and Russia combined, and a combination of China and India.

Table 1 Characteristics of Countries

Country	I	II	III	IV	V
1 GDP per capita, \$ (Y/N)	22,779	18,222	13,783	17,778	229
2 Population, mill. (N)	122	270	373	360	1,862
3 GDP, \$ bill. (Y)	2,779	4,920	5,141	2,800	557
4 Emission X, mill. tons (X)	238	1,320	815	1,263	722
5 Emission X, adjusted for subsidy or tax mill. tons (X*)	267	1,439	1,044	900	600
6 Emissions per capita, tons (X/N)	2.0	4.9	2.2	3.5	0.4
7 Emissions/GDP, tons per \$ mill. (X/Y)	86	268	159	451	1,296

Notes: Lines 1. through 4. derived from *World Resources* (1992-93);
Note that line 4. refers to carbon dioxide emissions from industrial sources.
Line 5: Author's estimate.

Damage Levels and Costs of Abatement

Table 2 translates these country characteristics into the resulting levels of pollution damage caused and suffered, as well as the level of abatement costs. The damage, for the purposes of this simulation exercise, is measured in terms of increase in mortality rates.

As shown in line 1 of Table 2, the damage per million tons of emission is assumed to be uniform for populations in all countries. Unfortunately, little information exists regarding the effect of global warming on worldwide mortality rates. Cline (1992, 116-117) refers to a study by Kalkstein, which suggests that the degree of global warming implied by a doubling of carbon emissions from current levels would raise the average death rate in US cities by 0.0397 persons per thousand. That figure will be used here as a guidepost. If assumed to apply worldwide and based on our initial global level of emissions of 4,358 million tons, this rate would imply an increase of 9^{-6} per thousand for each increase of one million tons of emission. This figure does not appear to allow for effects on mortality rates from carbon emissions other than those caused by higher average temperatures. In view of this and the fact that changes in the quality as well as length of life may be involved (including effects on morbidity rates), the damage level assumed here is taken to be somewhat larger, i.e. an increase of one million tons of carbon emission results in an increase in the worldwide mortality rate of 2^{-5} per thousand. Even then, the postulated damage level represents a minuscule increase in current world mortality rates. Although this figure is somewhat conjectural, its absolute level is not crucial to the main focus of the analysis, since it is applied uniformly and concerns the relative impact of alternative international arrangements on the various countries depending on their specific characteristics. This chapter focuses on a pure global externality and does not address damages other than increased mortality rates (such as crop damage, flood damage, and so forth), which are likely to be regionally differentiated.[1] These effects and their possible differential impact are disregarded, since it is the purpose of this chapter to focus on the role of population size and the mortality effect of global warming as a pure social bad.

In line 2, the death rate effect is then translated into the resulting increase in the number of deaths per million tons of X, based on the aggregate country populations given in Table 1. Line 3 separates out that part of line 2 that is suffered domestically by the originating country. Line 4(a) shows the values of life assumed to be posited by the various countries. Beginning with a commonly used level for country II, the remainder are then adjusted to fall as per capita income declines. Line 4(b) gives a uniform (weighted) average value that will be used for the larger part of the analysis. Line 5 shows the total or global damage per ton of X using the average value of life, line 6 shows the domestically suffered share thereof, while line 7 shows the share suffered externally by the rest of the world.

As emissions are reduced, so is the resulting damage, but to arrive at the net gain, the abatement cost must be deducted. That cost reflects the loss of consumer surplus due to reduced output of the damage-generating activity as well as the loss of surplus in the consumption of substitute products, the prices of which rise. For purposes of this exercise, the marginal abatement cost as shown in line 8 is taken to be a linear function of the cutback in emissions. The model chosen here assumes, as do most studies of carbon-abatement costs (see, for instance, Nordhaus, 1991b), that the marginal cost of reducing emissions by one unit from initial levels is zero and that thereafter marginal cost of abatement rises with the degree of abatement. Furthermore, it is assumed here that over the range of abatement that is relevant to this study, the relationship is linear, i.e.

---

1 See Cline (1992b) for an excellent discussion of the various damages from global warming.

Table 2 Assumptions Underlying Benefit and Cost Estimates

	Country	I	II	III	IV	V
Externalities						
1	Global damage per mill. tons X (as increase in mortality rates/1,000)	2^{-5}	2^{-5}	2^{-5}	2^{-5}	2^{-5}
2	Global damage per mill. tons X (as no. of deaths)	59.74	59.74	59.74	59.74	59.74
3	Domestic damage per mill. tons X (as no. of deaths)	2.44	5.4	7.46	7.2	37.24
4	Value of life in $ thousand					
	(a) Country values (v)	1,250	1,000	750	425	100
	(b) Weighted av. of country values (v*)	350	350	350	350	350
5	Global damage in $ per ton of X (using v*)	20.91	20.91	20.91	20.91	20.91
6	Domestic damage suffered in $ per ton of X (using v*)	0.85	1.89	2.61	2.52	13.03
7	Outside damage suffered in $ per ton of X (using v*)	20.06	19.02	18.3	18.39	7.88
Cost of Abatement						
(MC = c ΔX)						
8	Value of c	0.519	0.094	0.152	0.098	0.171

Notes: For the derivation of values shown in lines 1, 4(a) and 8, see the text.

MC = c ΔX. The value of parameter c varies among countries, and is derived here from the OECD-GREEN study (1992), also reported in Cline (1992, Table 4.2 and accompanying text), which suggests that for the world as a whole a 17 percent worldwide reduction from baseline emissions by the year 2000 would require a unit carbon tax of $21 per ton. Since the "countries" in the model each begin with different emissions-to-GDP ratios, it follows that the marginal cost of attaining a given cutback in emissions will vary widely, as is evident in looking at the differing values of parameter c. Beginning with a parameter value of 0.519 for country I, which has a relatively low carbon intensity (as shown on line 7 of Table 1), its value moves inversely with each country's initial level of emissions.

The analysis is thus based on certain key parameters, values for which are not readily available in forms that lend themselves to this model. However, their relative values are in line with such information as is available and are believed not to be entirely unreasonable. The model is also a static, single-year construction and does not allow for future growth paths in emissions. This simplified approach was taken to focus the analysis on the implications for relative national and international welfare of measures to deal with the pure global externalities involved.

ALTERNATIVE INTERNATIONAL ARRANGEMENTS

As noted at the outset, it is clear that unilateral action is not likely to solve the problem of global externalities, since the gains to be had by each country abating independently are quite small relative to the large gains accruing to the rest of the world. Yet by the same token, the gains to be had by one country from concerted action with the rest of the world will be relatively substantial. There are therefore potentially substantial gains for all to be derived by cooperative policies to curb the offending effluent. Table 3 lists various possibilities, beginning with the status quo and considering various forms of unilateral and multilateral action.

The A. cases feature no international cooperation, and each country pursues its own unilateral policy. A.1 is listed as the initial position or status quo.[2] This situation is modified in A.2, where a single country responds unilaterally to the emission damage that its own action imposes on its own population, while continuing to disregard the damage that it causes abroad. It thus cuts back its emissions to a level at which its marginal abatement cost equals its marginal domestic benefit from reduced emission. No responses come from other countries and hence no reduction occurs in spill-in damage. The policy is domestically efficient, but without response from other countries it will miss efficiency from a global perspective. Case A.3 rather unrealistically postulates a situation in which a single country unilaterally cuts back emission to a point at which its marginal abatement cost equals the marginal global benefit, now including damage avoided abroad as well as at home. In the absence of response from other countries, the outcome will be slightly improved from a global point of view, but still will fall far short of a globally efficient solution. Moreover, the outcome will be much less favorable for the individual country engaging in such unilateral action -- since its internal benefits will be increased only modestly, while its abatement cost increases substantially.

---

[2] As pointed out to me by Hirofumi Shibata, this initial position might be thought of as an imperfect Nash equilibrium from which countries might move as a matter of international cooperation.

Table 3 Alternative International Arrangements

A	No Cooperation
	1 Status Quo
	2 Unilateral: Domestic Efficiency Only
	3 Unilateral: Worldwide Efficiency
B	Cooperation: No Adjustment for Equity
	1 Multilateral Domestic Efficiency
	2 Multilateral Global Efficiency
C	Cooperation: Equity with De Novo Entitlements
	1 Equal Emissions Per Capita
	2 Equal Emissions to GDP Ratio
D	Cooperation: Equity with Existing Entitlements
	1 Equal Percentage Reductions in Emissions
	2 Equal Reduction in Emissions Per Capita
	3 Equal Percentage Reductions in Adjusted Emissions
E	Cooperation: Equity with Efficiency
	1 Tradable Permits
	2 Uniform International Eco-Tax, with Redistribution

As we turn from unilateral action to cooperation, countries under the B. cases are assumed to agree to follow a common pattern of abatement. Cases B.1 and B.2 postulate patterns similar to those followed unilaterally under A.2 and A.3. The level of abatement will be the same for any one country as before, but its resulting benefits will be larger, as all countries follow the same policy and the level of spill-in damage is reduced. Under pattern B.2, global pollution will be reduced to the efficient level.

While agreements under B. were directed at various efficiency goals, those shown under C. and D. aim at equitable solutions. For comparability, it is assumed that countries agree to reduce their collective emissions by the same amount as would ensue under B.2, the global efficiency case. Various possible equity rules are shown reflecting different concepts of equity. The view of equity in the C. cases begins with a *de novo* arrangement, which seeks distribution of emission entitlements that is independent of that prevailing in the status quo or baseline case A.1. Under C.1 the aim is to equalize the level of per capita emission in all countries, while C.2 aims at equalizing the ratio of emissions to GDP. The rather drastic rearrangements of emission assignments called for under

C.1 and C.2 would require cutbacks in baseline emissions for most countries, while assigning additional emission rights to others. In the absence of trading rights, the outcomes shown in Table 4 do not count the latter as gains because the countries would have been free to use higher emission levels in the baseline situation, prior to entering the agreement.

The D. cases reflect a less drastic view of equity and recognize initially existing levels of emission as shown in A.1, and then aim to set equitable cutbacks relative to those bases. Case D.1 calls for equal percentage reductions from the existing level, while D.2 calls for equal reductions in emissions per capita. A further consideration is introduced in case D.3, which allows for existing taxes on and subsidies to the polluting activity, as specified in line 5, Table 1. The adjustment then calls for equal percentage reductions from that level of effluent that would hold in the absence of tax or subsidy. In this way the percentage reduction may be less for those countries currently taxing the pollutant and more for those countries that currently subsidize the activity in question.
None of the equity-based cooperative solutions in sections C. and D., except for D.1, are likely to meet the standard of global efficiency that was found in case B.2., since abatements are not distributed in line with marginal abatement cost. Only case D.1, requiring equal percentage reductions in emissions, meets the global efficiency test, and this because of the assumption made of equal abatement-cost elasticities across countries.

Consequently and finally, section E. aims to meet standards of both equity and worldwide efficiency through cooperative agreements. Case E.1 shows an agreement based on the use of tradable permits, while E.2 makes use of an international eco-tax accompanied by a redistribution of the revenue. Both permits and revenue are assumed to be distributed on a per capita basis. The efficient overall level of permits is set so as to equate marginal abatement costs with marginal abatement benefits. Similarly, the tax per unit of emission would be set such that tax per unit of emission, together with marginal abatement cost, matches the marginal abatement benefit. The two measures are thus equivalent, with both yielding a globally efficient level of pollution and a similar pattern of net benefit distribution among countries. As under B.2 above, a globally efficient outcome is reached, but the net benefit distribution is not the same as under B.2. Other distributions of net benefits under the E. rules may be obtained by varying the way in which permits or revenue are distributed among countries.[3] In line

[3] For a more detailed discussion of other equity concepts applicable in this context, see Barrett (1992).

with the Coase Theorem, an efficient outcome results -- whatever the distribution of property rights.[4]

SIMULATION OF ALTERNATIVE AGREEMENTS

I now apply the various agreements shown in Table 3 to determine the outcome for each of the five countries, with the results shown in Table 4. These include the abatement costs suffered as well as the benefits enjoyed. The calculation applies the assumption of a uniform damage valuation as shown in line 4(b) of Table 2. In order to afford a basis of comparability, all cases from C.1 through E.2 are standardized by providing for the same globally efficient level of emission level as was shown to prevail in case B.2. Certain general conclusions may be drawn.

Status Quo

The initial situation shown under A.1 illustrates how the various country characteristics of Table 1 contribute to explain the locations from which global pollution damage originates, as well as those in which the resulting burden is suffered.

To begin with, note (lines (f) through (h)) that by far the larger share of total damage caused (around 80 percent in this illustration) is directed into spill-out damage, with only a small share directed into self-inflicted domestic damage. The only exception is country V, due to its large share in total population. Since total spill-out damage equals total spill-in damage, the same picture emerges if we consider the form in which any one country suffers its damage, with much the larger part imposed from the outside. This relationship follows from our case of pure global externalities and is a major reason why cooperative solutions are essential.

Consider now the distribution of damage caused by country of origin, as shown in lines (b) plus (d) of case A.1. A country's emission or contribution to global pollution (X) may be expressed as $X = e * y * N$, where e is the emission intensity (or emissions to GDP ratio), y is per capita GDP and N is population. If there existed a systematic relationship between these characteristics, say y and e, general conclusions regarding the contribution to pollution could be drawn. But

---

[4] However, as applies to the Coase Theorem in general, the efficient outcome is reached on the assumption of perfect markets. The real-world bargain for international abatement agreements will involve a limited number of parties and bargaining strategies may well yield inefficient outcomes. See also note 5.

Table 4 International Distribution of Net Benefits with Alternative International Arrangements (in $ billion)

	Country	I	II	III	IV	V
A	<u>No Cooperation</u>					
1	<u>Status Quo</u>					
(a)	Emissions, mill. tons	238	1,320	815	1,263	722
(b)	Spill-out damage	4.77	25.11	14.91	23.3	5.69
(c)	Spill-in damage	3.5	5.74	9.25	7.8	47.38
(d)	Own internal damage	0.2	2.49	2.13	3.18	9.41
(e)	Net domestic damage	3.7	8.23	11.38	10.98	56.79
(f)	Total spill-out damage = total spill-in damage =	73.71				
(g)	Total own internal damage =	17.41				
(h)	Global damage =	91.12				
2	<u>Unilateral Domestic Efficiency</u>					
(a)	Reduction in emissions, mill. tons	1.64	20.11	17.17	25.71	76.2
(b)	Reduction in spill-out damage	0.03	0.38	0.31	0.47	0.60
(c)	Reduction in spill-in damage	-	-	-	-	-
(d)	Reduction in own internal damage	-	0.08	0.04	0.06	0.99
(e)	Cost of reduction	-	0.02	0.02	0.03	0.50
(f)	Domestic net benefits	-	0.06	0.02	0.03	0.49
(g)	Global net benefits	0.03	0.44	0.33	0.50	1.09
3	<u>Unilateral Worldwide Efficiency</u>					
(a)	Reduction in emissions, mill. tons	40.3	222.4	137.6	213.4	122.3
(b)	Reduction in spill-out damage	0.81	4.23	2.52	3.92	0.96
(c)	Reduction in spill-in damage	-	-	-	-	-
(d)	Reduction in own internal damage	0.03	0.42	0.36	0.54	1.59
(e)	Cost of reduction	0.42	2.32	1.44	2.23	1.28
(f)	Domestic net benefits	(0.39)	(1.90)	(1.08)	(1.69)	0.31
(g)	Global net benefits	0.42	2.32	1.44	2.23	1.27

Table 4 (Continued)

B **Cooperation: No Adjustment for Equity**

1 **Multilateral National Efficiency**

(a)	Reduction in emissions, mill. tons	1.64	20.11	17.17	25.71	76.2
(b)	Reduction in spill-out damage	0.03	0.38	0.31	0.47	0.60
(c)	Reduction in spill-in damage	0.12	0.23	0.32	0.29	0.84
(d)	Reduction in own internal damage	-	0.08	0.04	0.06	0.99
(e)	Cost of reduction	-	0.02	0.02	0.03	0.50
(f)	Domestic net benefits	0.12	0.29	0.34	0.32	1.33
(g)	Global net benefits =	2.4				
(h)	Global reduction in emissions =	140.8 mill. tons				

2 **Multilateral International Efficiency**

(a)	Reduction in emissions, mill. tons	40.3	222.4	137.6	213.4	122.3
(b)	Reduction in spill-out damage	0.81	4.23	2.52	3.92	0.96
(c)	Reduction in spill-in damage	0.59	0.97	1.56	1.32	8.00
(d)	Reduction in own internal damage	0.03	0.42	0.36	0.54	1.59
(e)	Cost of reduction	0.42	2.32	1.44	2.23	1.28
(f)	Domestic net benefits	0.20	(0.93)	0.48	(0.37)	8.31
(g)	Global net benefits =	7.69				
(h)	Global reduction in emissions =	736 mill. tons				

C **Cooperation: Equity with *De Novo* Entitlements**

1 **Equal Emissions Per Capita**

(a)	Reduction in emissions, mill. tons	0	528.4	0	207.6	0
(b)	Reduction in spill-out damage	0	10.05	0	3.82	0
(c)	Reduction in spill-in damage	0.63	0.39	1.92	1.33	9.59
(d)	Reduction in own internal damage	0	1.0	0	0.52	0
(e)	Cost of reduction	0	13.12	0	2.11	0
(f)	Domestic net benefits	0.63	(11.73)	1.92	(0.37)	8.31
(g)	Global net benefits =	0.15				
(h)	Global reduction in emissions =	736 mill. tons				

Table 4 (Continued)

2	<u>**Equal Emissions to GDP**</u>					
(a)	Reduction in emissions, mill. tons	0	0	0	221.2	514.8
(b)	Reduction in spill-out damage	0	0	0	4.07	4.06
(c)	Reduction in spill-in damage	0.63	1.39	1.92	1.33	9.59
(d)	Reduction in own internal damage	0	0	0	0.22	6.71
(e)	Cost of reduction	0	0	0	2.40	22.66
(f)	Domestic net benefits	0.63	1.39	1.92	(0.88)	(13.07)
(g)	Global net benefits =	(10.01)				
(h)	Global reduction in emissions = 736 mill. tons					

D <u>**Cooperation: Equity from Baseline Entitlements**</u>

1	<u>**Equal Percentage Reduction in Emissions**</u>					
(a)	Reduction in emissions, mill. tons	40.3	222.4	137.6	213.4	122.3
(b)	Reduction in spill-out damage	0.81	4.23	2.52	3.92	0.96
(c)	Reduction in spill-in damage	0.59	0.97	1.56	1.32	8.00
(d)	Reduction in own internal damage	0.03	0.42	0.36	0.54	1.59
(e)	Cost of reduction	0.42	2.32	1.44	2.23	1.28
(f)	Domestic net benefits	0.20	(0.93)	0.48	(0.37)	8.31
(g)	Global net benefits =	7.69				
(h)	Global reduction in emissions = 736 mill. tons					
2	<u>**Equal Reduction in CO_2 Emissions Per Capita**</u>					
(a)	Reduction in emissions, mill. tons	30.1	66.5	91.9	88.7	458.8
(b)	Reduction in spill- out damage	0.60	1.26	1.68	1.63	3.62
(c)	Reduction in spill-in damage	0.60	1.26	1.68	1.63	3.62
(d)	Reduction in own internal damage	0.03	0.13	0.24	0.22	5.98
(e)	Cost of reduction	0.23	0.21	0.64	0.39	18.01
(f)	Domestic net benefits	0.40	1.16	1.28	1.46	(8.41)
(g)	Global net benefits =	(4.1)				
(h)	Global reduction in emissions = 736 mill. tons					

Table 4 (Concluded)

3	**Equal Percentage Reduction in Adjusted Emissions**					
(a)	Reduction in emissions, mill. tons	4.2	60.1	0	475.1	196.7
(b)	Reduction in spill-out damage	0.08	1.21	0	9.53	3.95
(c)	Reduction in spill-in damage	0.62	1.28	1.92	0.66	7.03
(d)	Reduction in own internal damage	-	0.11	0	1.20	2.56
(e)	Cost of reduction	0.01	0.17	0	1.11	3.31
(f)	Domestic net benefits	0.62	1.22	1.92	0.75	6.28
(g)	Global net benefits =	10.79				
(h)	Global reduction in emissions = 736 mill. tons					

E **Cooperation: Equity with Efficiency**

1	**Equal Per Capita Tradable Permits**					
(a)	Reduction in emissions, mill. tons	40.3	222.4	137.6	213.4	122.3
(b)	Reduction in spill-out damage	0.81	4.23	2.52	3.92	0.96
(c)	Reduction in spill-in damage	0.59	0.97	1.56	1.32	8.00
(d)	Reduction in own internal damage	0.03	0.42	0.36	0.54	1.59
(e)	Cost of reduction	0.42	2.32	1.44	2.23	1.28
(f)	Payments made () or received	(1.04)	(16.12)	(4.71)	(12.83)	34.70
(g)	Domestic net benefits	(0.84)	(17.05)	(4.23)	(13.20)	43.01
(h)	Global net benefits =	7.69				
(i)	Global reduction in emissions: 736 mill. tons					
2	**Uniform International Eco-Tax with Equal Per Capita Redistribution of Revenue**					
(a)	Reduction in emissions, mill. tons	40.3	222.4	137.6	213.4	122.3
(b)	Reduction in spill-out damage	0.81	4.23	2.52	3.92	0.96
(c)	Reduction in spill-in damage	0.59	0.97	1.56	1.32	8.00
(d)	Reduction in own internal damage	0.03	0.42	0.36	0.54	1.59
(e)	Cost of reduction	0.42	2.32	1.44	2.23	1.28
(f)	Tax paid	(4.13)	(22.96)	(14.15)	(21.94)	(12.56)
(g)	Revenue redistribution	3.09	6.82	9.44	9.11	47.28
(h)	Domestic net benefits	(0.84)	(17.05)	(4.23)	(13.20)	43.01
(i)	Global net benefits =	7.69				
(j)	Global reduction in emissions = 736 mill. tons					

as shown in Table 1, such is not the case. A high level of e is found in low-income country V, due to heavy reliance on carbon-intensive energy, but a high e is also found in high-income country II, due now to heavy use of energy-intensive technologies. Yet, though e is much lower in II than in V, and its N is lower, country II generates heavier emissions than V, due to its much higher y.

Matters are easier when it comes to the distribution of pollution damage suffered. With per capita damage from global pollution the same throughout, the global damage suffered by any one country is simply a function of that uniform damage and population. Highly populated countries suffer the larger damage.

Unilateral Action

A single country acting in isolation to internalize its domestic damage from its own emission will have only little effect on domestic or global pollution (lines (d) and (b) in Case A.2). Only countries with large populations, such as V, can reap significant gains by acting alone to achieve national efficiency.

While isolated action to secure domestic efficiency will do little to reduce domestic damage, the domestic outcome is even worse if the country in isolation acts to pursue worldwide efficiency, as in case A.3. This follows because it would require a country to cut back its emissions by much more than it would have to in pursuit of domestic efficiency. Marginal costs would now have to be equalized with marginal global benefits, rather than merely with marginal domestic benefits. This cutback would involve higher abatement costs without the benefits of reduced spill-ins, leaving a net loss for all but country V (line (e) of A.3).

Cooperative Action: Efficiency Based

If all countries would agree to cut back their emissions to their domestically efficient levels, but without further equity adjustments, each country would make modest gains (line (f) of B.1) and, together with the benefits these cutbacks confer on the rest of the world, limited global net benefits would result here amounting to about three percent of status quo global damage.

A cooperative agreement (as in case B.2) for each country to cut its emissions back to internationally efficient levels (thus internalizing domestic plus external damages) delivers a maximum level of global net benefits. Even though the outcome of this exercise is more instructive when applied to distributional results and relative positions, note that this efficient action reduces emissions by 17 percent from the aggregate status quo level. More important, it is shown that net benefits are unevenly distributed (case B.2 line (f)). The highly populated, low-income country V makes large gains from the more extensive reduction in spill-in damages, but those countries with high emission intensities (II and IV) suffer net losses.

Cooperative Action: Equity Based

Turning now to equity-based cooperation, note again that the various approaches are standardized so as to yield the same level of worldwide emission obtained in the worldwide efficiency case B.2.

De novo adjustment

Begin with a view of equity that sets permissible levels of emission independent of status quo levels. Case C.1 requires adjustments so as to equalize per capita emissions, but without increasing emission rights to any one country. It is not surprising that the result would be highly redistributive. Countries with initially high emissions per capita (such as II and IV) would have to make substantial cutbacks, while countries with low initial emissions per capita (such as V) would be spared. As noted before, they might in fact be allowed to expand emissions above the baseline level, but they would not find this advantageous. The large cutbacks imposed on high-emission countries, in turn, would provide large benefits to the populous countries by way of reduced spill-in damages. Note that the gain in global net benefits here is but two percent of what is achieved in the B.2 solution, in which required cutbacks are distributed efficiently, in line with abatement costs.

If, instead of per capita-based equalization, the *de novo* entitlement approach aims to equalize the emissions-to-GDP ratio (as under case C.2), only countries IV and V with their high emissions intensities are losers under this plan. The redistributive effect now reverses that under C.1, being favorable to high-income and unfavorable to low-income countries. Furthermore, substantial global net losses follow (as compared with the small net gain in C.1).

Adjustments from baseline

We now consider a less redistributive view of equity, which assigns the required emission reductions by recognizing emission levels at the status quo and from there seeking equitable cutbacks. One such agreement might call for an equal percentage reduction in emissions (as in case D.1). The results in this model parallel those reached in multilateral international efficiency case B.2. This follows from the assumption of uniform marginal abatement cost elasticities across countries. Country V, with a large population, will make substantial net gains because of the large spill-in benefits that accrue from the rest of the world, while country II, with high per capita emissions, loses most; gains and losses for other countries are relatively small.

A second possible adjustment from the baseline position might be a requirement for equal reduction in emissions per capita shown (as case D.2). The brunt of the reduction is now borne by low-income, highly populated country V, which experiences a net loss, while other high-income countries enjoy modest gains. The latter are not large enough to offset the loss and a global net loss ensues.

The situation changes appreciably if allowance is made for the fact that countries may be subsidizing or taxing the pollutant in the initial situation. It is postulated here (Table 1, line 5) that country II imposes a low tax, while countries I and III impose high taxes, with countries IV and V subsidizing the pollutant in the initial position. Accordingly, if equal percentage reductions are required in that level of emissions which would arise in the absence of the tax or subsidy (case D.3), the reduction from current levels would be less for the taxing countries and more for the subsidizing countries than would hold if no such allowance is made. All countries now gain and the level of global net benefits increases by some 40 percent over the internationally efficient outcome of case B.2, in which efficiency is evaluated from an initial situation in which taxes and subsidies are in place.

Efficiency with Equity

So far, most of the equity-based arrangements (cases C.1 through D.3) fall short of the maximum overall level of net benefits obtained under the internationally efficient standard given in case B.2. It remains to consider arrangements that combine that efficient outcome with some postulated equity standard.

Tradable permits

One such approach would provide for a system of emission permits that are tradable. This is illustrated in case E.1. Here, permits are distributed on an equal per capita basis, with the level of total permits set so as to yield the globally efficient effluent level. Countries that have relatively high emissions per capita may then purchase emission permits from those with low emissions per capita if their marginal benefits exceed their marginal costs. By this means, the emission abatements get to be distributed in an efficient fashion so as to achieve the efficient case B.2 result.[5]

---

[5] The outcome shown is that which would result in a competitive market process. But in a realistic setting, the number of countries involved in the trading of rights will be limited, and the outcome will depend upon the bargaining strategies used. The pattern of net gain or loss which countries with various characteristics derive from alternative agreements in turn determine their strategy in the bargaining process by which agreements may be reached. This game-theoretic aspect of the problem is not dealt with

International eco-tax

Another arrangement yielding an identical result would apply an international eco-tax at a uniform rate per unit of emission. This is illustrated in case E.2, with the tax rate set to reduce emissions by the same amount as under case E.1. The revenue collected in this way, shown in line E.2 (f), is distributed on a per capita basis, thus yielding the same pattern of net benefit distribution as in case E.1.

Thus, approaches E.1 and E.2 offer different techniques to reach identical results. Both achieve the same efficient level of abatement also reached under B.2. Moreover, E.1 and E.2 yield similar patterns of net benefits -- highly redistributive from high- to low-income countries and differing from the pattern resulting under B.2.

Ranking of Outcomes

The main purpose of this exercise is to compare the effects of various international arrangements for dealing with a pure global externality and to determine how the effects depend on the differing characteristics of the participating countries. For this purpose it is useful to note how the various countries would rank the different agreements with respect to resulting net benefits.

As shown in Table 5, no country would prefer the status quo, since unilateral national efficiency is obviously better. All would prefer, if they were to correctly assess the resulting costs and benefits, some form of cooperative agreement. Specifically, there should be no difficulty in moving from the status quo to multilateral domestic efficiency, since all countries would gain. This, however, leaves a poor outcome regarding global efficiency; further moves become more complex, however, since it is evident that there is much incompatibility between the rankings of country V and those of the high-income countries I, II and III. The more redistributive plans (such as that employing equal per capita tradable permits) rank high for the low- and low for the high-income countries, as does that plan calling for equal per capita emissions. The reverse holds for C.2, which calls for equal emissions to GDP, a plan ranked highest by the high-income countries and lowest by the low-income countries. It is also noteworthy that those plans that meet standards of international efficiency (B.2 and D.1) are also more acceptable to the low-income countries. It is also interesting to note that those high-income countries with relatively low emission-to-GDP intensity (such as countries I and III) will be

---

here. For treatment of this topic, see Barrett (1991, 1992); Grubb (1989); Nitze (1990), Sebenius (1991), and Schelling (1991).

more likely to line up with the low-income, highly populated country in accepting case C.1, which calls for equal per capita emissions.

A review of the ranking patterns across countries, giving equal weight to each, makes it clear that agreement D.3 might have the best chance of acceptance, since it is ranked quite high by all participants. This plan calls for equal percentage cutbacks, but does so from a base that allows for the existence of taxes on and subsidies to effluent-creating activities. Finally, note that plans E.1 and E.2, which combine efficiency and distributional concerns, are given top ranking by the low-income country V, while given a bottom or next-to-last ranking by the others.

Differentials in Valuations

The simulations have been based on the assumption that all countries assign the same damage cost to a unit of emission. Possible differentials in evaluation between countries are disregarded at the outset but are considered below. In our illustration this took the form of an accepted weighted-average valuation placed on the saving of life, although underlying that uniform value it was recognized that there might be a wide range of national valuations. These may reflect differences in underlying preferences as well as differences in per capita income. Regarding the latter, it is reasonable to suppose that high-income countries may, for purposes of their domestic policies on resource allocation (e.g. health expenditures) implicitly or explicitly place higher dollar values on life than do low-income countries. But in addressing the equity of international agreements, the latter countries would hardly be willing to accept higher values to be placed on outside lives and lower values to be placed on their own lives. In dealing with pure global externalities, these differentials in national valuations (whether they be of lives saved or any other damage averted) raise difficult philosophic questions. Indeed, the very concept of efficiency in the international context becomes problematical when the value of global externalities runs into differing national value systems.

Now replace the uniform (weighted average) valuation used in Table 4 by differential evaluations: the globally efficient level of emission cutback will be unaffected, as will be the overall global net benefit to be derived. The distribution of gains in terms of lives saved would also be unaffected. This follows from my reasonable assumption that, where a uniform value is used, it should be set as an average weighted by population. At the same time, there would be a major change in the distribution of net benefits among countries in dollar terms. While space considerations prohibit a full examination of this effect, high-income countries (such as I, II and III) would now gain a much larger share of the net benefits, while low-income country V would gain much less. This outcome is illustrated in Table 6, which compares the results of average with differential values for the case of equal percentage reduction in emissions (case

D.1). Whereas in the weighted-average case large net benefits existed for the low- and small benefits or net losses for the high-income countries, the case of differential valuation shows uniformly low but positive net benefits.

Table 5 Ranking of International Arrangements for Emissions Abatement[a]

	Country	I	II	III	IV	V
A.1	Status Quo	7	6	6	5	7
A.2	Unilateral Domestic Efficiency	6	5	5	4	5
A.3	Unilateral International Efficiency	8*	8*	7*	8*	6
B.1	Multilateral Domestic Efficiency	5	4	4	3	4
B.2	Multilateral International Efficiency	4	7*	3	6*	2
C.1	Equal Per Capita Emissions	1	9*	1	6*	2
C.2	Equal Emissions to GDP	1	1	1	7*	9*
D.1	Equal Percentage Reduction in Emissions	4	7*	3	6*	2
D.2	Equal Reduction in Per Capita Emissions	3	3	2	1	8*
D.3	Equal Percentage Reduction in Adjusted Emissions	2	2	3	2	3
E.1	Equal Per Capita Tradable Permits	9*	10*	8*	8*	1
E.2	Uniform International Eco-Tax with Revenue, Redistribution	9*	10*	8*	8*	1

[a]Rankings based on net domestic benefits shown on Table 4. Numbers with asterisk (*) indicate that net benefits are negative in those cases.

Differences in valuation assume further importance if the study departs from the purely global case and considers activities that generate damages that are spatially limited or damages that result from the location of movable by-products such as toxic substances. In that case, the switch from uniform (weighted average) valuation to differential valuations may well affect the outcome in real as well as monetary terms.

Since the damages that arise from current emissions may be felt largely by future generations, a further valuation problem arises in the choice of an appropriate discount rate and its treatment where countries differ in time preference. Although international financial markets may be sufficiently integrated to generate similar interest rates on funds that readily flow from one country to another, capital markets as a whole are far from perfect and a prevailing interest rate applicable to organized and sophisticated international financial institutions cannot be accepted as the international social rate of time preference. Consumers in low-income countries have demonstrably higher rates of time preference than do consumers in countries where per capita income is fifty times higher. Is the present value of a future flow of costs and benefits experienced within one country, then, to be different from that of the same flow experienced by another?

Table 6 Uniform and Differential Valuations of Damages (Case D.1,Table 4)

Country	I	II	III	IV	V
Weighted average valuation					
Domestic net benefits	0.20	(0.93)	0.48	(0.37)	8.31
Global net benefits =7.69					
National valuations					
Net national benefits	1.80	1.66	2.73	0.04	1.46
Global net benefits =7.69					

CONCLUSION

Spillovers across country borders, particularly in the form of pure global externalities, raise complex economic and philosophical issues for the application of cost-benefit analysis underlying international agreements. Such is the case on both efficiency and equity grounds.

Allowance for domestic effects only, while disregarding outside effects, leads to sub-optimal results. Not only will the level of emissions remain excessive from a global perspective, but any one country, by suffering spill-ins, will be left in a worse position than would be achieved under a cooperative agreement to account for external effects. Yet, cost-benefit analysis not infrequently holds that zero weights be assigned to outside effects. At best, this may be defended by arguing that outside responses are not within the acting government's control, but such reasoning stands in sharp contrast to the central stance of international trade theory. From Ricardo on, that view has based its case for free trade on worldwide gains, not the isolated domestic benefits of any one country only. This efficiency-focused reasoning, generally accepted with regard to commodity flows, should hold equally for the flow of cross-border externalities.

Here as elsewhere, policy adjustments pose distributional as well as efficiency issues. Starting out from an initial baseline or status quo position, the various policies, as shown in Table 4, result in net gains for some and net losses for other countries. The required abatement costs may fall short or exceed the value of damage reduction. In those cases in which a global net gain results, compensation could be paid to the losers, thus meeting the Hicks-Kaldor test. But the question remains whether this suffices and whether the adjustment should not be followed by actual compensation payment. A good case can be made for such a requirement in the context of most cost-benefit analyses, but the problem as posed here differs. Various equity rules have been designed to reflect what are considered to be equitable changes from the baseline position. By thus determining the desirable pattern of net benefits that is to result, I have already accounted for the compensation issue, setting the equity rule by aiming at net results.

Finally, it should be stressed that cooperation is required to obtain global efficiency and to do so in what is considered to be an equitable fashion. In the usual context of fiscal analysis (where spillover effects are excluded), a case is frequently made for intergovernmental competition, as distinct from cooperation, as a means of securing efficient policy. Residents and capital, by threatening to move, may force policy into a pattern that suits their preferences. Whatever the merit of this argument, it surely does not hold for my case of truly global pollution damages, since the damage suffered will be independent of its place of origin. Intergovernmental competition as an efficiency-inducing device becomes inoperative whenever effects are global in nature.

References

Barrett, S. (1991). "International Environmental Agreements as Games." In R. Pethig (ed.), *Conflicts and Cooperation in Managing Environmental Resources*. Berlin: Sprenger-Verlag.

Barrett, S. (1992). "Negotiating a Framework Convention on Climate Change: Economic Considerations." In *Convention on Climate Change: Economic Aspects of Negotiations*. Paris: OECD.

Cline, W.R. (1992). *The Economics of Global Warming*. Washington, DC: Institute for International Economics.

Grubb, M. (1989). *The Greenhouse Effect: Negotiating Targets*. London: Royal Institute for International Affairs.

Nitze, W.A. (1990). *The Greenhouse Effect: Formulating a Convention*. London: Royal Institute for International Affairs.

Nordhaus, W.D. (1991a). "The Cost of Slowing Climate Change: A Survey." *Energy Journal* 12/1, 37-65.

Nordhaus, W.D. (1991b). "Economic Approaches to Global Warming." In R. Dornbusch and J.M. Poterba (eds.), *Global Warming: Economic Policy Responses*. Cambridge, MA: MIT Press.

OECD. (1992). "The Economic Costs of Reducing CO_2 Emissions." *Economic Studies* 19. Paris.

Schelling, T.C. (1991). "Economic Responses to Global Warming: Prospects for Cooperative Approaches." In R. Dornbusch and J.M. Poterba (eds.), *op. cit.*

Sebenius, J.K. (1991). "Crafting a Winning Coalition: Negotiating a Regime to Control Global Warming." In *Greenhouse Warming: Negotiating a Global Regime*. Washington, DC: World Resources Institute.

Chapter 13

INTERNATIONAL COORDINATION OF ENVIRONMENTAL POLICIES AND STABILITY OF GLOBAL ENVIRONMENTAL AGREEMENTS

Carlo Carraro (University of Venice and FEEM)

and

Domenico Siniscalco (University of Torino and FEEM)*

INTRODUCTION

Over the last twenty years, scientists have stressed the importance of some new phenomena related to the use of the environment: climate change, ozone-layer depletion, acid depositions, freshwater pollution, marine pollution, deforestation and the loss of biological diversity (e.g. Cf. UNEP, 1991; World Bank, 1992; Worldwatch, 1994). Many of these problems have been recently discovered (e.g. the ozone-layer depletion); some have recently acquired importance due to their scale and potential impact (e.g. climate change, deforestation, or water pollution).

Most of the new environmental phenomena mentioned above have an intrinsic international dimension due to transnational or global spillovers. The depletion of the ozone layer and climate change depend on global aggregate emissions. Acid depositions and marine pollution usually cross the borders of the polluting country. Even deforestation is an international problem, as forests are a sink for CO_2 and provide important support for bio-diversity.

The international dimension of the environment is a source of substantial interdependence among nations: each country benefits from using the environment as a receptacle of emissions and is damaged by environmental deterioration. While the benefit for any country is primarily related to domestic

* The authors are grateful to Andrea Beltratti, Lans Bovenberg, Stephen Smith, Henry Tulkens, and an anonymous referee for helpful comments on a previous version of this paper. The usual disclaimer applies.

pollution alone, the environmental damage is related to both domestic and foreign emissions. This problem is not new to economists, and has been analysed in the area of externalities and public goods. What is new is the context in which these problems take place. Currently, the atmosphere and the waters are managed as global common-property goods, and no institution exists that possesses powers to regulate their use by means of supranational legislation, economic instruments, or by imposing a system of global property rights. This fact requires a shift in economic analysis, from a literature on government intervention to a literature on international policy coordination and voluntary agreements among sovereign countries. (see Carraro and Siniscalco, 1992a).

Environmental agreements among sovereign countries have been widely advocated and debated in the last few years, following the Montreal Protocol on CFCs, to protect the ozone layer, and the UN Conference on the Environment and Development, organised in Rio de Janeiro in 1992. Starting from the effort to design the international environmental agreements, this chapter provides some tools to understand the underlying economic theory.

The material is divided into three sections: the next section recalls some recent attempts to design global environmental agreements; the subsequent section introduces the analytical tools to characterise the agreements under review; the following two sections discuss the issues of transfers and negotiation linkages to increase the stability (rather than the profitability) of environmental agreements; finally, the last section draws some policy implications from the proposed analysis.

SOME RECENT EXPERIENCE IN ENVIRONMENTAL POLICY COORDINATION

The recent history of international agreements to protect the global environment has been marked by different attempts to achieve cooperation among countries.

The first attempt was to design worldwide agreements to cut emissions by bargaining solely on emissions. The result of these attempts was usually frustrating. The conventions, whenever they are signed by a great number of countries, are rather empty in terms of quantitative targets and/or deadlines. Precise commitments, on the other hand, are signed by small groups of "like-minded" countries. Examples include the Climate Change Convention, which is widely signed but notoriously empty, and by the EC countries' commitment on CO_2 emissions, which is rather binding and precise, but confined to a small number of countries. The early outcome of the Montreal Protocol provides another example of a small group of signatories, *vis à vis* a global phenomenon, with many other countries enjoying their positions of free-riders.

The dissatisfaction with such outcomes, and in particular with small coalitions, was followed by attempts to expand the agreements by bribing reluctant countries by means of transfers (as in the follow-up of the Montreal Protocol, or in the plans for financial implementation of the Agenda XXI), again undermined by several problems of free-riding and instability.

Finally, the most recent negotiating experience has been to try to link environmental protection to other international agreements: on technological cooperation (as in the case of the Climate Change Convention) and trade (as in the environmental clause in GATT).

With some amount of simplification, therefore, one can see a trend in the actual practise of environmental negotiations. When countries were bargaining on emissions only, attempts at full cooperation often failed, giving rise to small coalitions. Attempts to expand these coalitions first relied on money transfers and then on linked negotiations, and the latter routes seem to be more promising.

PROFITABILITY AND STABILITY OF INTERNATIONAL ENVIRONMENTAL AGREEMENTS

We believe that recent advances in the theory of non-cooperative formation of coalitions can be used to interpret the actual trends in environmental negotiations. In this section, in particular, we reorganise some recent results on coalitions formation, using the notions of profitability and stability.

Consider n countries ($n \geq 2$) that interact in a common environment and bargain over emission control of a specific pollutant. In order to simplify the presentation of the results, assume that these countries are symmetric,[1] i.e. the welfare function is not country specific. Let $W(x_1 \ldots x_n)$ be a country's welfare function, where x_i denotes a vector containing country i's emissions and all other economic variables affecting abatement costs and the environmental damage perceived in each country. The function W(.) captures countries' interaction in a global environment, as welfare depends on all countries' emissions as well as on other transnational variables (e.g. trade policy variables). Let $P_i(j)$ denote the value of country i's welfare when it decides to cooperate, and j-1 other countries also cooperate; moreover $Q_i(j)$ is the value of its welfare when country i does not join the coalition formed by j cooperating countries.

---

[1] The following analysis could be carried out even under the assumption of asymmetric countries without changing the main conclusions that will be derived. We choose the symmetric case because it helps us to emphasise the distinction between the problem of coalition profitability and the one of coalition stability.

As the focus of this section is to analyse the formation of a coalition, the only variable that defines the value of the welfare function is the number of cooperating countries.[2] However, it is implicit that all other relevant variables, including emissions and policy decisions in other countries, enter country i's welfare function.[3]

Let J stand for the set of cooperating countries, and let J^o denote the set of countries that play non-cooperatively. Hence $P_i(0)$, $i=1,2,..,n$ is a country's non-cooperative payoff (the non-cooperative Nash equilibrium payoff), whereas $P_i(n)$ is country i's payoff when all countries decide to cooperate.

Profitability

A first necessary condition must be met if an environmental coalition is to be formed: the welfare of each country signing the cooperative agreement must be larger than its welfare under non-cooperation. We call this requirement profitability of the coalition. A coalition is profitable if country i gains from joining the coalition, with respect to its position when no countries cooperate. Formally, a coalition formed by j players is *profitable* if:

$$(1) \qquad P(j) > P(0)$$

for all countries belonging to J. In the case of symmetric countries this condition is fairly trivial: it is either satisfied or it is not, for all countries involved in the negotiating process. As discussed in Cesar and De Zeeuw (1993), and in the section on stability and negotiation linkages below, this condition becomes much more relevant when countries are asymmetric.

Stability

Of course, profitability may not be a sufficient condition to induce many countries to sign a cooperative agreement. As stressed earlier, the main problem preventing the formation of any coalition is the possibility of free-riding by some countries. Free-riding can be explained as follows: since one country can profit from the reduction of emissions made by cooperating countries, that one country has an incentive to let the others sign the cooperative agreement, thus enjoining a

---

[2] For the case of asymmetric countries, it would be insufficient to indicate the number of cooperating countries, but it would be necessary to specify which countries enter the coalition.

[3] Examples of specific welfare functions used to analyse the problem of environmental cooperation can be found in Barrett (1992), Carraro and Siniscalco (1992b) and Heal (1994).

cleaner environment at no cost. If all countries are symmetric and choose non-cooperatively whether or not to adopt the cooperative strategy, no cooperation may take place. In other words, the (non-cooperative) "metagame" in which countries choose between cooperation and non-cooperation can be represented as a Prisoner's Dilemma, as in the early literature on the "tragedy of commons."

This theoretical conclusion is not consistent with the existence of small coalitions and the current trend of environmental negotiations described in the first section. It is therefore necessary to check more carefully whether the features of the problem of environmental cooperation actually coincide with those of a Prisoners' dilemma, as argued above. As a first step, assume that each country acts unilaterally and non cooperatively. In particular, assume that each country, in deciding whether or not to defect from the coalition, considers only its own actions, taking as given the actions of the other players.[4] Under these assumptions, do some countries adopt the cooperative strategy (at the non-cooperative Nash equilibrium) ?

The problem can also be stated more formally. A country chooses the cooperative strategy if P(j), the country's payoff for belonging to the j-coalition, is larger than Q(j-1), the country's payoff when it rejects the coalition, and lets other j-1 countries sign the cooperative agreement. Hence, Q(j-1) - P(j) is a country's incentive to defect from a coalition formed by j players, whereas P(j+1) - Q(j) is the incentive for a non-cooperating country to join a j-coalition (which, consequently, becomes a (j+1)-coalition).

As a consequence, j^*, the number of countries that unilaterally and non-cooperatively adopts the cooperative strategy, is defined by $P(j) - Q(j-1) > 0$ and $P(j+1) - Q(j) < 0$. Thus, a coalition formed by j players is *stable* iff: (i) there is no incentive to free-ride, i.e.:

(2a) $\quad Q(j-1) - P(j) < 0 \qquad$ for all countries belonging to J

(ii) there is no incentive to broaden the coalition, i.e.:

(2b) $\quad P(j+1) - Q(j) < 0 \qquad$ for all countries belonging to J^o.

This definition corresponds to that of cartel stability presented in the oligopoly literature (see D'Aspremont and Gabszewicz, 1986. A similar definition is also used in Barrett, 1992).

---

4 More formally, the equilibrium of the non-cooperative game, in which one country decides whether or not to adopt the cooperative strategy, must be a Nash equilibrium.

Small stable coalitions

It has been shown that under fairly general conditions stable coalitions do exist (see Donsimoni et al., 1986). However, this does not satisfactorily address the problem of protecting international commons because, as has been demonstrated both in the oligopoly and in the environmental literature (see, for example, D'Aspremont et al., 1983; D'Aspremont and Gabszewicz, 1986; Hoel, 1991; Barrett, 1992; Carraro and Siniscalco, 1992b), stable coalitions are generally formed by $j \leq n$ players, where j is a small number, regardless of n.[5] If stable coalitions are small, and countries are symmetric, the impact of their emission reduction on total emissions is negligible.

The existence of small stable coalitions leads to the following question: can the j cooperating countries expand the coalition through self-financed welfare transfers to the remaining players?

STABILITY AND TRANSFERS

The answer provided by the literature and by the practise of international agreements focuses on transfers as means to bribe non-signatory countries. Notice that we are not referring to the possibility of using transfers or side-payments to make the agreement profitable to all countries. This latter issue, which is also known as burden sharing and which becomes relevant when countries are asymmetric, is discussed in Chander and Tulkens (1993). Here we start from the necessary condition that the agreement is profitable (to all countries because players are symmetric), and we look at the possibility that transfers increase the stability of the agreement.

In this context, which is discussed in detail in Carraro and Siniscalco (1993), a non-trivial analysis of transfers implies imposing constraints on the amount of resources to be transferred: if the transfers were unconstrained, then all non-signatories could be bribed, but the mechanism would not be credible. Therefore, we assume the following:

---

[5] More satisfactory results are presented in Heal (1994), which introduces a fixed cost of forming the coalition, and in Bauer (1993), which considers the possibility that a bloc of countries bargains with other blocs of countries. However, the results in this latter paper are based on a conjectural-variations equilibrium concept that does not coincide with the fixed point defined by the best-reply mapping (i.e. it is not a Nash equilibrium).

a. transfers must be *self-financed*, i.e. the total transfer T must be lower than the gain that the committed countries obtain from expanding the coalition;

b. the move to a larger coalition must be *Pareto-improving*, i.e. all countries must increase their welfare *vis à vis* the situation preceding the coalition expansion, and *vis à vis* non-cooperation (the larger coalition must be profitable);

c. committed countries choose the transfer in order to *maximise the number of signatories* (given the above two constraints).[6]

Under these conditions, however, a negative result emerges. Self-financed transfers cannot induce free-riders to sign the environmental agreement unless some degrees of commitment constrain the strategic choices of cooperating countries.[7] This result is not trivial. As previously stated, if all countries were committed to cooperation, obviously no free-riding would occur (and the appropriate analytical tool would be cooperative game theory, as the problem is no longer stability but burden-sharing). However, there are several less demanding forms of commitment which, if associated with appropriate welfare transfers, can lead to large stable coalitions.

In particular, three types of partial commitment can be proposed that could serve as possible blueprints for environmental cooperation (of course, other types of institutional mechanisms could be proposed as well): *stable coalition commitment*, when only the j countries belonging to the stable coalition commit to cooperation; *sequential commitment*, when the j countries are committed to cooperation and any new signatory, as soon as it enters the expanded coalition, must commit to cooperation as well; *external commitment*, when a subset of non-cooperating countries commits to transfer welfare in order to induce the remaining non-signatories to cooperate, and to guarantee the stability of the resulting coalition.

Assuming these alternative commitment schemes and the above restrictions on transfers, Carraro and Siniscalco (1992b, 1993) analyse the formal

---

6 Notice that transfers enter a country's payoff in an additive way. Otherwise, each cooperating country would maximise the payoff with respect to both emissions and transfers. This would be the case if transfers were carried out using policy instruments that interact with other economic variables, and with emissions in particular (see Carraro and Siniscalco, 1994).

7 Carraro and Siniscalco (1993) prove the following proposition: if no countries can commit to the cooperative strategy, no self-financed transfer from the j cooperating countries to the other non-cooperating countries can successfully enlarge the original coalition. This proposition holds strictly only in the case of symmetric countries.

conditions to expand coalitions, and carry out a few examples with symmetric countries.

A general conclusion emerges from such analysis: both the existence of stable coalitions, and the possibility of expanding them, depend on the pattern of interdependence among countries, as described by *the slope of their best-reply function*. The reason that lies behind this result is intuitively simple. The best-reply functions reflect, *inter alia*, the marginal damage produced by the increase in foreign countries' emissions when the domestic country reduces its own emissions. Suppose countries bargain over emission levels in order to achieve an optimal aggregate outcome, taking into account reciprocal externalities. As is well known, cooperation among all countries is profitable and optimal, but it is intrinsically undermined by free-riding, as reflected by the best-reply function of a country that does not belong to the partial coalition. If this best-reply function is negatively sloped, the non-cooperating country will expand its emissions if the coalition restricts them, offsetting the effort of the cooperating countries. If, on the contrary, the best-reply functions are orthogonal or near-orthogonal, the free-riders will simply enjoy the cleaner environment without paying for it, but will not offset the emission reduction by the cooperating countries (technically, there is no "leakage").

As a consequence, the case in which best-reply functions are orthogonal or near orthogonal is the one in which the cooperating effort of some countries provides the expected environmental benefit, thus making cooperation more likely than in the case in which the increase of emissions in the non-cooperating countries, which take place when the best-reply functions are negatively sloped, increases their economic gains and reduces the environmental benefit.

Some reflections on the empirical nature of the benefit and damage functions seem to show that, in most cases of transnational or global pollution, the best-reply functions should be near-orthogonal and that the emission reduction by some countries should not be offset by greater emissions elsewhere (think of CFC, acid rain, and CO_2, whenever the so-called "carbon leakage" is not too high). This issue needs further empirical investigation in order to design operational schemes (see Larsen and Shah, 1994 for a first attempt). However, if it proves correct, the analysis of cooperation carried out thus far may become relevant for policy discussion.

STABILITY AND NEGOTIATION LINKAGES

Now consider an alternative route to achieve a large environmental agreement, namely the link between separate negotiations. This mechanism, too, offers interesting insights for the protection of the global commons.

Suppose that negotiations on environmental policy do not lead to a stable coalition, or that, if a stable coalition exists, it is too small to attain a satisfactory protection of the environment. Moreover, assume that no country can commit to cooperation, which implies that the strategic behaviour proposed in the previous section cannot be used to expand the coalition (all countries decide unilaterally and non-cooperatively).

Linking two negotiations refers to the following: two negotiations are said to be linked when the signing of the first agreement is conditional to the signing of the other one, and vice versa. Notice that the linkage changes the rules of the game, its strategy space and payoff functions *vis à vis* the case of two separate negotiations. In particular, each government has a more complex strategy space because its decision concerns whether or not to cooperate on environmental protection, and whether or not to link environmental negotiations to other negotiations.

The idea of linking two negotiations has been proposed in the environmental literature for the following reason (see Cesar and De Zeeuw, 1993): if countries are asymmetric and one country profits from cooperation on one issue and the other country profits from cooperation on another issue, the linkage of the two negotiations provides positive gains from cooperation to both countries. The condition that must be met is that the payoffs resulting from the two negotiations must be "almost offsetting" (see Cesar and De Zeeuw, 1993). Hence the linkage is a "profitability tool."

In our model, countries are symmetric, which implies that the above argument does not apply. Instead, we propose linking two negotiations in order to stabilise the resulting coalition, i.e. in order to offset the incentive to free-ride on the environmental agreement. Hence, the linkage becomes a "stability tool."

The mechanism can be explained through the following example: suppose the environmental negotiation is linked to the negotiation on R&D cooperation, which involves an *excludable positive externality* and increases the joint coalition welfare. In this way, the incentive to free-ride on the benefit of a cleaner environment (which is a public good fully appropriable by all players) is offset by the incentive to appropriate the benefit stemming from the positive R&D externality (which is a partially excludable club good and therefore fully appropriable only by the signatory countries). The latter incentive can stabilise the joint agreement, thus increasing its profitability because countries can reap both the R&D cooperation and the environmental benefit (this second benefit would be lost without the linkage).[8]

In a more formal way, let $P^1(j^*_1)$ be the payoff of the j^*_1 countries that join the first (environmental) agreement, whereas $Q^1(j^*_1)$ is the payoff of the non-signatories. Similarly, define $P^2(j^*_2)$, and $Q^2(j^*_2)$ for the second agreement. Assume now that the two negotiations are linked. The payoff of the countries signing the joint agreement is $P^u(j^*_u)$, whereas the non-signatories achieve $Q^u(j^*_u)$. Notice that, given the stability condition proposed above, j^*_k, $k=1,2,u$, is defined by $P^k(j^*_k) - Q^k(j^*_k-1) = 0$, $k=1,2,u$, i.e. j^*_k defines the number of countries belonging to the stable coalition. Moreover, $P^k(j^*_k)\geq 0$, $k=1,2,u$ (see Definition 1). Finally, let J^1, J^2, J^u be the sets of signatory countries in the three cases described above, whereas J^1_o, J^2_o, J^u_o are the sets of non-signatories. As we are analysing the possibility of expanding the environmental coalition, we consider only the case $J^1 \in J^u$.

Notice that linking two negotiations increases the dimension of the stable environmental coalition if $j^*_u > j^*_1$. The move to a larger stable coalition is profitable for the j^*_1 countries belonging to the stable environmental coalition if:

(3a) $P^u(j^*_u) \geq P^1(j^*_1) + P^2(j^*_2)$ when $J^1 \in J^u$, $J^1 \in J^2$

(3b) $P^u(j^*_u) \geq P^1(j^*_1)$ when $J^1 \in J^u$

The definition specifies the incentives to link the two agreements. The linkage can occur when it increases the welfare of the signatories, *vis à vis* the separate negotiations. Notice that the linkage may not be Pareto optimal. Countries that do not sign the joint agreement may lose, compared to the situation in which they belong to one of the two separate negotiations. The goal of the proposed mechanism is to expand the environmental coalition, $J^1 \in J^u$.
Hence, the welfare of countries that do not belong to the environmental agreement, but belong to the other one, does not decrease when the joint coalition is formed if:

(4a) $P^u(j^*_u) \geq P^2(j^*_2)$ when $J^2 \in J^u$

(4b) $Q^u(j^*_u) \geq P^2(j^*_2)$ when country $i \in J^2$, $i \notin J^u$

Furthermore, in order to achieve Pareto optimality, the joint coalition must not decrease the welfare of those countries that did not belong to any coalition. Let Q_o be the payoff of these countries before the joint coalition is formed. Moreover, J_o denotes the set containing them. The following condition

---

[8] The optimality of linking R&D cooperation and environmental cooperation is discussed in Carraro and Siniscalco (1994).

must be satisfied: linking two negotiations is Pareto optimal if conditions (3) and (4) hold and if:

(5a) $P^u(j^*_u) \geq Q_o$ when $J_o \in J^u$
(5b) $Q^u(j^*_u) \geq Q_o$ when country $i \in J_o$, $i \notin J^u$

Notice that this definition has an interesting implication: if n, the number of negotiating countries, is the dimension of J^2 and J^u (i.e. all countries sign the second and the joint agreement), and if $P^u(n) \geq P^1(j^*_1) + P^2(n)$, $j^*_1 < n$, then the linkage of the two negotiations expands the environmental coalition and is Pareto optimal.

Therefore, the linkage of two negotiations can lead a large number of countries to sign the cooperative agreement, thus ensuring environmental protection while guaranteeing Pareto optimality. Notice that this result can be achieved even in the case in which countries' reaction functions are negatively sloped, so that the environmental coalition (the one resulting from bargaining on emissions only) would be unstable for all j.

POLICY RECOMMENDATIONS

The above theoretical results may be used to provide some policy recommendations. In particular, they suggest a broader approach to environmental negotiations that accounts for all links between the environment and other economic variables. Three remarks can be emphasised:

a. The basic statement from which our analysis has started is that free-riding undermines the stability of negotiations to protect the global environment, and that no supranational authority exists to regulate countries' behaviour. Of course, the free-riding problem is relevant if policies to achieve environmental protection are costly. This is a stylised fact generally accepted both in the theoretical and the empirical literature (see Boero, Clark and Winters, 1993 and Grubb et al., 1993, for a survey).

b. Another crucial dimension of the analysis is time. Even if the above theoretical models are static, they suggest that cooperation can hardly be achieved through the simultaneous acceptance of the environmental agreement by all countries. Hence the need for a sequence of policy strategies (transfers, linkages, etc.) that broaden the environmental coalition step by step;

c. The above analysis also sheds some light on the role of environmental policy and of its traditional tools (taxes, command and control, etc.). In particular, the general remark is that no environmental policy should exist as such. Policymakers should, rather, use all policy tools (industrial and trade policy tools, for example) to control economic phenomena that may not be called environmental, but whose consequences on the environment may be relevant.

The remarks above lead to the following policy recommendations. Suppose a stable coalition, however small, exists. First, the policy evaluation concerns the costs and benefits of forming the coalition. If economic costs are low (because environmental innovation, tax-recycling and other no-regret policies are implemented), then environmental benefits may be such to provide countries forming the original coalition with the incentives (resources) to persuade other countries to enter the coalition. Policymakers of countries belonging to the original coalition may therefore attempt to induce other countries to join the coalition by firmly and credibly committing themselves to carry out the environmental strategy. Under this condition, a new larger stable coalition may emerge. This coalition can then try to become even larger by committing to cooperation and by transferring resources to other countries.

Another possible sequence of actions is the following. Suppose a group of countries decides to link negotiations on environmental issues with other negotiations. If the possibility of exploiting the benefits of R&D cooperation or those of a free trade agreement is linked to the acceptance of measures to reduce emissions, countries outside the "linked coalition"[9] achieve only the environmental benefit (which is not excludable). These countries may therefore want to enter the coalition, if environmental benefits are not sufficient to offset the competitiveness losses arising from the exclusion from the R&D or trade agreement.

Notice, that countries in the stable "linked coalition" may not want to accept other countries in their coalition. New countries, by reducing the competitiveness gains of countries in the original coalition, may reduce their welfare too, if the additional environmental benefits are not large (this point is formally discussed in Carraro and Siniscalco, 1994).

References

Barrett, S. (1992). "Self-enforcing International Environmental Agreements." *CSERGE Working paper* GEC 92-34. London.

Bauer, A. (1993). "International Cooperation over Environmental Goods." Paper presented at the Oslo Seminar on Environmental Economics, September 15-17.

---

[9] We use the name "linked coalition" to denote the stable coalition emerging from negotiations in which environmental issues are linked to other economic issues.

Boero, G., R. Clark, and A. Winters. (1993). "Controlling Greenhouse Gases: a Survey of Global Macroeconomic Studies." *Discussion Paper No. 93-18.* University of Birmingham.

Carraro, C. and D. Siniscalco. (1992a). "The International Dimension of Environmental Policy." *European Economic Review*, 379-387

Carraro, C. and D. Siniscalco. (1992b). "Transfers and Commitments in Environmental Negotiations." Paper prepared for the ESF Task Force 3. Forthcoming in K.G. Maler (ed.), *Transnational Environmental Problems: An Economic Perspective*. Dordrecht: Kluwer.

Carraro, C. and D. Siniscalco. (1993). "Strategies for the International Protection of the Environment." *Journal of Public Economics* 52, 309-328.

Carraro, C. and D. Siniscalco. (1994). "R&D Cooperation and the Stability of International Environmental Agreements." Paper presented at the CEPR workshop on "Environmental Policy, International Agreements and International Trade." London: November 11-12.

Cesar, H. and A. De Zeeuw. (1993). "Issue Linkage in Global Environmental Problems." Paper presented at the FEEM-IMOP workshop on "Market Structure and Environmental Policy." Athens: January 14-16, 1994.

Chander, P. and H. Tulkens. (1993). "Strategically Stable Cost-Sharing in an Economic-Ecological Negotiations Process." Forthcoming in K.G. Maler (ed.), *Transnational Environmental Problems: An Economic Perspective*. Dordrecht: Kluwer.

D'Aspremont, C.A. and J. J. Gabszewicz. (1986). "On the Stability of Collusion." In G.F. Matthewson and J.E. Stiglitz (eds.), *New Developments in the Analysis of Market Structure*. New York: MacMillan Press, 243-264.

D'Aspremont, C.A. et al. (1983). "On the Stability of Collusive Price Leadership." *Canadian Journal of Economics* 16, 17-25.

Donsimoni, M.P., N.S. Economides and H.M. Polemarchakis. (1986). "Stable Cartels." *International Economic Review* 27, 317-327.

Grubb, M., J. Edmonds, P. ten Brick and M. Morrison. (1993). "The Costs of Limiting Fossil-Fuel CO_2 Emissions: A Survey and Analysis." *Annual Review of Energy and Environment* 18, 397-478.

Heal, G. (1994). "The Formation of Environmental Coalitions." In Carraro, C. (ed.), *Trade, Innovation, Environment*. Dordrecht: Kluwer.

Hoel, M. (1991). "Global Environmental Problems: The Effects of Unilateral Actions Taken by One Country." *Journal of Environmental Economics and Management* 20/1, 55-70.

Larsen, B. and A. Shah. (1994). "Global Climate Change, Energy Subsidies and National Carbon Taxes." This volume.

UNEP. (1991). "The State of the World Environment 1991." New York: UNEP.

World Bank. (1992). *World Development Report 1992: Development and the Environment*. Oxford University Press.

Worldwatch. (1994). *State of the World 1994: A Worldwatch Institute Report on Progress Toward a Sustainable Society*. London: Earthscan.

PART VI

EFFICIENCY AND DISTRIBUTION ISSUES

Chapter 14

ENVIRONMENTAL TAXATION AND THE "DOUBLE DIVIDEND": A READER'S GUIDE

Lawrence H. Goulder (Stanford University)*

INTRODUCTION

Economists have long favored the use of taxes as instruments of environmental protection. Many economic analysts assert that in situations involving serious externalities taxes are the most effective mechanism for "getting the prices right" -- that is, for helping prices closely approximate marginal social costs. The notion that taxes can improve welfare outcomes by internalizing externalities traces back at least as far as Pigou (1938) and is a central tenet of environmental economics.

Real-world economies obtain public revenues not only through environmental (corrective) taxes but also through distortionary taxes such as income, payroll, and sales taxes. What constitutes getting the prices right (or merely closer to right) is more complicated in a second-best setting where both types of taxes are present than in a world with environmental taxes alone. This is the case for at least two reasons. First, taxes interact. In particular, the *gross costs*[1] of newly imposed environmental taxes depend on the marginal rates of

---

* The author is grateful to Lans Bovenberg, Sijbren Cnossen, Bill Hogan, Robert Kohn, Jenny Ligthart, Robert McMillan, Ian Parry, Jim Poterba, Robert Repetto, Ronnie Schöb, Jon Strand, Jim Sweeney, and two anonymous referees for helpful comments, and to Jesse David and Steven Weinberg for excellent research assistance. Financial support from the National Science Foundation (Grant No. SBR 9310362), Resources for the Future, and the IBM Corporation is gratefully acknowledged.

[1] "Gross costs" denotes the welfare sacrifices associated with environmental tax policies, abstracting from the welfare benefits associated with the change in environmental quality. The modifier "gross" is used to make clear that the (usually beneficial) environment-related welfare impacts are being disregarded. Under this method of accounting, the overall efficiency change from a policy initiative is the welfare benefit stemming from the change in environmental quality (the gross benefits) minus the gross costs. A more precise definition of costs will be offered in the first section. In this

pre-existing distortionary taxes.[2] Second, the presence of other taxes introduces the possibility of "swapping" an environmental tax for an existing tax. Consider a tax reform in which an environmental tax is introduced and its revenues are used to finance reductions in the income tax. The overall gross cost of this revenue-neutral package depends not only on the costs attributable to the environmental tax itself but also on certain efficiency benefits (avoided costs) associated with the reduction in income tax rates.[3]

These issues have come to life in recent analyses and policy debates surrounding the carbon tax option. Initial work on carbon taxes tended to ignore other distortionary taxes; most initial studies assumed that the revenues from this tax would be returned to the economy in a lump-sum fashion.[4] Subsequent analyses pointed out, however, that the revenues could be used to finance reductions in ordinary, distortionary taxes.[5] Several analysts have indicated that this could significantly reduce the costs of the carbon tax. Indeed, some have suggested that these costs could be zero or negative when opportunities to "recycle" the revenues through cuts in distortionary taxes are taken into account. However, some recent theoretical and empirical work points out an effect that works in the opposite direction, revealing ways that existing distortionary taxes may interact with the carbon tax and thereby enlarge the carbon tax's costs. While recognizing that recycling the revenues can reduce the costs of a carbon

---

chapter, "costs" always means gross costs unless stated otherwise.

[2] Partial equilibrium analysis suggests that the incremental efficiency cost of a new tax in a given market is higher, the larger are pre-existing taxes in that market. This follows from the basic notion in public economics that the costs of a given tax tend to rise with the square of the overall tax rate. However, as emphasized later in this chapter, tax interactions extend across markets: pre-existing taxes in other markets significantly affect the costs of a new tax in a given market.

[3] This idea was advanced several decades ago by Tullock (1967) and Kneese and Bower (1968), and somewhat more recently by Nichols (1984). Terkla (1984) appears to have been the first to perform a numerical assessment of the efficiency benefits associated with devoting environmental tax revenues to cuts in existing taxes.

[4] For example, in its Model Comparison Project undertaken in 1990-2, the OECD commissioned six models to investigate the costs of reducing CO_2 emissions through carbon taxes. In this investigation, costs were evaluated assuming lump-sum replacement of revenues. Only two of the six models were capable of assuming alternative forms of revenue-replacement. On this, see OECD (1992).

[5] For general discussions, see, for example, Pearce (1991), Poterba (1991), Oates (1991), Pezzey (1992), and Repetto *et al.* (1992). For numerical assessments, see Shackleton *et al.* (1992) and Gaskins and Weyant (1994) for results from several models.

tax, this work shows that for any given method of recycling, pre-existing taxes augment the costs. Much of this work indicates that this *tax interaction effect* is larger than the *revenue-recycling effect*, so that, overall, a revenue-neutral carbon tax is likely to involve positive costs in a second-best setting.

The term "double dividend" relates directly to these discussions. Pearce (1991) noted that swaps of environmental taxes for distortionary taxes may produce a double dividend by not only (1) discouraging environmentally damaging activities but also (2) reducing the distortionary cost of the tax system. The double-dividend concept is relevant to many important ideas in second-best environmental taxation. Unfortunately, the term is used in different ways: the dividend given by (2) above, in particular, can have very different interpretations. This has led to some confusion. This chapter will distinguish the different notions of "double dividend" and analyze the theoretical and empirical support for each. The main motivation here is not to develop a taxonomy but rather to clarify key issues relevant to the formulation of environmental policy in second-best economies. The discussion is intended to help delineate the circumstances under which the substitution of "green" or environmentally motivated taxes for typical existing taxes is likely to be an efficiency-improving venture.

Before launching into the specifics of the double-dividend issue, it may be useful to view the issue in a broader context. There is widespread agreement as to the ability of environmental taxes to confer the first dividend above (environmental improvement), although the magnitude of this dividend usually is highly uncertain. On the other hand, there is much debate as to what kind of additional (second) dividend, if any, might be offered by environmental taxes. The preoccupation with the possibility of a second dividend, in my view, reflects the uncertainties about the magnitudes of the first. Much of the controversy about the second dividend is in terms of whether environmental taxes can be introduced in a way that is costless. The no-cost idea is highly attractive to policymakers, who are interested in "green tax swaps" but are frustrated by the uncertainties as to the values of the environmental benefits that would result from such swaps. Under these conditions, the no-cost idea is especially appealing. If revenue-neutral environmental tax policies are costless, then the burden of proof facing the policymaker is much reduced: to justify the environmental tax on benefit-cost grounds, it suffices to know the *sign* of the environmental benefits -- to know that they are positive. If costs are zero (or negative), this guarantees positive net benefits. On the other hand, if one cannot be assured that the costs are zero, then before one can recommend an environmental tax swap on efficiency grounds one has to be involved in the messy business of comparing (uncertain) environmental benefits with the economic costs.

Thus, the debate about the double dividend reflects the desire to be able to make safe judgments about environmental reforms in the presence of uncertainty. In my view, it makes a difference how this debate is resolved, because this could influence what policymakers view as the minimal information requirements for fruitful environmental tax reform.

The rest of this chapter is organized as follows. The next section distinguishes three interpretations of the double-dividend concept and comments on the implications of each. The stronger interpretations are most relevant to policy evaluation, and subsequent sections of the chapter deal primarily with them. Then the theoretical evidence for or against the stronger double dividend notions is examined, followed by a presentation of numerical results. The subsequent section discusses the relationship between this issue and optimal environmental taxation. The final section offers conclusions.

THREE DOUBLE-DIVIDEND PROPOSITIONS

One can distinguish several double-dividend claims that might be made about a given environmental tax initiative such as a carbon tax. Here are three:[6]

> *Weak Form:* By using revenues from the environmental tax to finance reductions in marginal rates of an existing distortionary tax, one achieves cost savings relative to the case where the tax revenues are returned to taxpayers in lump-sum fashion.
>
> *Intermediate Form:* It is possible to find a distortionary tax such that the revenue-neutral substitution of the environmental tax for this tax involves a zero or negative gross cost.
>
> *Strong Form:* The revenue-neutral substitution of the environmental tax for *typical* or *representative* distortionary taxes involves a zero or negative gross cost.

---

[6] Other notions have appeared. In European discussions, reduced unemployment and increased profits are often referred to as the potential extra dividends (in addition to improved environmental quality) from environmental taxes. See, for example, Bovenberg and van der Ploeg (1993), Carraro, Galeotti, and Gallo (1994), and Sörenson, Pedersen and Nielsen (1994). In this chapter, the extra dividend is more fundamental: it relates to gross welfare costs rather than specific economic variables (such as employment or profits) that contribute to welfare.

The Weak Form

These hypotheses differ in terms of what they propose about the costs of revenue-neutral environmental tax policies. Let "gross cost" refer to the reduction in individual welfare (in wealth equivalents) of a given tax initiative, abstracting from the welfare effect from changes in environmental quality.[7] Let $C(t_E, \Delta T_L)$ denote the gross cost of the new environmental tax t_E in combination with lump-sum tax reductions ΔT_L sufficient to make the policy revenue-neutral. Similarly, let $C(t_E, \Delta t_X)$ denote the gross cost of the new tax t_E accompanied by cuts in the distortionary tax Δt_X sufficient to achieve revenue-neutrality. The first proposition asserts that:

$$C(t_E, \Delta t_X) \; < \; C(t_E, \Delta T_L). \tag{1}$$

The gross cost is lower when revenues are replaced through cuts in the distortionary tax than when revenues are replaced lump-sum. Under this proposition, the "second" dividend is the lower distortionary cost in the former case (left-hand side) relative to the cost in the latter case (right-hand side).

The weak double-dividend notion is relatively uncontroversial. This is fortunate, because the key assumption on which it depends is relatively innocuous. The weak double-dividend claim can be shown to be equivalent to the claim that replacing, at the margin, a lump-sum tax by a distortionary tax entails a positive welfare cost (apart from environmental considerations).[8] This latter

---

[7] I emphasize efficiency issues here, and thus I define gross costs in welfare terms. These costs are most easily defined with respect to a representative individual whose overall utility U can be defined by $U(u_1(E), u_2(G))$, where u_1 is utility from environmental quality (E) and u_2 is utility from "ordinary" goods and services (G). Policy initiatives generally lead to changes in both u_1 and u_2. Gross benefits are the wealth equivalent (as measured, for example, by the equivalent variation) of the change in u_1; gross costs are the negative of the wealth equivalent of the change in u_2.

[8] Consider the two post-reform situations associated with (1). One (associated with the left-hand side) involves a reduction in a distortionary tax; the other (associated with the right-hand side) involves a lump-sum tax reduction. The two post-reform situations are alike in other important respects: they involve the same tax revenue and all other tax rates (including t_E) are the same. One can write the levels of welfare associated with the two post-reform situations as $W(t_E, t_X', T_L)$ and $W(t_E, t_X, T_L')$, where $t_X' = t_X + \Delta t_X$ and $T_L' = T_L + \Delta T_L$. (Note that $\Delta t_X < 0$, $\Delta T_L < 0$.) The weak double-dividend assertion is equivalent to the assertion that $W(t_E, t_X', T_L) > W(t_E, t_X, T_L')$. This, however, is equivalent to the assertion that the gross efficiency cost of raising the distortionary tax from t_X' to t_X, where the change is financed through a reduction in lump-sum taxes from T_L to T_L', is positive. Thus, the weak double-dividend claim is upheld so long as the distortionary tax considered has a positive gross efficiency cost or excess burden.

claim is not difficult to uphold, since the idea that swapping a distortionary tax for a lump-sum tax has a positive welfare cost is part of the usual definition of "distortionary." So long as the tax t_X deserves its title as a distortionary tax, it will have a positive efficiency cost and the weak double-dividend claim will hold.[9]

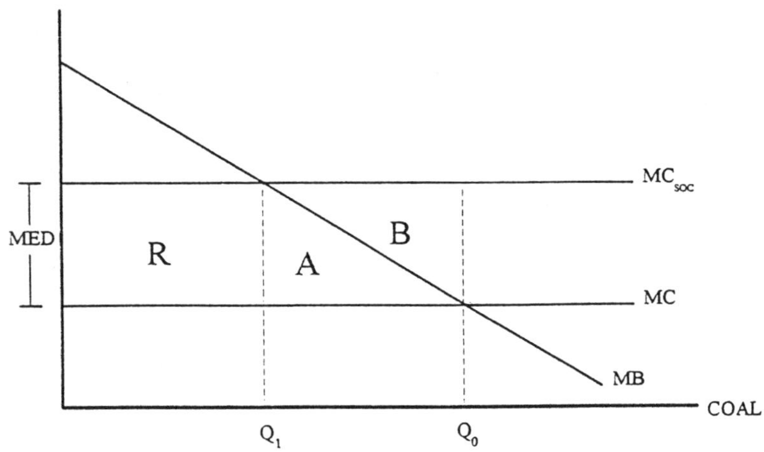

Figure 1 Partial equilibrium analysis of pollution taxation

---

[9] Two clarifications are in order here. First, it should be kept in mind that this discussion only considers *efficiency* costs and benefits from environmental tax initiatives: distributional concerns are ignored. When distributional impacts are taken into account, it is no longer guaranteed that the social costs of a revenue-neutral environmental tax initiative are lower when revenues are recycled through cuts in distortionary taxes than when they are recycled in lump-sum fashion. For an exploration of these issues, see Proost and van Regemorter (1994).

Second, it is interesting to consider whether the weak double-dividend notion is upheld in the case where t_X is negative in the *status quo ante*. Starting from a negative value (or subsidy), a further reduction in this tax (financed through lump-sum taxes) may be efficiency-worsening. Under these circumstances, if a new environmental tax is employed to finance reductions in a "distortionary" tax whose value is already negative, the weak double-dividend will not result. This does not contradict the claims in the text. In the situation described here, the key requirement of the weak double-dividend notion -- that the tax t_X have a positive marginal excess burden -- is missing.

The weak form can be related to the elements of Figure 1. The figure offers the typical partial equilibrium and first-best framework for analyzing welfare effects of an environmental tax.[10] MC denotes the private marginal costs of producing the given commodity, which in this example is coal. MC_{soc} represents the social marginal cost curve, incorporating the marginal external damage (marginal external cost), MED. MB represents the marginal benefit (demand) curve. If a tax is imposed on coal equal to the marginal external damage, social and private marginal costs become aligned. The usual textbook analysis regards the welfare gain as area B.[11] This is the value of the environmental improvement (A+B) minus the gross costs of the tax (A).[12]

This simple analysis suggests that the tax revenue R is transferred costlessly from those who pay the coal tax to the government and back to the private sector,[13] with no efficiency consequences. The weak notion of the double dividend recognizes the fact that this "recycling" of the revenues (R) may indeed have efficiency consequences. The assertion is that when revenues are used to cut existing distortionary taxes, in particular, they help reduce the overall (gross) distortionary costs of the tax system. *If* the gross costs under *lump-sum* replacement are given by area A, then under *distortionary-tax* replacement they are less than A.[14]

---

[10] In this example, the tax is a strict Pigovian tax in that it applies to a commodity with which pollution is associated rather than directly to pollution emissions. The basic lessons apply to emissions taxes as well.

[11] In the case with non-constant private marginal costs or non-constant marginal external costs, the presentation is slightly more complicated, but the results are essentially the same.

[12] The environmental economics literature often refers to *abatement costs*. These are usually defined in a way that corresponds, in Figure 1, to the entire area below the MB curve (including the area below the MC curve) over interval from Q_0 to Q_1. Gross costs are smaller than abatement costs because they net out the avoided expenditure on polluting inputs. In Figure 1, this avoided expenditure is the area under the MC curve from Q_0 to Q_1.

[13] In the simplest case, the revenues are returned to the private sector in a lump-sum fashion.

[14] Of course, the government could waste the revenues -- for example, by applying them to government projects with benefit-cost ratios below one. In this case the gross costs will be greater than A. In this chapter I do not consider which use of revenues is most likely; I leave that to political scientists.

It is important to recognize, however, that other important cost considerations may be obscured by the first-best, partial equilibrium framework embodied in Figure 1. In particular, when other taxes (such as income taxes) are present, the area A in Figure 1 is not a good indicator of the gross costs of the tax on coal. The weak double-dividend notion correctly claims that rebating revenues through cuts in distortionary taxes reduces gross costs relative to their level under lump-sum replacement, but the reference level from which the reduction occurs is generally quite different from that suggested by area A in the diagram. We return to this issue below.

The Stronger Forms

The intermediate and strong double-dividend notions involve assertions about the *sign* of the gross cost of a revenue-neutral policy in which an environmental tax replaces (some of) an existing distortionary tax. The assertion is:

$$C(t_E, \Delta t_X) < 0. \tag{2}$$

The assertion here is that swapping an environmental tax for a distortionary tax involves a negative overall gross cost.

The intermediate and strong double-dividend notions differ in the strength of the claim regarding the extensiveness of the class of distortionary taxes for which (2) holds. The intermediate notion affirms that there exists at least one distortionary tax t_X for which (2) applies; the strong notion claims that (2) holds for typical or representative existing taxes. For any given distortionary tax (t_X) involved in the tax swap, condition (2) is stronger than (1) assuming $C(t_E, \Delta T_L) > 0$.

One can decompose the overall gross cost $C(t_E, \Delta t_X)$ into that which is "directly attributable" to the environmental tax and that which is "directly attributable" to the reduction in the distortionary tax.[15] The intermediate and strong double-dividend notions can be interpreted as claims that the first cost -- the cost resulting from the environmental tax -- is smaller in absolute magnitude than the second cost -- the cost associated with the cut in the distortionary tax. In other words, when the taxes are scaled to imply the same revenue impact, the

---

[15] Since taxes interact, the attribution of the cost of the revenue-neutral policy change into "direct" effects of t_E and "direct" effects of Δt_X is not automatic. A reasonable decomposition is to split $C(t_E, \Delta t_X)$ into $C(t_E, \Delta T_{L1})$ and $C(\Delta t_X, \Delta T_{L2})$, where ΔT_{L1} and ΔT_{L2} are the lump-sum tax changes necessary to make the component changes t_E and Δt_X revenue-neutral.

environmental tax introduces a smaller cost than does an equal-revenue change in the distortionary tax.

It is important to keep in mind what is meant by "costs." As mentioned earlier, costs (C) are a measure of the policy-induced changes in individual welfare (abstracting from welfare effects associated with policy-related changes in environmental quality). In the related theoretical work, welfare depends directly on individual consumption of goods and services and enjoyment of leisure. Measured this way, the economic cost can differ in sign and magnitude from changes in important macroeconomic variables such as GNP or the growth of GNP. The question whether a given revenue-neutral tax swap entails positive costs is different from the question whether the swap entails a reduction in GNP or its growth rate.[16]

Relative to the strong version, the intermediate double-dividend version makes a more modest claim about the circumstances in which gross costs will be zero or negative. It asserts the existence of at least one particularly burdensome distortionary tax whose replacement by an environmental taxes would involve negative costs. It should be noted that the case for removing or reducing an unusually costly existing tax can be made without introducing environmental considerations. Even without bringing in the environmental dimension, there would be strong reasons for replacing this tax with other taxes on narrow (non-environmental) efficiency grounds.[17] Correspondingly, although the presence of this highly inefficient tax may supply a target for general tax reform, by itself it does not offer support for introducing the new environmental tax.

As mentioned in the introduction, the validity of the strong form of the double dividend would significantly reduce information hurdles associated with the evaluation of environmental taxes. If the strong double dividend resulted, it would only be necessary to establish the positive sign of environmental benefits to justify a given revenue-neutral environmental tax on benefit-cost grounds. In light of the substantial uncertainties surrounding the magnitudes of environmental benefits from green taxes, the appeal of the strong form is quite understandable.

---

16 GNP is an imperfect indicator of gross welfare for several reasons. One is that it only reflects the marginal welfare contributions of goods and services (to the extent that these are reflected in market prices); inframarginal surpluses are not incorporated. A second reason is that it disregards welfare contributions from non-marketed goods and services such as leisure.

17 If efficiency were the only consideration, optimal tax reform would achieve equality in the marginal efficiency costs of all taxes with non-zero rates. Of course, efficiency is not the sole policy objective. A highly inefficient tax may be worth keeping if it has important virtues along other dimensions such as distribution.

Much of the debate about the potential for environmental tax reform revolves around whether the strong form holds. The remainder of this chapter concentrates on this double-dividend notion.

DOES THE STRONG FORM HOLD? THEORETICAL CONSIDERATIONS

A Key Result

This section examines the theoretical evidence related to the strong interpretation of the double-dividend idea. An important theoretical investigation of this issue was provided by Bovenberg and de Mooij (1992, 1994a). These authors develop a simple general equilibrium model with one primary factor of production -- labor -- and three produced commodities -- a "clean" consumption good, a "dirty" consumption good, and a public good (nonrival in consumption). Production exhibits constant returns to scale, and the rates of transformation between the three produced commodities are constant and equal to unity. Environmental quality is negatively related to aggregate production of the dirty consumption good. A representative household derives utility from leisure, from consumption of the three produced goods, and from environmental quality. The two private goods and leisure are weakly separable from the public good and environmental quality in the utility function.

Bovenberg and de Mooij begin with the situation where the only tax is a labor income tax with a constant marginal tax rate. They then consider the effects of a revenue-neutral policy change in which a tax is imposed on the dirty consumption commodity and the revenues are devoted to a reduction in the labor tax rate. The commodity tax is non-infinitesimal.[18] The strong double-dividend claim is that this policy would yield an increase in non-environment-related welfare, that is, in the utility from the composite of consumption and leisure enjoyed by the representative household.[19] These authors find that this claim is substantiated if and only if the uncompensated wage elasticity of labor supply is negative, that is, if and only if the labor-supply function is backward-bending. Most empirical studies of labor supply yield positive values for the

---

[18] Alternatively, the tax is incremental but superimposed on an existing non-infinitesimal tax on the dirty consumption commodity. The significance of the "non-infinitesimal" assumption will be discussed later.

[19] The separability assumption enables the authors to evaluate utility from the consumption-leisure composite apart from the utility from public goods or environmental quality.

uncompensated elasticity;[20] thus, the Bovenberg-de Mooij results tend to reject the double-dividend proposition in its strong form.[21]

What lies behind the Bovenberg-de Mooij result? There are two key components. First, the tax on the environmentally harmful consumption good lowers the after-tax wage and generates distortions in the *labor* market, and these labor-market distortions are at least as great in magnitude as the labor-market distortions from a labor tax increment of equal revenue yield. Hence, the revenue-neutral swap in which the environmental tax replaces (some of) the labor tax leads to no reduction (and usually an increase) in labor-market distortions. Second, the tax on the environmentally damaging commodity induces changes in the commodity market -- "distorting" the choice among alternative commodities.[22] These two distortionary effects -- in labor and commodity markets -- imply that, apart from environmental considerations, the revenue-neutral combination of an environmental tax and reduction in labor tax involves a reduction in the non-environmental component of welfare. In fact, the distortions in the commodity and labor markets are connected. To the extent that the environmentally motivated commodity tax leads households to substitute other commodities for the taxed commodity, there is a reduction in the gross revenue yield of the tax. This *tax base erosion effect* limits the extent to which the environmental tax can finance a reduction in the labor tax, and augments the overall gross cost of the tax initiative.

The intuition for this result is as follows.[23] Consider a static model in which there is one labor market (no labor heterogeneity). Let Case 1 refer to the situation where there is just one produced commodity. Suppose that initially there is a tax on labor but no commodity tax. Under these circumstances the commodity tax (even an infinitesmal one) produces a non-infinitesimal excess burden. This occurs because, under the circumstances just described, the introduction of a commodity tax is formally identical to an increase in the labor

---

[20] See, for example, Hausman (1985).

[21] At the same time, Bovenberg and de Mooij's analysis supports the weak notion of double dividend. Their paper concludes with "Hence, there exists a 'double dividend' in the sense that a cost reduction can be achieved by using revenues from pollution taxes to cut distortionary taxes rather than returning these revenues in a lump-sum fashion."

[22] It should be kept in mind that we are concerned with "distortions" in the *gross* sense -- abstracting from environmental considerations. Clearly, environmental taxes can improve overall resource allocation -- reduce overall distortions -- when the environmental dimension is considered. We return to this issue below.

[23] See Poterba (1993) for a complementary discussion.

tax. It reduces the real wage in precisely the same way that an increase in a labor tax would.[24] Now suppose, in contrast to this first case, that there are many distinct produced commodities. As before, suppose that the initial situation involves only a tax on labor. Now consider (Case 2) the effect of introducing a uniform but small tax on *all* commodities, and (Case 3) the effect of introducing a small tax on just *one* of the commodities. Case 2 is formally similar to Case 1; it will generate non-infinitesimal excess burdens because this tax is equivalent to an increase in the wage tax.[25] Case 3 is relevant to the imposition of an environmental tax on one commodity. Bovenberg and de Mooij show analytically that, for a given revenue yield, the excess burden in Case 3 is non-infinitesimal and in fact is larger than in Case 1 or 2. The reason is that Case 3 generates the same or larger labor-market distortion as in the other cases and a larger distortion in the commodity markets.[26]

These results indicate that there are *two* important omissions in Figure 1. They work in opposite directions. First (as emphasized by the weak double-dividend claim), the revenues R can be used to reduce gross distortionary costs. This suggests that the partial equilibrium analysis would overstate the cost of the environmental tax initiative. At the same time, the presence of other (labor) taxes

---

[24] The after-tax real wage is (1) the after-tax nominal wage divided by (2) the gross-of-tax price of consumption. Income taxes directly affect the after-tax real wage by reducing (1), whereas energy taxes directly influence this wage by raising (2).

[25] A uniform commodity tax does not avoid distorting the commodity market in all circumstances. A sufficient set of conditions for its optimality is that commodity consumption be separable from leisure in utility and that the utility function be homothetic. See Bovenberg and de Mooij (1994a).

[26] The gross cost is greater in Case 3 than in Case 2 under the stated assumption that the environmental tax is non-infinitesimal. With an infinitesimal environmental tax, the commodity market gross distortion is negligible, and the net distortionary impact in the labor market is zero (the effect of the new environmental tax just compensates for the effect of the reduction in the labor tax). With the "large" environmental tax, in contrast, the commodity market distortion is non-incremental. So long as there is some substitutability across commodities, the tax on the environmental commodity leads to substitution away from this commodity in consumption. This implies a loss of tax base (relative to the case of no substitution) and, importantly, a reduction in the absolute magnitude of the cut in labor taxes that can be financed by the environmental tax. Because of this tax base erosion effect, the environmental tax is unable to finance a reduction in labor taxes large enough to compensate for the implicit increase in labor taxes represented by the environmental tax itself. Thus, when environmental taxes are "large," the revenue-neutral combination of environmental tax and labor tax cut involves a gross distortionary cost for two reasons: (1) labor market distortions are enlarged, and (2) a non-negligible distortion is generated in the commodity market.

implies that, for any given use of the tax revenues, the gross distortion from the environmental tax will be larger than implied by Figure 1. This implies the opposite bias from the partial equilibrium analysis. Bovenberg and de Mooij show that the latter effect is larger in magnitude than the former: overall, the presence of prior taxes implies higher gross costs from the environmental tax -- even when revenues are recycled through cuts in the distortionary tax. These two effects are schematized in Figure 2. Adopting terminology similar to that introduced by Parry (1994) in a recent paper, I call the former effect the *revenue-recycling effect* and the latter the *tax-interaction effect*.[27] (The tax-base erosion effect, mentioned three paragraphs above, contributes to the more general tax interaction effect.) Using a different analytical approach, Parry obtains results that conform to those of Bovenberg and de Mooij, showing that the tax interaction effect is of greater magnitude than the revenue-recycling effect under plausible values for parameters.[28,29]

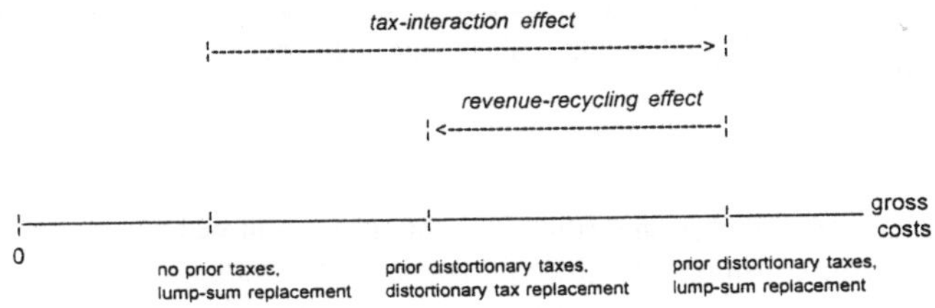

Figure 2 Gross costs of environmental tax under alternative initial conditions and recycling methods

---

27 Parry terms these the "revenue effect" and "interdependency effect."

28 In Parry's analysis, the revenue-recycling effect is greater than the tax-interaction effect only when two conditions simultaneously hold: (1) output from the polluting industry (which produces the good, subject to the environmental tax) is a weaker-than-average substitute for leisure, and (2) the tax elasticity of emissions is well below unity.

29 Figure 2 shows that to uphold the strong double-dividend claim, the magnitude of the revenue-recycling effect must in fact be *larger* than that of the tax-interaction effect -- by enough to bring the gross costs to zero. In Bovenberg and de Mooij and Parry's analyses, since the tax-interaction effect is larger than the revenue-recycling effect, environmental taxes not only involve positive gross costs, but larger gross costs than would result from the same tax in a first-best setting in which there are no pre-existing taxes and in which revenues are returned lump-sum.

Earlier analyses of revenue-neutral environmental taxes emphasized the revenue-recycling effect and did not recognize the tax-interaction effect. These analyses suggested that the strong double dividend was inevitable. An important contribution of Bovenberg-de Mooij, Parry, and others is to call attention to the tax interaction effect and its implications for the costs of revenue-neutral environmental tax reforms. A related and important lesson from this work is that partial equilibrium analyses of the gross costs of environmental taxes can be highly misleading. This research shows that environmental taxes importantly affect resource use in markets other than the market in which the tax is applied -- with significant implications for gross costs. These general equilibrium impacts are especially important if there are prior taxes in "other" markets such as the labor market.[30]

How General Are These Results?

The analytical results from Bovenberg-de Mooij and Parry stem from simplified models. What are the key simplifying assumptions of these models, and to what extent are these simplifications critical to the results?

Intermediate inputs

One simplification of the models is that they disregard intermediate inputs. However, in Bovenberg and Goulder (1994a), the Bovenberg-de Mooij model is extended to incorporate intermediate inputs, and the results above prevail in the extended model. Bovenberg and Goulder analyze the effects of environmentally motivated taxes on "dirty" intermediate goods and on "dirty" consumption goods. In each case, the strong double-dividend claim fails to materialize when the uncompensated wage elasticity of labor supply is positive. The economic basis for the result in the case of the intermediate good tax is the same as that provided above for the commodity (consumer good) tax. Similar results are obtained in Bovenberg and de Mooij (1994b).

Capital

Another simplification in the above models is their static nature. How would consideration of capital affect the results? Including capital introduces the possibility that a revenue-neutral environmental tax could shift the burden of taxation from one factor to another. This has efficiency implications to the extent that, in the initial tax system, the marginal efficiency costs of capital taxation and

---

[30] On the other hand, the partial equilibrium welfare analysis is apt for a first-best world with no prior taxes.

of labor taxation are different.[31] To assess the importance of including capital in the analysis, two considerations are important. First, one needs to consider the extent to which the marginal efficiency costs of capital taxation differ from those of labor taxation under the *status quo ante*. On what do these marginal efficiency costs depend? While labor taxes distort the labor-leisure margin -- the margin of choice between enjoying leisure and enjoying consumption (by earning labor income) -- capital taxes distort the intertemporal margin -- the margin of choice between consuming today and consuming in the future (by saving). For a given labor tax, the distortion in the labor-leisure margin is greater the larger is the (compensated) wage elasticity of labor supply. For a given capital income tax, the distortion along the intertemporal dimension is greater the larger is the intertemporal elasticity of substitution in consumption. Thus, the relative marginal efficiency costs of labor and capital income taxes depend on these elasticities and on the magnitudes of labor and capital income tax rates.

The second consideration is whether the burden or incidence of the environmental tax falls largely on labor or on capital. As is well known,[32] taxes on consumption commodities (such as a consumer gasoline tax) are implicitly taxes on labor.[33] On the other hand, the burden of taxes on intermediate inputs (such as a tax on fossil fuels) could potentially fall primarily on capital. This can occur because such taxes can raise the cost of producing capital goods and thereby lower the rate of return. If this effect is large enough, the burden of the tax will fall primarily on capital.[34]

---

[31] The discussion here is not formal. To my knowledge, no one has yet examined the double-dividend issue analytically in a dynamic, closed-model with capital and labor. Bovenberg and van der Ploeg (1994b) develop an analytical model for a small open economy that employs labor and capital. The efficiency issues I shall discuss here apply to an economy that is "large" in the sense that it influences the return to capital; these issues are largely absent in the Bovenberg-van der Ploeg model of a small, open economy, where the return to capital is given by the world market and the burden of domestically imposed green taxes is fully borne by labor. Ligthart and van der Ploeg (1994) develop a static model with labor and a fixed factor, capital. The issues I shall discuss here are only of interest when capital is not fixed in supply.

[32] For a general discussion, see Auerbach and Kotlikoff (1987).

[33] Indeed, this phenomenon underlies the Bovenberg-de Mooij result. The environmental (commodity) tax is an implicit labor tax that produces both labor and commodity market distortions.

[34] Specifically, capital income will fall by a larger percentage than labor income. Labor income falls because the tax raises costs of production, leading to an increase in the prices of consumer goods and a reduction in the real wage.

These considerations indicate that the gross costs of a revenue-neutral environmental tax will be lower to the extent that:

(1) in the initial tax system, the difference in marginal efficiency costs (marginal excess burdens) is large,

(2) the burden of the environmental tax falls primarily on the factor with relatively low marginal efficiency cost,

(3) the base of the environmental tax is relatively broad, so that the gross distortions it generates in intermediate good and consumer good markets are small, and

(4) revenues from the tax are devoted to reducing tax rates on the factor with relatively high marginal efficiency cost.

These considerations raise the possibility of a zero or negative cost tax swap. For example, if the marginal excess burden of capital taxes is higher than that for labor, and if the burden of a new environmental tax falls primarily on labor, then introducing this new tax in combination with a revenue-preserving cut in capital taxes can involve zero or negative gross costs.

Would this result, if it occurred, resurrect the double-dividend in its strong form? Or would it only confirm the intermediate double dividend claim? The issue is semantic: the answer depends on whether one classifies a (pure) tax on capital as a typical or representative tax. A more clear-cut confirmation of the strong double-dividend notion would occur if the gross costs were negative in a case where revenues are returned through a general income tax cut (that is, through cuts in the marginal rates applicable to capital and labor). Returning the revenues through a general income tax cut "dilutes" the gain associated with revenue-replacement, because it involves a reduction in the relatively efficient factor tax as well as the more inefficient one. This method of revenue-replacement does not exploit the potential contribution of item (4) above; but foregoing (4) enables the policy to test unequivocally the strong double-dividend claim. The above considerations indicate that if the capital tax is extremely costly relative to the labor tax, the gross costs can be negative even if the potential contribution of (4) is not exploited.

Thus, once capital is considered, the theory gives more scope for the possibility of the strong double dividend. Is it likely empirically? Most empirical studies for the US indicate that capital taxes have higher marginal excess burdens than do labor taxes.[35] Under these circumstances, the prospects for a strong double dividend become more favorable to the extent that:

(1') in the initial tax system, there is a large difference in the marginal excess burden (MEB) per dollar of revenue for capital taxation, as compared with the MEB per dollar for labor taxation,

(2') the burden of the environmental tax falls primarily on labor, and

(3') the base of the environmental tax is broad.

Are these conditions met in the real world? Condition (2') indicates that the strong double-dividend claim could potentially be upheld for an environmentally motivated commodity tax, because such taxes tend to be borne largely by labor. An example of an environmentally motivated commodity tax is a tax on consumer purchases of gasoline. In a world where capital taxes have much higher excess burdens than labor taxes, the gasoline tax (with revenues returned through cuts in income taxes or, even better, through cuts in capital taxes) is appealing because it helps shift the burden of taxation from capital to labor. Intuitively, this consumer-level tax functions like a consumption tax and shares the appeal of a consumption tax in reducing intertemporal distortions, or distortions in capital markets. On the other hand, a tax on consumers' gasoline purchases is obviously much narrower than a general consumption tax. As indicated by condition (3'), the narrowness of the base is a liability. To the extent that substitution from gasoline is easy, a given gasoline tax will generate less revenue and introduce larger gross distortions in commodity markets, worsening the prospects for zero or negative gross costs.[36,37]

---

35 See Ballard, Shoven, and Whalley (1985), and Jorgenson and Yun (1990).

36 Paradoxically, the very feature -- a narrow base -- that contributes to larger gross costs is the feature that often makes environmental taxes attractive in terms of overall efficiency. The narrower these taxes are, the closer they often are to the source of the externality (for example, certain types of emissions). Hence, even though the narrower base tends to amplify gross costs, it also tends to expand gross benefits and overall efficiency (gross benefits minus gross costs). To the extent that the objective is overall efficiency, large gross costs should not stand in the way of environmental taxes.

37 This implicitly assumes that a uniform commodity tax is optimal (see note 25). Optimal commodity tax principles indicate that, other things equal, there is an efficiency rationale for relatively higher taxes on commodities that are highly complementary to leisure. See, for example, Atkinson and Stiglitz (1980), chapter 12.

The prospects for the strong claim are worse in the case of a tax on intermediate inputs. Certain intermediate input taxes -- for example, a tax on fossil fuel inputs -- may have a strong justification when environmental benefits are taken into account. But the double-dividend issue relates only to the cost side, and intermediate input taxes do not have much promise on this side of the ledger. The reason is that these taxes are often borne to a significant degree by capital. To the extent that fossil fuels are employed more intensively in the production of capital goods than in the economy in general, an environmentally motivated tax on fossil fuels will raise the relative price of these goods and be borne more by capital. In the data set employed in my general equilibrium energy-economy-environment model of the US (see Goulder, 1994a), capital goods production is highly energy intensive. Consequently, in policy simulations the burden of fossil fuel (intermediate input) taxes falls primarily on capital, virtually ruling out the possibility of the double dividend in the strong form.[38]

The new dimension that arises when capital is considered is that of shifting the burden of taxes between factors and thereby exploiting differences in the marginal excess burdens of factor taxes. This dimension has force only insofar as the existing tax system is inefficient in the sense that the marginal excess burdens of different taxes are not equal. Arguably, the case for reducing these inefficiencies is independent of the case for a new environmental tax. The possibility of a double dividend in the strong form is more difficult if one considers only situations where the "rest of the tax system" is optimal in some sense. I'll return to this issue later in the chapter.

Pre-existing subsidies

The issue just raised also is relevant to evaluating the strong double dividend claim in the presence of prior subsidies: there may be independent reasons for revoking the subsidies. Still, it is worthwhile noting that prior subsidies clearly offer scope for the strong double dividend. For example, a carbon tax can serve to undo prior subsidies on fossil fuels and thereby reduce or eliminate the gross distortionary costs in fuels markets. Shah and Larsen (1992) emphasize this point in considering the potential for carbon taxes in developing countries.

Emissions taxes

The analytical models discussed above consider taxes on commodities (consumer goods or intermediate inputs) whose production or use generates pollution. Other things equal (abstracting from monitoring costs, etc.), it is most efficient to tax

---

[38] Many of these issues are examined in detail in Goulder (1994b), which compares the costs of a fossil fuels tax (Btu tax) and gasoline tax.

the source of the externality, that is, emissions. Relative to commodity taxes, emissions taxes can yield higher *overall* efficiency gains. But the results concerning the double dividend are not changed. One can think of emissions as another intermediate input. As with taxes imposed on an ordinary intermediate input, emissions taxes alter the mix of all intermediate inputs and distort factor markets. The same mechanisms as discussed earlier indicate that emissions taxes will involve larger gross costs per dollar of revenue than the income (labor) taxes they replace. Whether they are more or less attractive in terms of gross costs than intermediate input taxes depends on the production technology. It may seem counterintuitive that emissions taxes do not automatically fare better than ordinary input taxes in terms of the double dividend. It helps to keep in mind that the issue here is gross costs per dollar of revenue, not overall gross costs. The same feature that makes emissions taxes attractive in terms of overall efficiency -- their ability to induce larger emissions reductions per dollar of revenue than ordinary intermediate input taxes -- curtails their attractiveness in terms of gross costs.

Exhaustible resources and decreasing returns

Until this point we have only considered taxes on produced goods. In assessing the gross costs, we needed to consider the distortions produced in the markets for the factors that produced these goods. In particular, in examining the costs of a tax on fossil fuels, we regarded fossil fuels as produced intermediate goods and attended to the related distortions in the markets for labor and capital.

Of course, fossil fuels in their original form are primary resources, not produced goods. They are exhaustible resources whose reserves are given by nature.[39] The essential scarcity of natural resource stocks gives rise to scarcity rents. A basic result from the literature on the taxation of natural resources is that a constant tax on these rents does not alter the intertemporal allocation of these resources and has no efficiency cost.[40] This suggests that an environmentally motivated tax on, say, fossil fuels, might not have a significant (gross) efficiency cost after all, given the exhaustible nature of these fuels. If this were the case, then swapping a tax on these fuels for ordinary income taxes might be a negative cost option, upholding the double-dividend claim in its strong form.

---

[39] *Known* reserves, of course, are endogenous, a function of exploration activity. This does not alter the main points discussed here.

[40] See, for example, Sweeney (1977, 1993), and Dasgupta and Heal (1979). It may be noted that under these circumstances the environmental tax does not yield any environmental benefits.

Should the previous results be discarded because they are derived from models that ignored exhaustible resources? Not necessarily. It should be recognized that in most cases, the environmentally motivated taxes under actual consideration are not taxes on scarcity rents. The base of a carbon tax, for example, is not the scarcity rent but rather the quantity of fuel purchased (since carbon combustion and the amount of CO_2 emitted are functions of the quantity of fuel consumed). In contrast with a tax on pure rent, the tax on fuel output affects intertemporal choice and introduces a gross efficiency cost.[41] Despite the exhaustible nature of fossil fuels, a carbon tax does not attain the non-distortionary ideal.

Of course, this does not entirely disprove the significance of exhaustibility to the double-dividend issue. Even if the environmental taxes in question are related to fuel output rather than rent, it is possible that the gross efficiency costs of these taxes might be considerably lower when these "outputs" are extracted exhaustible resources. To continue with the fossil fuels example, *extracted* fossil fuels are the product of several primary factor inputs: natural resource stocks (fuels in the ground), capital and labor. The gross efficiency cost from taxes on these fuels will be a weighted average of the gross efficiency cost generated by the tax in the exhaustible resource market, the capital market, and the labor market.[42] To discern this efficiency cost, one would need to know the extent to which the tax is borne by each of these factors and the marginal efficiency cost associated with reductions in scarcity rents, the return to capital, and the wage. This information is not easy to obtain, to say the least. Ascertaining the marginal efficiency cost in the resource market alone is exceedingly difficult. Despite these difficulties, further empirical work aimed at gauging these magnitudes might be of considerable value.

Involuntary unemployment

The above models assume all markets clear. However, introducing involuntary unemployment in the models by way of a fixed real wage above the market-clearing value would not change the basic results, assuming no other structural

---

[41] This is the case because, in general, the present value of the tax payment per unit of fuel is not constant through time. On this see, for example, Sweeney (1977).

[42] Whether the strong double dividend arises depends on whether the government has already taxed fixed-factor rents. In a model with three factors of production -- labor, resources, and a fixed factor, Bovenberg and van der Ploeg (1993) show analytically that a tax on a "dirty" input fails to yield a double dividend (in the strong sense) if all the rents from the fixed factor are already taxed. On the other hand, if fixed factor rents are not taxed, the same type of tax offers the double dividend. In the latter case the input tax proxies for the tax on fixed factor rents, and enjoys much of its efficiency potential.

changes. In the original models, introducing the (revenue-neutral) environmental tax reduces labor productivity and shifts leftward the labor-demand function. The same phenomenon occurs in a model in which the real wage is fixed above the original market-clearing level; the result is higher involuntary unemployment. The gross costs of this revenue-neutral policy are positive for the same reasons as in the market-clearing models.[43] However, the results differ when a further structural change is introduced. Bovenberg and van der Ploeg (1993) develop a model with involuntary unemployment (resulting from a fixed real wage) and three factors of production: labor, resources, and a fixed factor. In this model, under some circumstances the revenue-neutral combination of an environmental tax and a labor tax cut boosts employment and raises the non-environmental component of welfare (that is, the gross welfare cost is negative). This occurs when the production share of the fixed factor is large and substitution between labor and resources is easy. Under these circumstances a large share of the tax burden from the environmental tax is borne by the fixed factor, with relatively little efficiency cost. Hence the revenue-preserving cut in the labor tax yields a gross efficiency gain that more than offsets the gross efficiency cost attributable to the environmental tax.[44]

The environment as a capital good

In the analytical models discussed so far, environmental quality, if it is represented at all, enters as an argument of the utility function. It is a nonrival consumption good. The models that incorporate environmental quality usually assume that utility from consumption of "ordinary" goods and services is separable from utility from environmental quality. This permits a clean separation of gross costs and environmental benefits in the welfare analysis.

Environmental quality may also function as a capital good. That is, a cleaner environment may contribute positively to output, holding other inputs constant. Accounting for the productive contribution of environmental quality is a key element in developing truly integrated economy-environment models. Such integration permits models to capture an important potential benefit from

---

43 Labor taxes introduce a wedge between the marginal product of labor and the marginal disutility of labor. With a binding minimum wage, this wedge is even larger. Thus a leftward shift in labor demand is likely to be more costly with involuntary unemployment than under a market-clearing wage.

44 The same qualitative result would obtain if there were no involuntary unemployment in this model. In this model, the environmental tax serves as an implicit way to tax the rents from the fixed factor in the absence of a more direct instrument to do so.

environmental taxes: namely, the avoidance of the adverse impact of environmental deterioration on production.[45]

Once the capital goods aspect of environmental quality is included in an economic model, it becomes more difficult to separate cleanly the gross costs and (environmental) benefits of tax policies. "Gross costs" signifies the effect of the environmental tax on welfare, abstracting from welfare changes linked to changes in environmental quality. In a model with environment as a capital good, it is difficult (though not impossible) to perform this abstraction. In a very recent paper, Bovenberg and de Mooij (1994b) develop a model in which environmental quality functions both as a capital good and consumption good. They show that, under certain circumstances, an environmental tax produces the two "dividends" of a cleaner environment and faster economic growth. In my view, their result, strictly speaking, does not provide support for the strong double-dividend notion because it involves *benefit*-side issues; this is not a case of negative gross costs. Rather, their result supports the notion that *when production-related environmental benefits are taken into account*, environmental taxes sometimes can lead to faster growth as well as a cleaner environment.[46]

Terms of trade effects

All of the analyses discussed above adopt a closed-economy framework. In open economies, another economic dimension -- changes in the terms of trade -- can affect welfare. In certain cases an open economy can employ environmental taxes as a means of improving its terms of trade. Specifically, if the economy in question is a net importer of the good on which the environmental tax is imposed, and if this economy is large enough to exert monopsony power (that is, large enough to influence world prices of this good), then imposing the environmental tax can improve this economy's terms of trade. By taxing this imported good, this economy reduces national (and thus global) demands for the good and induces a reduction in the good's before-tax price in world markets. This shifts some of the burden of the domestic environmental tax onto foreigners,

---

45 This is an important dimension of the environmental benefits from a carbon tax. To the extent that a carbon tax slows the accumulation of greenhouse gases and reduces global warming, it avoids productivity losses in agriculture and other climate-sensitive sectors.

46 Consideration of the capital good aspect of environmental quality seems to be very important in the *overall* benefit-cost evaluation of environmental tax options and may help justify the replacement of ordinary taxes by environmental taxes in several instances. But the strength of the case for the strong double-dividend claim, as defined here, seems largely independent of whether the capital good function of environmental quality is accounted for.

assuming that the revenues from the tax are devoted to cuts in taxes paid by domestic consumers or domestically owned firms. If the terms of trade gains are large enough, they could offset the other gross costs associated with the domestically imposed environmental tax and produce the strong double dividend from the point of view of the domestic economy.[47]

For the US, an environmentally motivated tax on imported oil could fall in this category.[48] On the other hand, it is difficult to make the case that the US would enjoy terms of trade gains from a (unilateral) carbon tax because this tax applies largely to coal, of which the US is a net exporter.

A Common Theme

A unifying theme emerges from these considerations. Although the above analysis reveals circumstances in which the strong double dividend can occur, with one exception (the terms-of-trade case) these circumstances involve an essential inefficiency in the existing tax system on non-environmental dimensions. That is, the strong double dividend can arise only if the initial tax system involves inefficiencies in the form of differences in the marginal excess burdens of various taxes. If the environmental tax shifts the tax burden from factors associated with high MEB's to factors with relatively low MEB's, it can thereby reduce the non-environmental inefficiencies of the tax system and (if this effect is large enough) yield the strong double dividend. Thus, the strong double-dividend presumes a need, based on non-environmental considerations, for factor tax reform. It is reasonable to ask why these inefficiencies cannot be addressed directly (through changes in factor tax rates) rather than through an environmental tax. It seems more natural to address these inefficiencies directly as part of general tax reform than indirectly as part of environmental policy.

One qualification to this last remark might be offered: there may be political opposition to reforms that directly alter the relative taxation of factors. Under these circumstances one could regard a revenue-neutral environmental tax policy as a more feasible alternative to direct reforms which are politically less palatable. The empirical support for this political argument is far from clear. Even if there were an empirical basis for the argument, one might feel

---

47 Terms-of-trade effects do not support the strong double-dividend claim in global terms.

48 The analysis is made more complicated by the existence of the OPEC cartel and the imperfectly competitive nature of the world oil market.

uncomfortable rallying behind it since it essentially exploits the inability of voters to recognize the true incidence of environmental taxes.

NUMERICAL FINDINGS

This section presents some numerical results applicable to the double-dividend issue. Here I consider only the results that bear on the strong double-dividend notion defined above: the issue under investigation is the gross costs (non-environment-related welfare costs) of revenue-neutral tax swaps. As mentioned earlier (see note 6) some other studies define reductions in unemployment or increased profits as additional "dividends" from green taxes. I do not report results from these studies here because I do not have the information necessary to combine the reported changes in unemployment, profits, and other economic variables into the more general welfare notion expressed by "gross costs."[49]

Table 1 presents results from five numerical models. These are the Goulder and Jorgenson-Wilcoxen intertemporal general equilibrium models of the US, the DRI and LINK econometric macroeconomic models of the US, and the Shah-Larsen partial equilibrium model, which has been applied to five countries, including the US[50] The results in Table 1 are for the revenue-neutral combination of an environmental tax (usually a carbon tax) and reduction in the personal income tax, except in cases where this combination was not available. I have focussed on the case where revenues are returned through cuts in the personal income tax because this tax seems "typical" and "representative;" hence this case is relevant to the strong double-dividend claim.

All welfare changes abstract from changes in welfare associated with improvements in environmental quality (reductions in greenhouse gas emissions). Thus they correspond to the gross-cost concept discussed above. In the Goulder model and the Jorgenson and Wilcoxen model, welfare changes are reported in terms of the equivalent variation; in the Shah and Larsen model, the changes are

---

49 Nor do I consider the evidence for the weak double-dividend notion. As mentioned earlier, there is virtually unanimous numerical support for this notion, so a weighing of this evidence seems unnecessary.

50 For a more detailed description of these models, see Goulder (1994a), Jorgenson and Wilcoxen (1990, 1994), Shackleton *et al.* (1992), and Shah and Larsen (1992). The Shah and Larsen model is by far the simplest of the five, in part because it takes pre-tax factor prices as given. However, the model addresses interactions between commodity and factor markets and thus incorporates some of the major efficiency connections discussed earlier.

Model	Reference	Country	Type of Environmental Tax	Form of Revenue Replacement	Results: Welfare Change	Results: Pct. Change in Real GNP, pd. 1	Results: Pct. Change in Real GNP, pd. 21
DRI	Shackleton et al. (1992)	U.S.	Phased-in Carbon Tax[a]	Personal Tax Cut	-0.39[b]	-0.00	-0.76
Goulder	Goulder (1994a)	U.S.	$25/ton Carbon Tax	Personal Tax Cut	-0.33[c]	-0.15	-0.22
"	Goulder (1994b)	U.S.	Fossil Fuel Btu Tax	Personal Tax Cut	-0.28[c]	-0.09	-0.18
Jorgenson-Wilcoxen	Shackleton et al. (1992)	U.S.	Phased-in Carbon Tax[a]	Labor Tax Cut	1.01[d]	0.03	-0.41
"	"	"	"	Capital Tax Cut	0.19[d]	0.20	0.95
LINK	Shackleton et al. (1992)	U.S.	Phased-in Carbon Tax[a]	Personal Tax Cut	-0.51[b]	0.00	-0.35
Shah-Larsen	Shah and Larsen	U.S.	$10/ton	Personal Tax Cut	-1049.[e]	-0.020[f]	
"	"	India	"	"	-129.	-0.060	
"	"	Indonesia	"	"	-4.	-0.005	
"	"	Japan	"	"	-269.	-0.008	
"	"	Pakistan	"	"	-23.	-0.070	

Notes: (a) Beginning at $15/ton in 1990 (period 1), growing at five percent annually to $39.80 per ton in 2010 (period 21), and remaining at that level thereafter. (b) Percentage change in the present value of consumption; the model does not allow for utility-based welfare measures. (c) Welfare cost per dollar of tax revenue, as measured by the equivalent variation. (d) Equivalent variation as a percentage of benchmark private wealth. (e) Compensating variation in levels (millions of U.S. dollars). (f) Percentage change in GDP (not GNP). This is a one-period model.

based on the compensating variation.[51] In the DRI and LINK macroeconomic models, the percentage change in aggregate real consumption substitutes for a utility-based welfare measure.[52]

In most cases, the revenue-neutral green tax swap involves a reduction in welfare, that is, it entails positive gross costs. This militates against the strong double-dividend claim. Results from the Jorgenson and Wilcoxen model, however, support the strong double-dividend notion.[53] To what might the differences in results be attributed?

A thorough examination of the differences in structure of these models, and an extensive test of how these differences account for differences in model outcomes, is beyond the scope of this chapter. However, one potential explanation lies in the differences between the Jorgenson and Wilcoxen model and the Goulder model in the marginal excess burden of capital taxation.[54] The interest elasticity of saving is higher in the Jorgenson and Wilcoxen model than in the Goulder model. In addition, the Jorgenson and Wilcoxen model assumes that capital is fully mobile across sectors, while the Goulder model includes adjustment costs, which limit the speed at which capital can be reallocated and lower the elasticity of capital demand. Thus, elasticities of capital supply and capital demand are higher in the Jorgenson and Wilcoxen model; correspondingly, the marginal excess burden of capital taxation is considerably higher in the Jorgenson and Wilcoxen model than in the Goulder model, and the

---

[51] The equivalent variation is the lump-sum change in wealth which, under the "business-as-usual" or base case, would leave the household as well off as in the policy-change case. Thus a positive equivalent variation indicates that the policy is welfare-improving. The compensating variation is the lump-sum change in wealth which, in the policy-change scenario, would cause the household to be as well off as in the base case. In reporting the Shah-Larsen results I adopt the convention of multiplying the compensating variation by -1, so that a positive number in the table signifies a welfare improvement here as well.

[52] The demand functions in these models are not derived from an explicit utility function. Hence they do not yield utility-based measures. A difficulty with using consumption as a proxy for welfare is that it disregards welfare changes linked to changes in leisure.

[53] It might also be noted that the Shah-Larsen model yields negative gross costs for four of five countries in simulations which combine a carbon tax and reduction in the corporate income tax. (Such results do not appear in Table 1.) Only in the case of the US is such a tax swap a gross-welfare-reducing venture.

[54] Of the five models in Table 1, these two are the most similar and allow for the most straightforward comparisons.

difference in the marginal excess burdens of capital and of labor is larger. In the Goulder model, the difference in MEB's per dollar is $0.10;[55] the difference in the Jorgenson and Wilcoxen model appears to be considerably higher.[56] As indicated in the previous section, a large deviation in the MEB's on capital and labor taxes works in favor of the strong double dividend (particularly if the burden of the environmental tax falls on labor). This helps explain why, in the Jorgenson and Wilcoxen model, a revenue-neutral combination of carbon tax and reduction in *capital* tax involves negative gross costs (that is, a positive change in gross welfare). It is more difficult to account for the fact that substituting a carbon tax for a *labor* tax involves negative gross costs in the model.

Table 1 also includes GNP impacts. These are included simply because they may be of interest. As mentioned above, conceptually they bear no systematic relationship to welfare impacts. Indeed, in some of the numerical simulations the GNP and welfare changes are of opposite sign.

Like the theoretical results, the numerical outcomes in Table 1 tend to weigh against the strong double dividend claim. But there is less than perfect agreement among the numerical results. Discerning the sources of differences in results across models is difficult and frustrating, in large part because of the lack of relevant information on simulation outcomes and parameters. Relatively few studies have performed the type of analysis that exposes the channels underlying the overall impacts. There is a need for more systematic sensitivity analysis, as well as closer investigations of how structural aspects of tax policies (type of tax base, narrowness of tax base, uniformity of tax rates, etc.) influence the outcomes. In addition, key behavioral parameters need to be reported. Serious attention to these issues will help explain differences in results and, one hopes, lead to a greater consensus on likely policy impacts.

---

55 The MEB's per dollar are $0.22 and $0.12 for the tax on individual capital and individual labor income, respectively.

56 This assumption is based on reported values in Jorgenson and Yun (1990). The Jorgenson and Yun and Jorgenson and Wilcoxen models have some similarities in structure and parameters, but there are important differences as well. Marginal excess burden numbers from the Jorgenson and Wilcoxen model were not yet available.

RELATIONSHIP TO OPTIMAL ENVIRONMENTAL TAXATION

The Optimal Tax Rate and the Strong Double Dividend Claim

The previous discussions concerned the effects of "incremental" tax reforms. How do these issues relate to the optimal levels of environmental taxes?

The basic partial equilibrium analysis of the optimal environmental tax invokes the "Pigovian" principle: the optimal environmental tax is at a rate equal to the marginal external costs, or marginal environmental damages (MED). This principle implicitly assumes that the gross marginal cost (or marginal abatement cost) associated with an environmental tax is equal to the tax rate. Hence if the tax is set equal to MED, gross costs and environmental benefits will be equated at the margin, assuring optimality.[57]

The foregoing discussion indicates, however, that in a second-best setting a given environmental tax may give rise to marginal gross costs that differ from the tax rate. As indicated above, the revenue-replacement effect tends to reduce the gross costs; other things equal, it tends to imply a higher optimal tax rate, that is, a rate above MED. On the other hand, the tax interaction effect tends to raise gross costs; other things equal, it implies an optimal rate below MED. A central question underlying the strong notion of double dividend is whether the revenue-replacement effect is strong enough to outweigh the tax-interaction effect. The answer to this question also determines the relationship between the optimal tax rate and MED in a second-best setting. The previous discussion indicated that, under plausible conditions, the tax-interaction effect is of greater magnitude than the revenue-recycling effect. This implies that, under these conditions, the optimal environmental tax rate is below the rate implied by the Pigovian rule.

This result is supported by analytical work by Bovenberg and van der Ploeg (1994a) and by Parry (1994).[58] From Bovenberg and van der Ploeg's

---

[57] Assuming the appropriate second-order conditions apply.

[58] The seminal contribution to the theory of integrating environmental and distortionary taxes was by Sandmo (1975), who analyzed the optimal configuration of commodity taxes when one of the commodities generates an externality.

Parry derives analytically the conditions under which the tax-interaction effect will be greater than the revenue-replacement effect. Conducting broad sensitivity analysis, he finds that under plausible parameter values the tax interaction effect is of greater magnitude. Parry suggests that in the US the optimal environmental tax rate is "typically" between 60 and 90 percent of the MED.

analysis, in a second-best setting the optimal environmental tax rate, t*, is given by the following:

$$(3) \qquad t^* = MED / \mu$$

where μ is the marginal cost of public funds.[59] In their model, μ exceeds unity if and only if the uncompensated wage elasticity of labor supply is positive. This is precisely the condition that denies the double dividend in its strong form.

Bovenberg and van der Ploeg offer the following intuition for their optimal tax formula. Environmental quality can be viewed as a good which the public sector provides by obtaining public revenues through the environmental tax. The higher the value of μ, the more costly it is to provide environmental quality. Thus, the higher is μ, the lower the optimal amount of environmental quality and the lower the value of t*, other things equal.[60]

---

[59] The marginal cost of public funds (MCPF) is the welfare cost of an incremental increase in government spending financed by taxes. For an explication of this concept and its relationship to the notion of marginal excess burden, see Ballard and Fullerton (1992). For an analysis of how aggregate activity influences the MCPF, see Ligthart and van der Ploeg (1994). Brendemoen and Vennemo (1993) develop and assess numerically a broader notion of the MCPF that includes the environmental benefits from public projects.

[60] The Bovenberg and van der Ploeg formula may appear to contradict an optimality requirement articulated by Lee and Misiolek (1986). In fact, the results are consistent. (For a close examination of this issue, see Schöb, 1994.) The Lee and Misiolek requirement for an optimal environmental tax is that the difference between marginal abatement costs and marginal environmental benefits per marginal dollar of revenues must equal the marginal excess burden of existing distortionary taxes. This can be written as:

$$\frac{\partial C/\partial t_E}{\partial R/\partial t_E} - \frac{\partial B/\partial t_E}{\partial R/\partial t_E} = \lambda(t_X)$$

where $C(t_E,t_X)$ is gross cost, B is total environmental benefit, R is revenue, λ is the marginal excess burden of other taxes, and t_E and t_X are the rates of the environmental and existing distortionary tax. Note that C is a function of both t_E and t_X. If t_X is zero, λ is zero and the equation above reduces to:

$$\partial C/\partial t_E = \partial B/\partial t_E$$

the Pigovian rule. Let t_E* represent the tax rate that satisfies the above equation. Now consider the case where there are pre-existing taxes ($t_X > 0$). How does t_E**, the optimum when other taxes are present, differ from t_E*? Oates (1991) points out that the

These issues are highly relevant to considerations of the optimal carbon tax. Nordhaus (1993) calculates the optimal carbon tax in the case where revenues are returned lump sum, and compares that with the optimal tax when revenues are returned through cuts in distortionary taxes. In the former case, the optimal tax begins at about $5 per ton, while in the latter case it begins at $59 per ton! This seems to contradict the notion, indicated by the optimal tax formula above, that second-best considerations should *reduce* the optimal tax rate. It turns out, however, that although Nordhaus's experiments attend to the revenue-replacement effect, they do not capture the tax-interaction effect because the model did not include pre-existing taxes. Including pre-existing taxes in the benchmark data would likely reverse the Nordhaus results. In this connection, simulations by Bovenberg and Goulder (1994b) indicate that the optimal carbon tax declines with the level of pre-existing taxes.

Caveats

These models indicate that in the presence of other taxes, the optimal environmental tax rate is lower than MED. Although these results seem intuitive, they are no more compelling than the case against the strong double-dividend idea. To the extent that more complex models might expand the possibilities for a strong double dividend, they also offer scope for an optimal environmental tax above MED.

It is also important to recognize that the optimal tax formulae by Bovenberg and van der Ploeg and Parry assume that the optimal environmental tax is part of an optimal tax system. These formulae do not address the *constrained* second-best problem: how best to establish environmental taxes when the rest of the tax system is suboptimal. Much depends on how the rest of the system departs from optimality. Simulation experiments by Bovenberg and Goulder (1994b) for the US economy indicate that the "suboptimality"[61] of the US tax system implies a constrained second-best optimal carbon tax which is lower than that suggested by the Bovenberg and van der Ploeg formula. The

---

tax elasticity at the Pigouvian optimum is relevant here. If the derivative $\partial C/\partial t_E$ were independent of t_X, one could determine from this elasticity whether t_E** is greater or less than t_E*. However, as discussed above, higher pre-existing taxes raise the value of $\partial C/\partial t_E$ evaluated at a given level of t_E. This means that in the presence of other taxes, a *lower* value of t_E may be required to satisfy the Lee and Misiolek requirement.

[61] The tax system is suboptimal in the sense that marginal gross efficiency costs of different taxes are not equal. To the extent that the tax system is meeting other objectives, this need not be suboptimal in a broader sense.

intuition is that a given environmental tax generates a larger gross cost in a suboptimal setting than in a setting where other taxes are set optimally.

"Is a Small Tax Efficiency-Improving?" and Related Theoretical Issues

The cost-side-oriented discussion of the double dividend is driven by the broader question as to whether the substitution of environmentally motivated or "green" taxes for ordinary taxes might be a welfare-improving venture. It is important to recognize that the difficulty of substantiating the strong double-dividend claim is not an indication that green tax swaps are ill-advised. Even if gross costs are positive, the environmental benefits could outweigh these costs, possibly by a substantial margin.

But in light of the uncertainties, are positive net benefits likely? There is a line of reasoning that suggests that policymakers can be assured that benefits will exceed costs provided that the environmental tax is sufficiently "small." The argument is that the gross costs of an incremental environmental tax (with revenues devoted to reducing ordinary taxes) is infinitesmal; in contrast, the gross benefits from an incremental environmental tax are non-incremental. This argument prompts two questions: (1) Does a "small" environmental tax indeed have a "small" gross cost? And (2) If so, is this of practical value to policymakers?

A result from Bovenberg and de Mooij (1992, 1994a) is relevant to the first question. In their framework, an *incremental* environmental tax, with revenues returned as a reduced labor tax, indeed has a zero gross cost. When the environmental tax is small, it yields the same labor-market distortion as the labor tax it replaces, and the gross distortion in the commodity market is infinitesmal. Hence the overall gross cost is zero. So long as the tax yields a non-incremental environmental benefit, it offers an overall welfare improvement.[62]

However, the claim that a small tax yields an overall efficiency improvement may not apply in more general circumstances such as the case where the initial tax system is highly inefficient. The issue is an example of the general second-best tax problem investigated by Lipsey and Lancaster (1956), Bertrand and Vanek (1971), Dixit (1975), and others: In a suboptimal tax system, under what circumstances does introduction of a tax in one market that helps bring prices closer to marginal cost in that market yield an overall efficiency improvement? Although a number of authors have shown particular conditions under which this will occur (see, for example, Dixit, 1975), this cannot be guaranteed for tax systems that simultaneously involve labor, capital, and

62 A similar result appears in Bovenberg and van der Ploeg (1994a).

commodity taxation. These notions are reinforced by results from simulations of a carbon tax by Bovenberg and Goulder (1994b) and by Parry (1994). In these simulations, an incremental carbon tax has incremental gross costs when the initial tax situation is efficient in the sense that marginal excess burdens of taxes are equated. However, the incremental tax involves large gross costs when the initial situation reflects inefficiencies of the US tax system.

This suggests that, in many cases, one cannot be assured that a small environmental tax will produce small gross costs. Moreover, even if this were the case, the knowledge that gross costs will be small seems to be of limited practical value. In my view, it cannot substitute for information about the magnitudes of environmental benefits. In the extreme case where analysts have virtually no information about the magnitudes of these benefits -- except perhaps that they are non-negative -- one cannot tell how small is small enough. No matter how small the tax, it's possible that the (small) gross costs exceed the (even smaller) gross benefits. Although it is useful to know that incremental taxes produce negligible costs, to be of practical value this knowledge needs to be coupled with information about benefits.

This discussion was not meant to lead to pessimism. Rather, it was intended to serve the constructive purpose of drawing attention to significant issues in policy evaluation. The key implication of this discussion is that in evaluating possible green tax swaps, information about environmental benefits is indispensable.[63] Much of the interest in the possibility of a double dividend was fueled by a hope that benefits considerations could be pushed aside. This seems to have been a false hope. The benefits side is critical, and further research to reduce the uncertainties about environmental benefits is likely to be of great value to policy analysts.

SUMMARY AND CONCLUSIONS

This chapter articulated different senses of the potential "double dividend" from environmental taxes and examined the theoretical and empirical support for each.

The weakest double-dividend claim is that returning tax revenues through cuts in distortionary taxes leads to cost savings relative to the case where revenues are returned lump sum. This claim is easily defended on theoretical grounds and (thankfully) receives wide support from numerical simulations.

---

[63] Ballard and Medema (1993) offer one of the first studies to incorporate environmental benefits in a computable general equilibrium assessment of environmental tax reforms.

The stronger versions contend that the costs of revenue-neutral swaps of environmental taxes for ordinary distortionary taxes involve zero or negative gross costs. The strongest version affirms that this is typically the case. An intermediate version asserts that this is the case in at least one instance. The existing theoretical analyses of this issue cast doubt on the strong double-dividend claim. At the same time, the theoretical case against the strong form is not air-tight. In simple models the strong version would result if the uncompensated wage elasticity of labor supply were negative. In more elaborate models, it could arise if the initial tax system were highly inefficient in factor markets (leading to significant differences across factors in marginal value products per dollar) and the environmental tax served to shift the burden of taxes to more efficient factors. These inefficiencies may offer more of a justification for direct attention to these inefficiencies than for indirect approaches through a revenue-neutral environmental tax.

Although the evidence is mixed, numerical results tend to militate against the strong double-dividend claim. There is a need for more systematic numerical investigations of revenue-neutral environmentally motivated tax policies. More extensive policy analyses and broader sensitivity analyses would help identify the channels driving the results and reveal the sources of differences in model outcomes. The extent to which the burden of environmentally motivated taxes is imposed on exhaustible resources (and on resource rents) remains a very interesting empirical issue, and greater attention to this issue in numerical models could generate important insights.

The difficulty of establishing the strong double dividend does not contradict the common numerical finding that an environmental tax can promote higher levels of national income when revenues are earmarked for capital formation. National income is distinct from welfare, and the double-dividend notions (at least as defined here) concern welfare costs.

The presence or absence of a strong double dividend is directly relevant to the issue of the optimal environmental tax. In simpler models, the conditions under which the strong double dividend claim is absent (present) are closely related to the conditions under which the second-best optimal environmental tax is lower than (greater than) the marginal environmental damages. Existing analytical models, in casting doubt on the strongest double-dividend claim, also indicate that the optimal environmental tax is lower than the rate suggested by the simple Pigovian rule.

The keen interest in the double dividend -- particularly in its strong form -- reflects the uncertainties faced by policymakers concerning environmental benefits. The strong double dividend notion, if true, would reduce significantly the amount of information that policymakers need to make a benefit-cost case for green tax swaps. The difficulty of establishing the strong claim implies, unfortunately, that assessing magnitudes of environmental benefits remains a crucial task in policy evaluation. This conclusion cannot be dodged by appeals to the net benefits from "small" environmental taxes. One never knows *a priori* how small is small enough, and, moreover, even small environmental taxes may have "large" gross costs when the existing tax system is suboptimal (in some sense).

The critical importance of attending to the environmental benefits may be the most fundamental message from this examination. Research that helps establish the environmental benefits associated with various environmental tax options will have considerable value for policy analysts. While research can reduce the uncertainties, they will remain a basic fact of life in environmental policymaking. Policymakers cannot afford to wait until circumstances permit riskless policy choices; they would wait forever. A better approach is to recognize the uncertainties and endorse environmental tax reforms when the prospects for gain make the risks worth taking.

References

Atkinson, A.B. and J.E. Stiglitz. (1980). *Lectures on Public Economics*. New York: McGraw-Hill.

Auerbach, A.J. and L.J. Kotlikoff. (1987). *Dynamic Fiscal Policy*. Cambridge, UK: Cambridge University Press.

Ballard, C.L. and S.G. Medema. (1993). "The Marginal Efficiency Effects of Taxes and Subsidies in the Presence of Externalities." *Journal of Public Economics* 52, 199-216.

Ballard, C.L. and D. Fullerton. (1992). "Distortionary Taxes and the Provision of Public Goods." *Journal of Economic Perspectives* 6/3, 117-131.

Ballard, C.L. J.B. Shoven and J. Whalley. (1985). "General Equilibrium Computations of the Marginal Welfare Costs of Taxes in the US." *American Economic Review* 75/1, 128-138.

Bertrand, T.J. and J. Vanek. (1971). "The Theory of Tariffs, Taxes, and Subsidies: Some Aspects of the Second Best." *American Economic Review* 61, 925-931.

Bovenberg, A.L. and R.A. de Mooij. (1994). "Environmental Taxation and Labor-Market Distortions." *European Journal of Political Economy* 10, 655-683.

Bovenberg, A.L. and R.A. de Mooij. (1994a). "Environmental Levies and Distortionary Taxation." *American Economic Review* (September), 1085-1089.

Bovenberg, A.L. and R.A. de Mooij. (1994b). "Environmental Tax Reform and Endogenous Growth." *Working Paper*. Tilburg University.

Bovenberg, A.L. and F. van der Ploeg. (1993). "Consequences of Environmental Tax Reform for Involuntary Unemployment and Welfare." *Working Paper*. Tilburg University.

Bovenberg, A.L. and F. van der Ploeg. (1994a). "Environmental Policy, Public Finance and the Labour Market in a Second-Best World." *Journal of Public Economics* 55, 349-390.

Bovenberg, A.L. and F. van der Ploeg. (1994b). "Green Policies in a Small Open Economy." *Scandinavian Journal of Economics* 96/3, 343-363.

Bovenberg, A.L. and L.H. Goulder. (1994a). "Integrating Environmental and Distortionary Taxes: General Equilibrium Analyses." *Working Paper*. Stanford University.

Bovenberg, A.L. and L.H. Goulder. (1994b). "Optimal Environmental Taxation in the Presence of Other Taxes: An Applied General Equilibrium Analysis." *Working Paper*. Stanford University.

Brendemoen, A. and H. Vennemo. (1993). "The Marginal Cost of Funds in the Presence of External Effects." *Discussion Paper* 99. Norway: Central Bureau of Statistics.

Carraro, C., M. Geleotti and M. Gallo. (1994). "Environmental Taxation and Unemployment: Some Evidence on the Double Dividend Hypothesis in Europe." *Working Paper*. GRETA Econometrics.

Dasgupta, P.S. and G.M. Heal. (1979). *Economic Theory and Exhaustible Resources*. Cambridge, UK: Cambridge University Press.

Deaton, A. (1987). "Econometric Issues for Tax Design in Developing Countries." In D. Newbery and N. Stern (eds.) *The Theory of Taxation for Developing Countries*. New York: Oxford University Press.

Diamond, P. and J. Mirrlees. (1971). "Optimal Taxation and Public Production I: Production Efficiency and II: Tax Rules." *American Economic Review* 61, 261-287.

Dixit, A. (1975). "Welfare Effects of Tax and Price Changes." *Journal of Public Economics* 4/2, 103-123.

Gaskins, D. and J. Weyant. (1994). *The Costs of Controlling Greenhouse Gas Emissions*. Stanford, CA: Stanford University Press.

Goulder, L.H. (1994a). "Effects of Carbon Taxes in an Economy with Prior Tax Distortions: An Intertemporal General Equilibrium Analysis." (Forthcoming in *Journal of Environmental Economics and Management*.)

Goulder, L.H. (1994b). "Energy Taxes: Traditional Efficiency Effects and Environmental Implications." In James M. Poterba (ed.), *Tax Policy and the Economy* 8. Cambridge, MA: MIT Press.

Hausman, J.A. (1985). "Taxes and Labor Supply." In A.J. Auerbach and M.S. Feldstein (eds.), *Handbook of Public Economics*, Vol. 1. Amsterdam: North-Holland.

Jorgenson, D.W. and P.J. Wilcoxen. (1990). "Environmental Regulation and US Economic Growth." *The Rand Journal of Economics* 21/2, 314-340.

Jorgenson, D.W. and P.J. Wilcoxen. (1994). "Reducing US Carbon Emissions: An Econometric General Equilibrium Assessment." In D. Gaskins and J. Weyant (eds.), *The Costs of Controlling Greenhouse Gas Emissions*. Stanford, CA: Stanford University Press.

Jorgenson, D.W. and K-Y. Yun. (1990). *Tax Policy and the Cost of Capital*. Oxford University Press.

Kneese, A.V. and B.T. Bower. (1968). *Managing Water Quality: Economics, Technology, Institutions*. Baltimore: Johns Hopkins Press.

Ligthart, J.E. and F. van der Ploeg. (1994). "Pollution, the Cost of Public Funds and Endogenous Growth." (Forthcoming in *Economic Letters*.)

Lee, D.R. and W.S. Misiolek. (1986). "Substituting Pollution Taxation for General Taxation: Some Implications for Efficiency in Pollution Taxation." *Journal of Environmental Economics and Management* 13, 338-347.

Lipsey, R.G. and K. Lancaster. (1956). "The General Theory of the Second Best." *Review of Economic Studies* 24, 11-32.

Nichols, A.L. (1984). *Targeting Economic Incentives for Environmental Protection*. Cambridge, MA: MIT Press.

Nordhaus, W.D. (1993). "Optimal Greenhouse-Gas Reductions and Tax Policy in the 'DICE' Model." *American Economic Review* 83/2, 313-317.

Oates, W.E. (1991). "Pollution Charges as a Source of Public Revenues." *Discussion Paper* QE92-05. Washington, DC: Resources for the Future.

OECD. (1992). *The Economic Costs of Reducing CO_2 Emissions*. Economic Studies 19. Paris.

Parry, I.W.H. (1994). "Pollution Taxes and Revenue Recycling." *Working Paper*. US Department of Agriculture.

Pearce, D.W. (1991). "The Role of Carbon Taxes in Adjusting to Global Warming." *Economic Journal* 101, 938-948.

Pezzey, J. (1992). "Some Interactions between Environmental Policy and Public Finance." *Working Paper*. University of Bristol.

Pigou, A.C. (1938). *The Economics of Welfare*, 4th ed. London: Weidenfeld and Nicolson.

Poterba, J.M. (1991). "Tax Policy to Combat Global Warming: On Designing a Carbon Tax." In R. Dornbusch and J.M. Poterba (eds.), *Global Warming: Economic Policy Responses*. Cambridge, MA: MIT Press.

Poterba, J.M. (1993). "Global Warming: A Public Finance Perspective." *Journal of Economic Perspectives* 7/4, 47-63.

Proost, S. and D. van Regemorter. (1994). "Testing the Double-Dividend Hypothesis for a Carbon Tax in a Small Open Economy." (Forthcoming in *International Tax and Public Finance*.)

Repetto, R., R.C. Dower, R. Jenkins et. al. (1992). *Green Fees: How a Tax Shift Can Work for the Environment and the Economy*. Washington, DC: World Resources Institute.

Sandmo, A. (1975). "Optimal Taxation in the Presence of Externalities." *Swedish Journal of Economics* 77.

Schöb, R. (1994). "Environmental Levies and Distortionary Taxation: Environmental View vs. Public Finance View." *Working Paper*. University of Munich.

Shackleton, R., M. Shelby, A. Cristofaro et. al. (1992). "The Efficiency Value of Carbon Tax Revenues." *Draft*. US Environmental Protection Agency.

Shah, A. and B. Larsen. (1992). "Carbon Taxes, the Greenhouse Effect and Developing Countries." *Working Paper* 957. Washington, DC: World Bank.

Sörenson, P.B., L.H. Pedersen and S.B. Nielsen. (1994). "Taxation, Pollution, Unemployment and Growth: Could There Be a 'Triple Dividend' from a Green Tax Reform?" *Working Paper*. Copenhagen Business School.

Sweeney, J.L. (1977). "Economics of Depletable Resources: Market Forces and Intertemporal Bias." *The Review of Economic Studies* 44, 125-142.

Sweeney, J.L. (1993). "Economic Theory of Depletable Resources: An Introduction." In A.V. Kneese and J.L. Sweeney (eds.), *Handbook of Natural Resource and Energy Economics*, Vol. 3. New York: North Holland.

Terkla, D. (1984). "The Efficiency Value of Effluent Tax Revenues." *Journal of Environmental Economics and Management* 11, 107-123.

Tullock, G. (1967). "Excess Benefit." *Water Resources Research* 3, 643-644.

Chapter 15

ENERGY LEVIES AND ENDOGENOUS TECHNOLOGY IN AN EMPIRICAL SIMULATION MODEL FOR THE NETHERLANDS

Frank A.G. den Butter, Rob B. Dellink

and

Marjan W. Hofkes (Free University, Amsterdam)*

INTRODUCTION

A major economic policy question today is to what extent (regulating) energy levies can contribute to a reduction of energy demand and, hence, of greenhouse gas emissions. Two different ways of modelling the effects of energy levies emerge from the literature. First, the analysis is conducted using highly aggregated, theoretical, general equilibrium models (see *e.g.* Bovenberg and van der Ploeg, 1992, 1993, 1994 and Bovenberg and de Mooij, 1992, 1993, 1994). These models focus mainly on the possibilities of obtaining a so-called double dividend, *i.e.* the possibility of improving the quality of the environment and yielding a less distortionary tax system at the same time by substituting energy taxes for ordinary income taxes. The second approach uses disaggregated empirical models, either applied general equilibrium models (*e.g.* the GREEN-model of the OECD, see OECD, 1994) or dynamic policy models in the Tinbergen tradition (*e.g.* the combined use of the SENECA energy model and the ATHENA multi-sector model by the Netherlands Central Planning Bureau (CPB, 1992) and the HERMES model by the EC (see Laroui and Velthuijsen, 1992a, 1992b)).

This chapter illustrates the economic effects of energy levies and the redistribution of their proceeds using a relatively simple and aggregated empirical macroeconomic model for the Netherlands, called the EnTech model. In particu-

* The authors would like to thank Lans Bovenberg, Larry Goulder, Dale Jorgenson, Ruud de Mooij and three anonymous referees for useful comments on a previous version of the paper. The usual disclaimer applies.

lar we test the double dividend hypothesis empirically and analyse the relationship between our results and the theoretical findings.

The main feature of the EnTech model is the endogeneity of (the direction of) technological change. Technology not only enhances the productivity of physical capital, but also creates spillovers to human capital. Furthermore, the EnTech model explicitly describes how an energy levy induces an increase in energy-saving technology. In this way the model reckons both with the externalities of investment in R&D and with price effects on the direction of technological progress. Note that our model is *not* an applied general equilibrium model, as markets do not clear in each period and prices are explicitly determined in behaviourial equations. Moreover, the parameter values of the model are based on empirical results from the literature and the model is calibrated such that a dynamic simulation over the past reproduces more or less the historical time path of the observed endogenous variables. The model may illustrate the dynamics of the introduction of an energy levy.

Simulation experiments assess the consequences of an energy levy for economic activity, employment and technological innovation. On the one hand, an energy levy may hurt the competitive position of a country (when introduced on a national level), having negative effects on economic activity and employment. On the other hand, it may also stimulate energy-saving technological innovation, which can improve the competitive position of the national industry and which may shift the direction of technological progress away from labour saving. Moreover, the proceeds of an energy levy can be used to decrease labour taxes. In this context the energy tax may yield an employment double dividend: it may reduce energy use and increase employment at the same time. In that case the loss of employment because of the deterioration of the competitive position will be more than fully compensated by employment gains induced by the decrease in labour-saving technological progress, by the substitution of energy for other production factors including labour, by the reduction of labour costs, and by the demand effects of the redistribution of the tax proceeds. The model makes it possible to distinguish between the effects of a levy on the industrial use of energy and a levy on household consumption of energy.

The following section briefly surveys the literature on theoretical and empirical models of energy taxes. The next section outlines the EnTech model. Then the results of our experiments are given. The last section concludes.

LITERATURE

Theoretical Literature

The effects of energy taxes on environmental quality, economic activity and employment have been investigated both by means of theoretical models and empirical models. In the theoretical literature general equilibrium models play a major role. In a number of papers Bovenberg and van der Ploeg (*e.g.* Bovenberg and van der Ploeg, 1992, 1993, 1994) analyse the effects of environmental policies in a second-best framework of a small open economy. It appears that the occurrence of an employment double dividend, *i.e.* a cleaner environment and increasing employment at the same time, depends strongly upon the existence of a fixed factor in production. In a model with only two production factors, labour and resources, where the price of resources is exogenously given by the world market price, labour, not surprisingly, always bears the whole tax burden. Replacing a labour tax with a resource tax in this case amounts to replacing an explicit tax on labour with an implicit tax on labour. However, when there is a fixed factor of production, part of the costs of an environmental tax can be shifted to this fixed factor. This suggests that an employment double dividend is more likely in the short run than it is in the long run, since in the short run more factors are fixed. It appears that the larger the share of the fixed factor is in production, the lower the profit tax and the larger the substitution possibilities are between labour and resources, the easier it is to shift the costs of an environmental tax to the fixed factor and the more likely an employment double dividend will be.

Bovenberg and de Mooij (*e.g.* Bovenberg and de Mooij, 1992, 1993) also analyse the effects of environmental taxes in a general equilibrium framework. They allow for the existence of income transfers provided by the government. In their model, an employment double dividend can appear even when there is perfect international capital mobility. In fact, workers are able to shift part of the burden of the environmental tax to transfer recipients and owners of financial wealth. In order to illustrate this mechanism, Bovenberg and de Mooij (1993) provide a numerical simulation, again using a model of a small open economy facing exogenous world market prices for commodities and capital. This model explicitly models two classes of households, differing with respect to their sources of income. The model is calibrated using Dutch data for 1989. It appears that a broad-based energy tax on both firms and households yields an employment double dividend, *i.e.* increased employment and reduced energy use at the same time, at the expense of a fall in non-labour income.

Goulder (1995) overviews different notions of the double dividend. He concludes that theoretical analyses tend to cast doubt on what he calls the strong double-dividend hypothesis, by which he denotes a revenue-neutral substitution of an environmental tax for a typical distortionary tax at zero or negative gross costs. However, he also states that a strong double dividend cannot be ruled out theoretically and that numerical evidence is mixed. Once capital is added next to labour, the possibility of a double dividend increases. Finally, it must be noted that the above-mentioned theoretical models do not account for endogenous technological progress.

Empirical Studies

A number of empirical exercises using large macro-economic models have been conducted in order to calculate the consequences of energy levies for energy use and greenhouse gas emissions and economic growth and employment. However, the outcomes of these exercises are difficult to compare because they relate to different types of measures, to various levels at which the measures are to be taken (worldwide, OECD, European Union (EU), national), and because the baselines used for the model simulations differ from each other.

The recent 'White Book' of the European Commission (EC, 1993) presents the results of simulation exercises on the effects of a reduction of the employer's contribution to social security financed by an energy levy and/or increasing the value-added tax, using three different models. All these models show moderate positive employment effects of a regulating energy levy. The Dutch Central Planning Bureau (CPB, 1992) calculates the effects of a 50% and a 100% levy, imposed at the OECD or national level.[1] According to these calculations, the competitive position of many energy-intensive industries is dramatically affected in the 'going-alone' scenario for the case of a general energy levy, while a broader geographical base (OECD) does not contribute sufficiently to the solution of this problem for industries in the exposed sector. These calculations predict a considerable decrease in employment for the case of a general levy, whereas the impact of a household levy on employment is virtually nil.

On balance, these models provide no uniform evidence with respect to the double dividend hypothesis. Yet, it is obvious that the energy levy is more efficient in reducing greenhouse gas emissions when it is effected at a worldwide

---

[1] Their study has been influential in the decision of the Dutch government not to impose a general energy levy unilaterally, but to restrict the scope of the levy to the 'household' sector (including small enterprises in the sheltered sector).

level rather than at a European or national level. In the latter case, the influence of industries closing down and/or moving abroad is larger so that the competitive position deteriorates. The resulting decrease of economic activity mitigates the possible positive effects on employment of the redistribution of the proceeds of the levy. Differences between the outcomes of these model exercises also emerge because of differences in the implementation of the measures in the model and because of differences in modelling the major mechanisms. For instance, the Central Planning Bureau explicitly reckons with the effects of industries moving abroad, using additional information not contained in the model, whereas other models, including ours, implicitly describe the moving abroad and closing down of industries by substitution in the production function, which mirrors shifts from energy-intensive industries to industries that use relatively less energy. One may question the adequacy of this implicit modelling of the 'moving-abroad' problem. Finally, the effect of higher energy prices on the direction of technological progress is ignored in most models, or it is treated in a superficial manner. One of the major features of our EnTech model is to endogenise this mechanism.

THE ENTECH MODEL

A major drawback for practical policy analysis is that most endogenous growth models are mainly theoretical. This chapter uses an empirical simulation model with endogenous technology and spillovers to human capital, called the EnTech model. The EnTech model is inspired by features of endogenous growth theory, although it is not an empirical counterpart of these theoretical models. It tries to intermediate between the formal endogenous growth models and the traditional macroeconomic models used in policy analysis in the Netherlands. At the core of the EnTech model is the production block, where investments in technology capital and in human capital play a major role. The external effects of R&D are modelled in such a way that R&D investments not only lead to more technology capital, but also have a positive impact on human capital through 'learning-by-doing' and 'learning-by-designing.'

In order to capture the main characteristics of the structure of production and the role of technology capital in our simulation model, we use a framework of nested CES functions, which allows for different elasticities of substitution between the various levels distinguished in the model. Whilst the manner in which the various production factors are combined in such a nested construction is rather *ad hoc*, its design is based on arguments and empirical evidence from the literature, and it is open to sensitivity analysis (see den Butter and Wollmer, 1992, for more details). For a similar approach, see Gelauff and Graafland (1994).

Our simulation model takes the view that prices are rigid and that markets can be in disequilibrium for a prolonged period. In line with the traditional dynamic policy models for the Netherlands, we assume that production is mainly demand determined. Therefore we derive the specification of the factor-demand equations, which determine the factor inputs in the production block, from cost minimisation, given production (or productive capacity) and given the nested structure of the production block. Thus, the derived factor-demand equations have total production as the scaling variable and the relative prices of all production factors as other explanatory variables. In order to make the specification of the factor-demand equations more realistic and in line with empirical specifications from the literature, we have included lagged reactions and additional explanatory variables in the factor-demand equations. In that case, combined factor demand will, through the production block, no longer automatically yield desired production. Consequently, we have to consider one production factor as residual in the model. Labour demand is taken as the residual factor demand, so that the model describes implicitly the effects of technology and the resulting spillovers on labour demand.

The advantage of our disequilibrium approach over a general equilibrium model is that the model describes the historical time path of major economic variables and that we are able to simulate the effects of (policy) shocks on the short- and medium term. Yet, these short- and medium-term simulations require a specific interpretation of the kind of technological progress that is endogenised in the model. As fundamental technological innovations will have long-term effects that are mainly beyond the time horizon of the model, the policy-induced increases in technology capital of this chapter paper relate primarily to a faster diffusion and implementation of existing best-practise techniques.

This chapter presents only a general outline of the model. For more details, see den Butter and Wollmer (1992) and den Butter, Dellink and Hofkes (1994). Figure 1 shows the structure of the production block. At the highest level of nesting, productive capacity (y_n) is a function of efficiency units of capital (including energy, K_c), and labour (L_c). L_c refers to efficiency units of labour (L), adjusted for changes in contractual working time. L is determined by the full capacity demand for labour (a^*) and the level of human capital (HC). K_c is determined by efficiency units of capital (excluding energy, K_T), combined with energy, also measured in efficiency units (E_c).

Technology capital (T_c and ($T_d + T_E + T_i$)), enters the production block in two related ways. First, firms accumulate knowledge either by undertaking research and development, which provides a domestically produced stock of technology capital (T_d), or by importing knowledge, providing a 'substitute' stock of foreign technology capital (T_i). This technology capital is combined to T_c, which may be viewed as being embodied in capital (K), or disembodied and

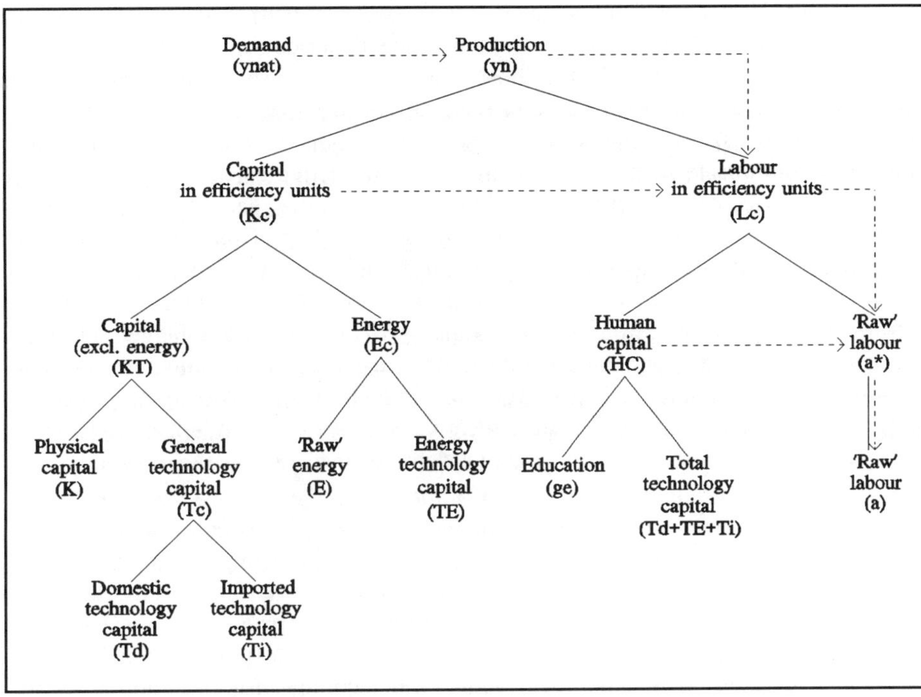

Figure 1 Structure of Production Block

increasing the productivity of existing capital, or both, and thereby determines the level of *efficiency units* of capital, K_T. Moreover, in order to endogenise energy-saving technology, this version of the model distinguishes between general technology capital and accumulated investments in energy-saving technology, which enhances the efficiency of the use of energy (T_E). Therefore, energy-saving technology capital, combined with input of 'raw' energy (E), yields efficiency units of energy (E_c).

In line with recent theories of endogenous growth, we assume spillovers of knowledge associated with research and development and the import of technological capital. This assumption serves to augment the human capital of workers as they work with the new technologies, *i.e.* through 'learning-by-doing' and 'learning-by-designing.' This augmentation of human capital works in a complementary manner with the education level of workers, which is itself determined by real government expenditures on education (g_e).

When all factor inputs (including labour supply) are given, y_n is endogenously determined by the production block. Hence, 'causality' runs from the lower elements in the tree of Figure 1 (indicated by solid lines) to the highest hierarchical level, where y_n is the final variable to be determined. However, as mentioned before, our model has productive capacity fully demand-determined and labour demand (*a*) to follow as a residual; in this case a positive investment impulse will, *ceteris paribus*, result in less labour demand, but also in a higher level of labour productivity. Then 'causality' runs from the left to the right in Figure 1 (indicated by broken lines), so that L_c, and hence *a*, result as final variables to be determined by the production structure. The model was calibrated on the basis of this specification.

Although the calibration of the model is based on actual annual time-series data (with reference period 1972-1990), we have deliberately not tried to estimate the full model using this data set. The two major reasons are that the model contains a number of unobservables and that the information contents of the data set are too poor to yield plausible parameter estimates in the behavioural equations of the model. Instead, our calibration procedure uses a much wider set of information, namely information contained in the body of empirical knowledge. Moreover, when no empirical results are available from the literature, we include parameter estimates (or 'guesstimates') of our own. Finally, we note that in each behavioural equation the constant term is determined by least-squares estimation. The Appendix shows for some major endogenous variables the historical time paths and a dynamic simulation with the baseline version of the model.

We extend the factor-demand equations, which are derived demand equations, in an *ad hoc* manner with adjustment lags and some additional explanatory variables suggested by the literature. The production block is calibrated in such a way that the (implicit) wage-cost elasticity of labour demand amounts to about -0.5, which conforms with labour-demand studies for the Netherlands. Government expenditure on education is exogenous to the model.

Although we aimed to keep the working of the model transparant, we needed some disaggregation of the energy block for the simulation exercises of this chapter. The industrial use of energy (*E*) is modelled in the same manner as the other factor-demand equations and depends on the price of energy, the prices of the other production factors and the volume of production as a scaling variable. In conformity with the literature (see, *e.g.*, Seale Jr. *et al*, 1991), we assume a long-run price elasticity of -0.75 and a mean adjustment lag of one year. We note that the (absolute) value of this elasticity is rather high, compared to empirical findings for the Netherlands. R&D expenditures on energy-saving technology depend on the relative price of energy in comparison with the price of R&D. Besides the industrial use of energy, the model also uses an equation for the demand for energy by households. This energy consumption depends on the price of energy paid by households -- with a long-run price elasticity of -0.5 -- and on disposable income.

The rest of the model has been kept as simple as possible and its specification and calibration is derived from policy models in the Netherlands. Desired productive capacity is determined by aggregate demand, which gives the basic version of the model a demand-oriented character. In order to describe the quality competition with other industrialised countries, we have given the import and export equations an indicator for the relative stand of technology (*riv*), which is endogeneous in the model. Hence, a relative increase in R&D expenditures in the Netherlands will, according to the model, enhance the volume of exports and lower the volume of imports, so that it fosters demand. Of course, part of this effect will also be achieved through price competition, which is modelled in the import and export equations in the usual manner.

Prices in the goods market are determined by a simple mark-up over total costs, consistent with monopolistic competition. Wages are determined by a wage-curve equation derived from Graafland (1992). This equation describes the wage negotiation process in the Dutch institutional setting. In the wage equation, the coefficient of labour productivity growth has been set to unity. This is in line with the findings of den Butter and van de Wijngaert (1992) that in the long run the 'wage space' (price inflation plus labour productivity growth) is fully used for wage increases. It does, however, imply a negative wage-price spiral in simulations where, as a result of some impulse, labour productivity decreases as com-pared to the baseline. Net government wages and social security benefits follow net market wages.

Finally, the model contains a number of technical equations, which endogenise the government budget. First, the financial deficit of the government is exogenous. Furthermore, the size of the wedge -- *i.e.* the difference between gross labour costs and net wages -- is determined endogenously to achieve *ex post* budget neutrality. All the additional revenues of the energy levy are redistributed to producers and/or consumers.

Note that our model, with various types of endogenous technology investments and spillovers, distinguishes three ways an energy levy reduces energy demand:

a. the income effect, when the energy levy leads to a decrease in demand and hence also to a decrease in economic activity;
b. substitution from energy to other factors of production;
c. more energy-saving technology induced by the higher price of energy leading to, *ceteris paribus*, a lower demand for 'raw' energy.

Most studies on the effects of energy prices on energy demand either disregard the last source or combine the latter two under the label of changes in energy efficiency.

SIMULATION RESULTS

In our simulations we consider only the *regulating energy levy*, as it is the major instrument in the policy proposals of, *e.g.*, the EU and the Dutch government, which aim at curbing energy consumption and greenhouse gas emissions.[2] The major aim of a regulating energy levy is to induce a decrease of energy demand, through substitution and technological change. Thus a regulating levy is not a Pigouvian tax, which is raised in order to internalise the costs of environmental damage, nor is it a general tax, which aims at raising government income. An energy levy is just one among many instruments of environmental policy, such as subsidies, regulations, information, emission standards and marketable emission permits (see *e.g.* Sterner, 1990, for a short review). In the case of regulating levies, the redistribution of the proceeds should foster fiscal neutrality. An open question in the proposals is whether there should be *ex ante* or *ex post* fiscal neutrality. The first approach implies the redistribution of the current proceeds of the levy only, whereas the second approach implies that all net revenues be redistributed: both the current proceeds and the endogenously determined and possibly negative indirect revenues. Since the latter approach is theoretically most elegant, we consider *ex post* fiscal neutrality in our model exercises.

The aim of our simulation exercises using the EnTech model is three-fold. First, and most important, they are meant to illustrate the working of our model and especially those mechanisms that are associated with endogeneity of (the direction of) technological progress. Second, the simulations may provide some quantitative indication of the effects of regulating energy levies. In this respect our calculations should be considered with caution because we run only very simple simulations and our model contains a number of assumptions on which little empirical evidence is available. For that reason we also conduct a sensitivity analysis, showing which parameter values and assumptions are crucial in the policy simulations and should be given empirical attention. Thus, we come to the third and implied aim of our policy simulations: they provide guidelines for further research on this subject.

---

[2] Most policy proposals consider an energy levy that is partly based on energy value and partly on the carbon contents of energy. In our simulation model we simply implement an energy levy as a rise in the price of energy.

The impulse-response tables of this chapter give the differences between the baseline projection and the projection from the impulse (1986-90) for certain variables of interest that are endogenous in the model and are relevant in the discussion on the double-dividend hypothesis. The simulations are run only for the last five years of the reference period to avoid too much influence of the volatile movements in energy prices on the results. Furthermore, the model is designed primarily to assess the short- and medium-term effects of technological progress; the structure of a macroeconomic model like our EnTech model is not suited for analysis of long-term effects.

The Effects of an Energy Levy

Table 1 reports the results of an energy levy of 50% on the energy price, where the proceeds are used to lower the employers' contributions to social security payments (*employers-scenario*). In the case of a *general levy*, *i.e.*, a levy on the energy prices of both consumers and producers, it appears that energy use decreases, while employment increases. In the short run the effects are smaller than in the longer run. This is due to the fact that in the short run the production structure is fixed mainly with respect to the composite capital input so that industrial energy use cannot decrease much. Also the consumers need some time to adjust their energy use. In the longer run substitution possibilities between energy and labour are larger. This effect, however, will be mitigated by shrinking production (relative to the reference path) and consequently decreasing use of all factors of production.[3]

In the first year, production growth is stable due to substitution by households from energy use to other goods. In subsequent years, however, this effect is dominated by increasing costs of production and hence a deterioration of the international competitive position, resulting in a slowdown in production growth. In the short run, (real) disposable income[4] increases due to the decreasing tax burden on labour. As a matter of fact, employees are able to shift part of

---

[3] We note that decreases in investments are induced only by the price and volume effects of the energy levy and not by the effect of reduced profitability, as capital income (or some other measure of profits) is not included as an explanatory variable in the investment equation. Experiments with a version of the model not reported here show that inclusion of profits as an explanatory variable in the investment equation does not alter the working of the model because the estimated coefficient of profits appears to be rather small.

[4] It must be noted that nominal disposable income is deflated with the production price index, which does not take account of increases in the household price of energy.

the (energy) tax burden to employers.[5] However, in the longer run labour taxes increase again due to decreasing proceeds from the energy levy and, thus, disposable income decreases. Total expenditure on R&D decreases in the longer run due to the slowdown of production growth; however, expenditure on energy-saving R&D increases.

Table 1 A regulating energy levy of 50%: employers scenario; demand-oriented model

		General levy			Household levy		
Year		1	3	5	1	3	5
Production	%	0.0	-0.5	-0.9	0.1	-0.0	0.0
Price of production	%	0.9	1.2	1.2	-0.0	-0.3	-0.6
Labour demand	1000 ly.	30.1	45.9	45.3	0.5	2.3	3.5
Unemployment	%-point	-0.5	-0.8	-0.8	0.0	-0.0	-0.1
Nominal wage costs	%	-0.8	-2.3	-3.3	-0.4	-0.9	-1.3
Nominal wedge	%	-8.1	-5.9	-6.2	-4.2	-2.8	-2.9
Disposable income workers	%	2.6	0.4	-0.3	1.4	0.5	0.4
Disp. income non-workers	%	2.0	-0.4	-1.2	1.4	0.5	0.3
Consumption	%	1.3	0.0	-1.1	0.9	0.4	0.2
Total energy use	%	- 11.3	- 19.6	- 22.1	-5.1	-8.7	-9.6
Research and development	%	-0.0	-0.7	-1.2	-0.0	-0.2	-0.3
of which: energy-saving R&D	%	*3.9*	*3.3*	*2.8*	*-0.0*	*-0.1*	*-0.2*

Concluding, both energy use and unemployment are reduced at the expense of lower disposable income in the longer run, especially of non-workers, and of a slowdown in production growth.[6]

---

[5] The way in which wage costs and gross wages are modelled implies that changes in labour taxes are, in the short run, for the most part passed on to employees; in the longer run also wage costs are flexible.

[6] Experiments (not reported here) over a longer time period (ten years) indicate that in the longer run wage costs decrease further, but that the increase in labour demand fades away due to shrinking production.

In the case of a *household levy*, *i.e.*, a levy on the energy price of households only, the effects on labour demand and energy use are much smaller than they are for the case of a *general levy*. Production growth is stable even in the longer run, due to the fact that the producers, who are confronted with lower demand, do not face higher energy prices under a *household levy*, but do take advantage of the lower taxes on labour. Labour costs decrease relative to the reference path, and substitution takes place from energy, capital and technology (including energy-saving technology) to labour.

On balance, the occurrence of an employment double dividend in our empirical model seems to be in line with the theoretical analysis of Bovenberg c.s. In the EnTech model capital is largely fixed in the short run. Consequently, it is possible to shift part of the burden from labour to capital. In the longer run this effect will die out. However, the existence of transfers and the fact that we do not impose income neutrality, still makes it possible to shift part of the burden away from labour, *i.e.*, from workers to non-workers. Furthermore, the substitution elasticities are such that labour and energy are rather good substitutes. Substitution to labour lowers the labour productivity, which in turn puts downward pressure on wages, leading to additional substitution to labour. Finally, the effects on employment are influenced by the endogeneity of technological progress. Higher energy prices induce an increase in energy-saving technological progress.

Alternative Scenarios

The *employers scenario* analysed in the previous section (see also Table 1) appeared to have a negative impact on total investments in technology. This section will discuss a scenario in which part of the proceeds of the levy are used to compensate for this endogenous fall in R&D expenditures, called the *R&D scenario*. Since total expenditures on R&D in the Netherlands are relatively small compared to the proceeds of the levy, only a minor part of the proceeds is needed for this purpose. Table 2 gives the results of the *R&D scenario*, *i.e.*, a regulating energy levy of 50%, where 99% of the proceeds are used to lower employers' contributions to social security, while one percent of the proceeds of the levy are used for higher R&D expenditures.

Under the *R&D scenario* with a *general levy*, a double dividend still occurs, though slightly weaker than it would be in the *employers-scenario* with a *general levy*. Furthermore, disposable income of both workers and non-workers is a bit higher. The enhanced R&D expenditures lead to higher spillovers to human capital, which results in a more efficient use of labour in the production process and consequently in a downward effect on the demand for labour. *Ceteris paribus*, this leads to a higher labour productivity, which exerts upward pressure

Table 2 A regulating energy levy of 50%: R&D scenario; demand-oriented model

		General levy			Household levy		
Year		1	3	5	1	3	5
Production	%	0.0	-0.5	-0.8	0.1	0.0	0.1
Price of production	%	0.9	1.2	1.1	-0.0	-0.4	-0.6
Labour demand	1000 ly.	28.4	41.5	39.3	-0.5	-0.3	0.0
Unemployment	%-point	-0.5	-0.7	-0.7	0.0	0.0	0.0
Nominal wage costs	%	-0.8	-2.2	-3.1	-0.4	-0.9	-1.2
Nominal wedge	%	-8.0	-5.7	-6.0	-4.2	-2.7	-2.8
Disposable income workers	%	2.6	0.4	-0.2	1.4	0.5	0.4
Disp. income non-workers	%	2.0	-0.4	-1.1	1.4	0.5	0.4
Consumption	%	1.3	0.1	-0.9	0.9	0.5	0.3
Total energy use	%	-11.3	-19.6	-22.1	-5.1	-8.7	-9.6
Research and development	%	1.9	2.8	3.4	1.2	2.0	2.5
of which: energy-saving R&D	*%*	*5.9*	*6.9*	*7.7*	*1.2*	*2.0*	*2.6*

on the wages and hence on disposable income and consumption. Finally, higher consumption boosts production, as the model is demand oriented.

Under the *R&D scenario* with a *household levy*, a double dividend no longer occurs. Neither labour demand nor unemployment change significantly. The negative effect on the demand for labour resulting from increased expenditures on R&D has approximately the same size as the positive substitution effects to labour due to lower labour costs and higher energy prices.

Finally, Table 3 presents the results of simulations for a *supply-oriented version* of the EnTech model, in order to assess to what extent the results are influenced by the demand-driven character of the model. In the *supply-oriented version*, production capacity is determined by the factor-demand equations instead of by aggregate demand. A labour-demand equation is added to the model; wages are still determined by the wage-curve equation (for a full description of the supply-oriented model, see den Butter and van Zijp, 1995). The rate of capacity utilisation now exerts a major influence on the determination of prices on the goods market.

Table 3 A regulating energy levy of 50%: employers scenario; supply-oriented model

		General levy			Household levy		
Year		1	3	5	1	3	5
Production	%	-0.2	-1.0	-1.4	0.1	0.0	0.1
Price of production	%	1.0	1.8	2.0	-0.1	-0.3	-0.6
Labour demand	1000 ly.	-2.6	2.0	10.1	0.1	2.3	4.9
Unemployment	%-point	0.1	-0.0	-0.2	0.0	-0.0	-0.1
Nominal wage costs	%	-0.5	-1.0	-1.6	-0.4	-0.9	-1.4
Nominal wedge	%	-6.4	-4.3	-4.9	-4.0	-3.0	-3.3
Disposable income workers	%	1.4	-0.7	-1.0	1.3	0.5	0.5
Disp. income non-workers	%	1.5	-0.8	-1.2	1.3	0.5	0.4
Consumption	%	0.9	-0.2	-0.9	0.8	0.4	0.2
Total energy use	%	-11.4	-19.5	-21.7	-5.1	-8.7	-9.6
Research and development	%	-0.0	-0.7	-1.1	-0.0	-0.2	-0.3
of which: energy-saving R&D	%	*3.8*	*3.3*	*2.8*	*0.0*	*-0.1*	*-0.2*

Table 3 indicates that in the *supply-oriented version* the price of production increases more than it does in the demand-oriented model. This is due to an increasing rate of capacity utilisation. As a result, downward pressure on actual demand is larger, while wage costs decrease less. Hence less substitution to labour takes place and a smaller increase in labour demand results. Only a very weak double dividend remains.

Sensitivity Analysis

The results obtained in the previous sections obviously depend upon the chosen parameter values of the model. This section will investigate how sensitive the outcomes of Table 1 are with respect to the values of some key parameters (see Table 4). By varying these parameter values we may also get more insight into the relative (empirical) importance of the various mechanisms at work in the model. The most important parameters in our simulation experiments are the price elasticity of energy use and the substitution elasticities between energy, on the one hand, and labour and capital, on the other hand.

First we consider the substitution possibilities between energy and capital. In our basic specification, the (direct) substitution elasticity between energy and capital is set at 0.33. When this substitution elasticity is doubled it appears that the effect of a general energy levy of 50% on labour demand is, according to the *employers scenario*, much smaller than it is in the basic version of the model. Now, less substitution will take place from energy to labour in favour of substitution to capital and R&D. Consequently, expenditures on energy-saving R&D increase.

The substitution possibilities between energy and labour can be decreased by decreasing the elasticity of substitution between capital and labour. Table 4 indicates that a decrease in the latter elasticity from 0.8 to 0.5 results in a smaller increase in labour demand, as less substitution to labour takes place.

The impact of the spillovers may be tested by lowering the weight of the spillovers in the CES equation for human capital from two-third to one-third, and by lowering the weight of human capital in the CES function for efficiency units of labour from 0.5 to 0.25. Now the effect of the energy levy on labour demand is smaller, since the influence of the fall in spillovers on labour demand is less. This shows that the spillovers from technological progress have a significant impact on the efficiency of labour, whilst the specification of the interaction between human capital and 'raw' labour is of minor importance.

One of the most distinguishing features of the EnTech model is its inclusion of endogenous technology. Yet, if investments in technology are assumed to be exogenous, the positive effect on labour demand appears to be of the same size as in the original specification. On the one hand, R&D expenditures (excl. energy-saving R&D) increase relative to the original specification, which lead, through increased spillovers to human capital, to a decrease in labour demand. On the other hand, however, R&D expenditures on energy-saving R&D decrease relative to the original specification, which counterbalances the above-mentioned effect through decreased spillovers to human capital and hence increased demand for labour.

As no decisive conclusion can be drawn from the literature with respect to the substitution elasticity between energy and energy-saving technology, this parameter is an obvious candidate for sensitivity analysis. Lowering this elasticity to about two-thirds of its value alters the results only negligibly.

Table 4 Sensitivity analysis on a levy of 50%: employers scenario; demand-oriented model (results in percentage change)

Year 5	energy use	energy R&D	pro-duction	labour demand	wage costs
General levy	*industrial*				
Original specification	-26.6	2.8	-0.9	5.2	-3.3
More subst. energy / capital	-26.2	4.3	-0.5	2.8	-2.7
Less subst. capital / labour	-25.9	4.3	-0.8	4.0	-3.0
Less effects of spillovers	-26.2	3.1	-0.6	3.4	-2.7
Exogenous technology	-26.5	-	-0.9	5.1	-3.3
Less substitution energy / energy technology	-26.6	2.8	-0.9	5.4	-3.4
Lower price elasticity industrial energy use	-19.1	3.0	-0.8	3.9	-2.9
Higher price elasticity household energy use	-26.6	2.8	-1.0	5.0	-3.2
Less effect of labour productivity on wages	-26.0	3.1	-0.9	4.8	-2.5
International energy levy	-26.0	3.2	-0.5	5.7	-3.0
Household levy	*household*				
Original specification	-17.8	-0.2	0.0	0.4	-1.3
Higher price elasticity household energy use	-25.6	-0.2	0.0	0.3	-1.2
Less effect of labour productivity on wages	-17.7	-0.1	0.0	0.4	-1.2

A lower price elasticity of industrial energy use mitigates, not surprisingly, the reduction in industrial energy use. Consequently, less substitution takes place between energy and the other factors of production, including labour. A higher price elasticity of household energy use increases household energy savings but has almost no (additional) effect on the production structure and thus on employment.

One of the major mechanisms of the model is the effect of the lower labour productivity on wage costs and hence on labour demand. To test for this mechanism, Table 4 gives the results of an alternative specification with a long-term effect of labour productivity on wage costs of 0.75 instead of 1.0. The table shows that only wage costs are affected significantly, which prompts the conclusion that factor substitution is driven mainly by higher energy prices and the wage-cost reduction through the redistributed revenues of the tax and only to a minor extent by the second-order effect that lower labour productivity leads to downward pressure on wages and hence boosts labour demand.

Apart from price- and substitution elasticities, the scope of the levy plays an important role. So, finally, the effects of an international energy levy are studied. Since the EnTech model is not designed for this purpose, this policy alternative is implemented in a rudimentary way by assuming that the effect of the energy levy on import prices is proportional to the effect on domestic prices. It appears that the results differ little from a national levy, where domestic prices increase only to a minor extent and hence the international competitive position does not deteriorate much.

CONCLUDING REMARKS

This chapter tests the double-dividend hypothesis using a relatively simple and aggregated empirical macroeconomic model for the Netherlands called the EnTech model. The EnTech model highlights especially the role of technological progress in production and endogenises both the speed and the direction of technological progress.

Simulation experiments with an energy levy of 50% show that at least in the short and medium term a so-called employment double dividend (*i.e.* reduced energy use and increased employment at the same time) can occur. A decrease in the wedge between labour costs borne by the employer and the net income received by the employee induces substitution effects to labour. These substitution effects are enhanced by the explicit modelling of technology spillovers to labour. A general levy yields stronger effects than does a levy on the energy use of households only. Under a general levy production may decrease, but the positive effects on energy use and employment, on the other hand, are also larger.

Exercises using the EnTech model illustrate how the relative importance of the economic mechanisms (which are relevant in the discussions on the effects of energy levies) depends upon the size of the substitution elasticities between energy, labour, technology capital and physical capital. Inspired by modern endogenous growth theory, the specification of the production block pays special attention to the spillovers from technology capital to human capital. These spillovers, which represent the positive externalities of R&D investments, are important for the impact of an energy levy on employment. The negative income effect of the energy levy results in a decrease in economic activity and hence in a decrease in employment, while the substitution effect enhances labour demand. Furthermore, in the EnTech model the fall in economic activity will lead to less investment in R&D, resulting in a lower level of spillovers to human capital and thus an upward pressure on labour demand. The fact that the relative increase in the energy price induced by the levy leads to a shift from labour-saving to energy-saving R&D further complicates the picture. The net effect on labor demand is ambiguous, the more as endogenising both the speed and the direction of technological progress further complicates the interaction of the various mechanisms at work. Given the selected parameter values of the basic version of our model, the higher investments in energy-saving R&D are more than offset by the decrease in total R&D expenditures. Therefore, we advocate that (a small) part of the proceeds of the energy levy be used for additional investments in R&D expenditures, although this way of redistribution has a negative effect on employment compared to other redistribution schemes.

The outcomes of the model, however, largely hinge on the parameter values we select. When calibrating the model we have tried to use as much information as is available from the literature. Yet, more empirical information on the structure of production and especially on the size of the externalities of R&D expenditures is essential for a proper quantitative assessment of the effects of an energy levy on energy use and employment if one believes (and we do) that the mechanisms modelled in this chapter are important. This information should be obtained from detailed (microeconomic) evidence on the key relations of the production structure: further estimation of the behaviourial equations of our model at the macro level (and using macro data) would not enhance its quality or alter its working very much.

References

Bovenberg, A.L. and R.A. de Mooij. (1992). "Environmental Taxation and Labour-Market Distortions." *CentER Discussion Paper* 9252. Tilburg University.

Bovenberg, A.L. and R.A. de Mooij. (1993). "Environmental Policy in a Small Open Economy with Distortionary Labor Taxes: a general equilibrium analysis." *OCFEB Research Memorandum* 9304. Rotterdam: Erasmus University.

Bovenberg, A.L. and R.A. de Mooij. (1994). "Environmental Levies and Distortionary Taxation." *American Economic Review* 94, 1085-1089.

Bovenberg, A.L. and F. van der Ploeg. (1992). "Environmental Policy, Public Finance and the Labour Market in a Second-Best World." *CentER Discussion Paper* 9243. Tilburg University.

Bovenberg, A.L. and F. van der Ploeg. (1993). "Optimal Taxation, Public Goods and Environmental Policy with Involuntary Unemployment." *CentER Discussion Paper* 9377. Tilburg University.

Bovenberg, A.L. and F. van der Ploeg. (1994). "Consequences of Environmental Tax Reform for Involuntary Unemployment and Welfare." *CentER Discussion Paper* 9408. Tilburg University.

Butter, F.A.G. den, R.B. Dellink and M.W. Hofkes. (1994). "Energy Levies and Endogenous Technology in an Empirical Simulation Model for the Netherlands." *VU Research Memorandum* 1994-16, FEWEC. Amsterdam: Free University.

Butter, F.A.G. den, and R.F. van de Wijngaert. (1992). "Wage Inflation and Labour Conflicts in the Netherlands: an Empirical Investigation using the Cointegration Approach." *European Journal of Political Economy* 8, 527-541.

Butter, F.A.G. den and F.J. Wollmer. (1992). "Endogenising Technical Progress in the Netherlands." *Beleidsstudies Technologie Economie* 22. The Hague: Ministry of Economic Affairs.

Butter, F.A.G. den, and R.W. van Zijp. (1995). "Technologie en werkgelegenheid." *Beleidsstudies Technologie Economie* 28. The Hague: Ministry of Economic Affairs.

CPB. (1992). "Economische Gevolgen op Lange Termijn van Heffingen op Energie." *CPB Working Document* 43. The Hague.

European Commission. (1993). *White Book*. Brussels.

Gelauff, G.M.M. and J.J. Graafland. (1994). "Modelling Welfare State Reform." *Contributions to Economic Analysis* 225, Amsterdam: North-Holland.

Goulder, L.H. (1995). "Environmental Taxation and the "Double Dividend:" A Reader's Guide." This volume.

Graafland, J.J. (1992). "From Phillips Curve to Wage Curve." *De Economist* 140, 501-514.

Laroui, F. and J.W. Velthuijsen. (1992a). "An Energy Tax in Europe." *SEO Report* 281. Amsterdam.

Laroui, F. and J.W. Velthuijsen. (1992b). "The Economic Consequences of an Energy Tax in Europe." *SEO Report* 282. Amsterdam.

OECD. (1994). "GREEN: the reference" *Working Paper* 143. Paris.

Seale Jr., J.L., W.E. Walker and I.-M. Kim. (1991). "The Demand for Energy." *Energy Economics*, 33-40.

Sterner, T. (1990). "An International Tax on Pollution and Natural Resource Depletion." *Energy Policy*, 300-302.

APPENDIX DYNAMIC BASELINE SIMULATIONS

This appendix shows graphs of historical time paths of major endogenous variables (volume indices, 1980=1) and dynamic baseline simulations according to the basic model.

Industrial production

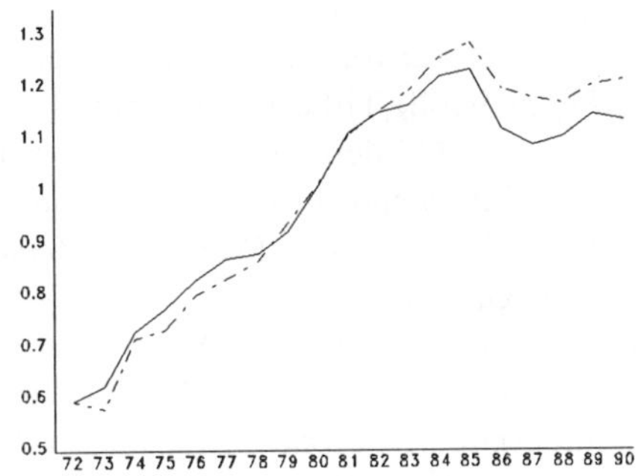

Consumption prices

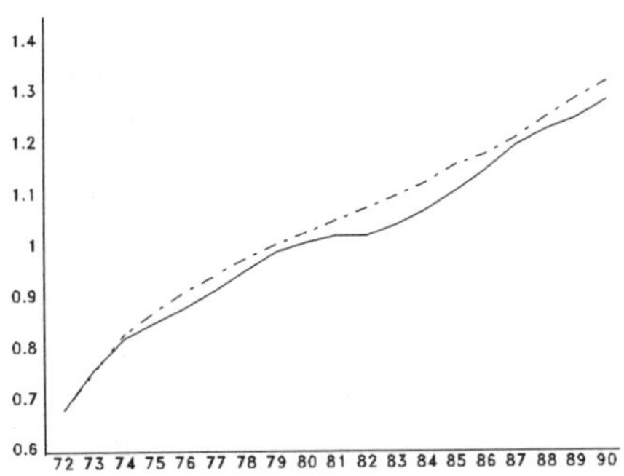

Efficiency units capital

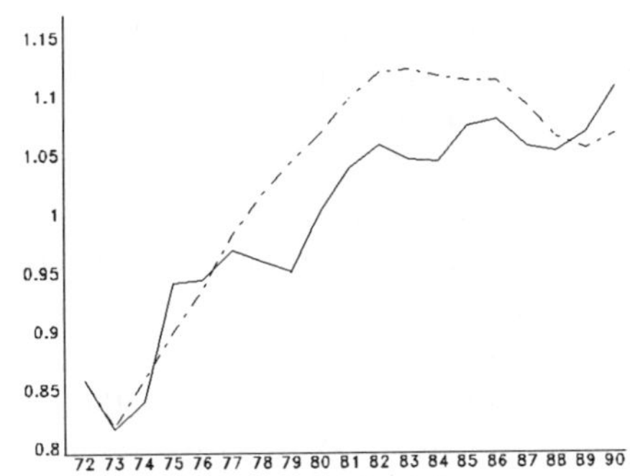

Efficiency units labour

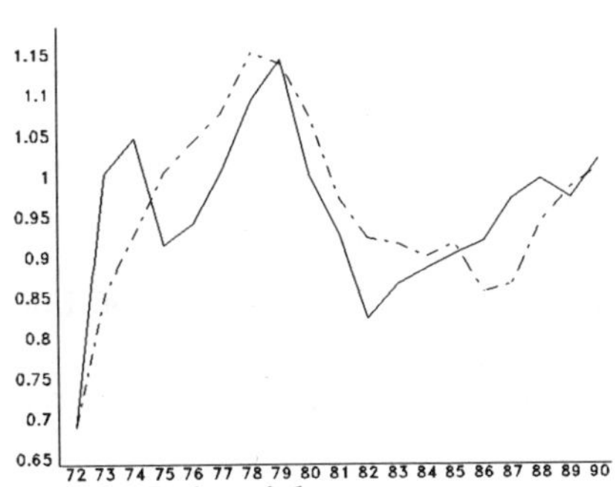

Industrial energy use

Household energy use

——— time series data -·-·-·- base line simulation

Chapter 16

WELFARE AND THE ENVIRONMENT: IMPLICATIONS OF A RECENT TAX REFORM IN NORWAY

Haakon Vennemo (Statistics, Norway)*

INTRODUCTION

Many industrialized countries have recently enacted tax reforms that aim to improve economic efficiency and welfare.

While higher economic efficiency obviously can improve welfare, it may also lead to more pollution and environmental damage. For instance, a tax reform that increases private income will probably increase the demand for private transportation and add to the problem of air pollution from private vehicles. A tax reform that expands production may also expand pollution from production. The environmental damage will cut into the welfare gain and possibly overturn it.

From a public economics perspective, the problem with a tax reform is that it may increase the importance of remaining distortions. Often, "distortions" are taken to be tax-induced distortions. But the same logic applies to environmental distortions. There is nothing to guarantee the success of a tax reform that obeys some principles from first- or second best, given environmental distortions in the economy. The outcome is purely an empirical matter.

The problem should also be of concern from a policy perspective. With higher environmental awareness in society, policymakers have to face critical examination of the environmental consequences of their actions. A policy that aims to increase production and growth is especially vulnerable to such criticism.

I submit that an evaluation of a tax reform would do well to consider the possible increase in environmental damage. This chapter carries out such an evaluation, using a framework that accounts for important linkages between the

---

* Thanks to Knut Alfsen and Brita Bye for comments on an earlier draft. This research was supported by the Research Council of Norway concerning the program "Economy and Ecology: Management Tools for a Sustainable Development."

economy and the environment, as well as welfare from environmental quality. I use the model DREAM (Dynamic Resource/Environment Applied Model), a computable general equilibrium (CGE) model that captures "producer-consumer," "producer-producer," "consumer-producer" and "consumer-consumer" externalities between the economy and the environment.

I analyze the tax reform carried out in Norway. The reform combines a cut in capital and labor income tax rates with a wider tax base, and includes the introduction of a tax on CO_2 emissions. It has many similarities to recent reforms in other countries. Indeed, countries seem to have inspired each other in their reform efforts. The analysis should therefore be of general interest.

The analysis is related to several strands of the literature. A vast literature uses CGE models to analyze tax reforms. Some recent examples are Jorgenson and Yun (1990) and Goulder and Thalmann (1993) on the US tax reform, and Holmøy and Vennemo (1995) on important parts of the Norwegian tax reform. None of these analyses incorporate environmental impacts. Researchers typically find an increase in efficiency and welfare that amounts to a small percentage of initial welfare, but a (much) larger percentage of tax revenue, real capital wealth or annual GDP.

Studies of tax reforms designed to reach environmental goals (especially emission limits) include Glomsrød, Vennemo and Johnsen (1992) and Brendemoen and Vennemo (1994) for Norway, Bergman (1991) for Sweden and the EMF-12 project (Gaskins and Weyant, 1993) for the US. Jorgenson and Wilcoxen (1993) survey this literature, which estimates costs of reaching the specified goals.

Ballard and Medema (1993) and Brendemoen and Vennemo (1993) use static models that include "producer-consumer" and "producer-producer" externalities (Ballard and Medema) and "producer-consumer" and "consumer-consumer" externalities (Brendemoen and Vennemo) to estimate the marginal cost of public funds.

THE MODEL

Our model is a growth model of the Cass-Koopmans type for an open economy. A small economy faces an exogenous interest rate and prices on competitive products. An infinitely lived consumer with perfect foresight maximizes utility from goods and leisure. There are nine industries. Six of them have competitive producers with perfect foresight. One of these produce tradables. This (large) sector determines the wage, which with the user cost of capital determines the output prices of non-tradables. Exogenous industries are production of petroleum and hydro-power[1] and a public sector.

Capital demand is determined such that the value of the marginal productivity of capital equals the user cost of capital. The user cost of capital includes self-fulfilling expectations of future investment goods' prices, taxes and cost-minimizing financial behavior by firms (see Sinn (1987) on the latter). Labor demand is such that the value of the marginal productivity of labor equals the price of effective labor input. There is an exogenous two percent trend increase in labor productivity.

Trade balances intertemporally. The annual trade balance reflects intertemporal optimization by producers and consumers, and changes with underlying economic conditions. This is similar to recent work in trade theory, see Obstfeld (1982) for an early contribution. A lump-sum tax clears the public budget. I impose an annual budget balance. This is an innocent assumption because of Ricardian equivalence.

The intertemporal balance-of-trade constraint equals the budget constraint of the representative household. In particular, household residual saving (saving over and above saving in real capital and domestic bonds) can occur only abroad. The long-run development of the current account therefore mirrors the long-run dynamic budget constraint of the consumer, and we can use the two terms interchangably. See Vennemo (1994) or Blanchard and Fisher (1989, ch. 2) for details.

The model tracks emissions to air of nine important pollutants, as well as road traffic volumes. These are chosen for their national importance and because they are related to aggregate economic entities. A set of detailed emission coefficients links emissions to material inputs, heating fuel and gasoline consumption. Traffic volumes depend on gasoline and auto-diesel consumption. The emissions to air and the traffic volumes form the impacts from the economy to the environment.

The models of environmental effects of macroeconomic policy developed by Brendemoen, Glomsrød and Aaserud (1992)[2] and Glomsrød, Nesbakken and Aaserud (1994) emphasize 11 external effects of economic activity; acidification of lakes, acidification of forests, health damage and annoyance from exposure to

---

1 Norway is a large exporter of petroleum, and is self-sufficient in hydro-power.

2 This paper is in Norwegian. Alfsen, Brendemoen and Glomsrød (1992) or Brendemoen and Vennemo (1993) are English papers with some of the same material.

NOx, SO_2, CO and particulate matter, noise, corrosion of building materials, traffic accidents, congestion and road depreciation.

This list was used to identify three links from the environment to the economy (see figure 1). One is a link from the environment to consumer welfare. This link comprises all the effects on the list except corrosion and road depreciation. A second link concerns labor productivity. I argue that noise, traffic accidents and reduced quality of air will increase the number of sick-leaves and reduce labor productivity.

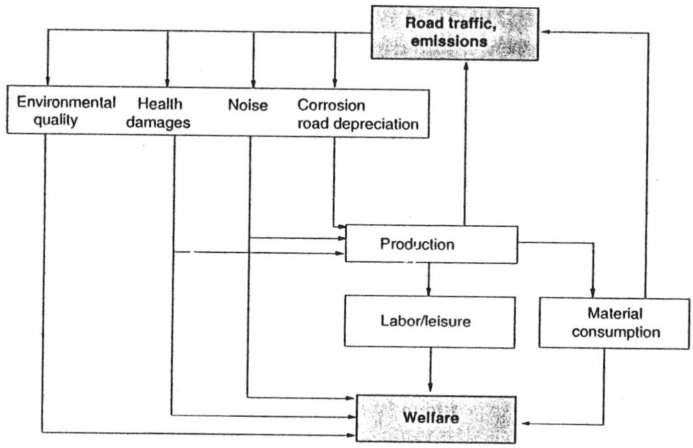

Figure 1 Overview of the model

A third link goes from environmental quality to the rate of capital depreciation. It has two motivations. One is the increase in corrosion caused by sulphur emissions, in particular. Another is the impact of traffic on road depreciation. Heavy traffic wears down the roads and increases the need for road maintenance.

The welfare function is a statistic of consumption of goods/services, leisure and environmental services over the entire infinite time span. The welfare function has the following properties: It rationalizes consumer behavior. It implies that environmental quality does not affect the choices made by the consumer. It implies that the marginal willingness to trade environmental goods of any period for consumption in the same period equals the base-year empirical estimate.[3] The willingness to trade environmental goods of any period for consumption of a later period equals the discounted value of the same parameter.

Changes in welfare are measured by equivalent changes in (human plus financial- and real capital) wealth, i.e. the welfare change from a price increase is measured by the equivalent change in wealth at the original set of prices. Changes in environmental quality are translated into monetary terms and integrated in the measurement. This is the traditional equivalent variation method in a dynamic context and extended to non-market goods. To arrive at a unit-free measure of welfare change, we divide equivalent variation by initial wealth.

Feedbacks from the model are limited to impacts of fossil fuel emissions and material inputs (for process emissions), and give only a rough indication of environmental effects of economic policy. The model's merit is the general equilibrium perspective on the link between the economy and the environment. Higher fossil fuel consumption can be expected to lead to higher emissions to air, and more traffic, which will create environmental externalities that will feed back into the economic model. Vennemo (1994) gives a detailed presentation and discussion of the model.

DATA

Parameters of the Model

The important parameters of the model are the substitution parameters in production and consumption, and the parameters reflecting environmental damage. These are compiled from econometric evidence and other empirical studies in Norway.

Output is produced in multi-level CES production functions. At the top level, material input and a capital-energy-labor composite combine into gross production. Material input is a fixed factor. This is a standard assumption in CGE models, and a reasonable approximation to the data of Norway (compare, e.g. Glomsrød, Vennemo and Johnsen, 1992, table 2).

---

3 This is a contested assumption. Many would argue that the willingness to pay for environmental services increases with income and with increases in the scarcity of the environment. Empirical studies, however, find small income effects (see, e.g., Kanninen and Kristöm, 1993).

The capital-energy-labor composite aggregates labor and a capital-energy composite, while energy aggregates fuel oil and hydro power, all in successive CES-nests. The elasticities of substitution, which differ among the "endogenous" industries, are derived from Alfsen, Bye and Holmøy (1995), and from Mysen (1991).

The elasticities of substitution are listed in Table 1. Most elasticities of substitution are below unity, showing an inelastic production structure.

Like technology, preferences are assumed to have a multi-level CES structure. The intertemporal elasticity of substitution is 0.5, a value broadly consistent with econometric evidence in Norway (Steffensen (1989), Biørn and Jansen (1982), Frisch (1959)). In the first stage of a three-stage budgeting procedure, the consumer spends total wealth on full consumption, i.e. consumption of goods and leisure. This aspect of her behavior can be described by an Euler-equation that relates the interest rate, the rate of pure time preference, growth in full consumption and the elasticity of marginal utility in a familiar textbook way.

Table 1 Elasticities of substitution

	Tradables	Petroleum refining	Construction	Wholesale and retail trade	Housing	Other services
Material input vs. labor-capital-energy	0.0	0.0	0.0	0.0	0.0	0.0
Labor vs. capital-energy	0.72	0.0	0.0	1.08	0.16	0.8
Capital vs. energy	0.52	0.0	0.02	0.7	0.16	0.67
Heating fuel vs. electricity	0.42	0.0	0.13	0.37	0.0	0.18

The consumer then spends full consumption on leisure and consumer goods. Time-series evidence in Norway points to a low wage elasticity of *labor force participation* (0.0 for men, 0.2 for unmarried women, Zakariassen (1994)). Cross-section studies find a larger, but similar response (0.2 for men, 0.4 for women; Dagsvik and Strøm, 1992) and a significantly larger uncompensated wage elasticity of *labor supply*: 0.3 for men, 0.9 for women. Dagsvik and Strøm (1992) report a low ($\approx$ 0.0) income effect on labor supply, indicating that uncompensated and compensated elasticities are close. Weighing up this information, I assume an uncompensated elasticity of 0.3 and a compensated elasticity of 0.4. This gives credit in a time-series model to the time-series information, while using the cross-section information as well. The time endowment and the elasticity of substitution between leisure and consumption are calibrated to get the labor-supply elasticities. Consumer expenditure is spread on each good in a Cobb-Douglas system.

The parameters describing the interaction between the economy and the environment are difficult to pin down. The worst difficulties are associated with the impacts from the environment to the economy. The model assumes a constant marginal impact on welfare over the relevant range (here the difference between the pre tax reform and post tax reform scenarios), and a quadratic impact on productivity and depreciation. The quadratic form is truncated at a chosen maximum. The technical reason for imposing a maximum is to facilitate the steady state. The economic reason is that a more than 100 percent decrease in productivity, or a higher than 100 percent depreciation is impossible. By all accounts the maximum occurs before that. The maximum productivity loss is set at approximately 15 percent. The maximum depreciation rates are three times the base-year rates. The maxima are only of academic interest in the present case.

The estimates of marginal economic impacts of environmental damage are based on Norwegian case studies. These combine production loss, avoidance costs and willingness to pay studies, with an emphasis on production loss. The impacts from traffic and the productivity loss from NOx emissions are the largest in value. NOx emissons are more severe than other emissions because of high concentration levels and because they occur in heavily populated areas. See Brendemoen, Glomsrød and Aaserud (1992) or Brendemoen and Vennemo (1993) for further discussion.

Gaskins and Weyant (1993) stress the importance of a sound baseline scenario for simulation output, as baseline input effects the marginal properties of a model. I use a baseline scenario aggregated from a similar scenario compiled by the Ministry of Finance for the last long-term projection of the Norwegian economy. The projection ends in 2030. From then on, the simulation assumes exogenous values consistent with a steady state.

The Norwegian Tax Reform

Taxes are high in Norway. In 1992, firms and households paid 46.8 percent of GDP in taxes. Only the Swedes and Danes pay more. Norwegians also receive more transfers than most others. The high reliance on taxation in public policy makes proper tax design important.

The recent reform in the Norwegian tax system emphasized efficiency over redistribution, indirect taxation over direct taxation and made some efforts in the direction of green taxes. The main ingredients of the reform are the following:

Reduce the marginal tax rate on capital income. The system of capital income taxation used to be progressive, with a top personal marginal tax rate of 43 percent until 1992. There is now a flat rate of 28 percent, and no double taxation of dividends or retained profits. The corporate wealth tax is eliminated. The personal wealth tax is reduced.

Increase the tax base for capital income. The old system of capital income taxation allowed firms to deduct a share of income if used for financial investments of specific kinds, and allowed lax rules for capital depreciation allowances. The tax reform tightened depreciation allowances and ruled out tax-deductible financial investment.

Reduce the marginal tax rate on labor income. The top marginal tax rate on labor income was 62 percent in 1989, and is 49.5 percent in 1994.

Reduce employers' contributions to social security. In Norway, as in some other countries, employers contribute to the funding of social security through a tax on labor. The rate of contribution was reduced from 15 percent to 13 percent in 1993.

Increase the VAT. To make up for some loss in revenue, the VAT increased from 20 percent to 22 percent in 1993.

Introduce a tax on CO_2. The government introduced a tax on CO_2 emissions in 1991. The tax treats emissions from different sources separately. The tax on gasoline is the highest, at Nkr 342 per ton CO_2 (around $50), while the tax on petrol-coke is the lowest, at Nkr 84 per ton CO_2.

BASELINE ESTIMATE OF WELFARE EFFECTS

This section studies the welfare potential of the tax reform package. The tax reform is assumed to be unexpected, and the model and data are as described above. Some of these assumptions will be relaxed later.

Prices and Quantities

One cannot explain the welfare impact without understanding the price and quantity impacts of the reform. The first column of Table 2 lays out the changes in factor prices.

Table 2 Changes in input prices and unit input demands. Percent. 2030

Input	Price	Quantity
Capital	9.9	-3.0
Heating fuel	4.2	-2.8
Labor	-5.8	2.7

The changes in capital taxation increase the user costs of capital for businesses. This is an intended effect of the reform. One of its goals was to increase marginal productivity, or rate of return to real capital, which formerly was known to yield too low a social return relative to financial capital. In our multi-period, rational-expectations setting, the marginal productivity of capital (in the restricted profit function) is equated to the user cost of capital. It is the increase in the user cost of capital that induces businesses to demand higher return to their capital.

The CO_2 tax increases the price of fuels. To understand this result, recall that the world market price of crude petroleum is fixed, which basically fixes the pre-tax price of fuels. The CO_2 tax is then added on in the purchasers' price. (Since petroleum refining is quite capital intensive, the pre-tax price actually increases to reflect the added capital cost of refining petroleum. This adds to the increase in the purchasers' price.)

To understand the impact on the price of labor, which is the wage plus other labor costs, bear in mind that the product price is fixed in the tradables industry. This fact caps unit costs of production in this industry; unit costs cannot increase in equilibrium. Since the user cost of capital and price of fuel do increase, the price of labor has to fall in order to save unit costs from inreasing. Being the sole immobile factor, labor "bears the burden" of the other factor taxes. The fall in the price of labor is divided between a reduction in the employers' contribution to social security, which, recall, is a component of the reform, and a fall in nominal wages paid to consumers and households.

The second column of Table 2 shows *unit* demand reactions to the input price changes. Own-price effects dominate, as demand for real capital and fuel both fall, while demand for labor increases. The economy restructures towards more labor intensive, energy efficient production. Table 2 reveals that quantities change less than prices. This is as expected, considering the rather inelastic technology of the model.

Given a set of unit demands, the labor supply determines the scale of production, measured as gross production. Gross production is constant, for instance, if labor supply increases 2.7 percent. Labor supply is influenced by the net real wage rate and by full consumption. The changes in labor supply and its determinants are given in Table 3.

Table 3 shows that labor supply increases, but not enough to maintain the original scale of production. To explain the increase in labor supply, note that the *net-of-tax* wage increases. The explanation here is that the reform lowers the income tax. The increase would have been larger but for the equilibrium fall in producer labor costs, however. A higher net-of-tax wage tends to increase labor supply.

Table 3 Change in labor supply and its determinants. Percent. 2030

Labor supply	1.3
Wage, net of tax	5.3
Price of consumer goods	2.8
Full consumption	0.8

The increase in labor supply is modified by an increase in the price level, which influences the *real* net-of-tax wage. The price level basically increases because the VAT increases, which is yet another piece of the reform. Net of VAT, prices of capital-intensive (sheltered) goods increase, but prices of labor-intensive goods fall.

The increase in labor supply is, in addition, modified by an increase in full consumption: that is, an increase in the index of consumer goods and leisure of the period. This effect can be explained the following way: The fact that full consumption increases in equilibrium tells us that if full consumption had been constant, the representative consumer would have been on a dynamic path that would lead her to work "too much" and consume "too little" in the sense that she would have saved more than she would ever have planned on spending. The intertemporal budget constraint would not have been binding. Such a path would not have been consistent with rational, forward-looking behavior.

Recall that the year-by-year change in consumption is governed by the Euler equation, which in this model does not allow sudden jumps in consumption and spending. The consumer must shift the consumption path as a whole, and similarly for labor supply in order for the budget constraint to bind. The increase in full consumption is obviously a sign that the reform makes her better off.

In summary, the reform makes energy and capital more expensive, inducing producers to shift out of these factors and into labor. Labor supply increases, but not enough to keep gross production from falling. The reform nonetheless makes the representative consumer more affluent.

Welfare

Table 4 lists the key factors that influence welfare.

The increase in welfare is equivalent to an increase in full wealth of 0.9 percent (at pre-reform prices). This rather small percentage which (considering the literature) should come as no surprise, translates into a monetary gain of 60 billion Nkr.

Table 4 Change in welfare and its determinants. Percent. 2030 except welfare, which is intertemporal

Welfare	0.9
Full consumption	0.8
Gasoline consumption	1.4
Fuel consumption	-3.6

To explain the increase in welfare, we inspect the change in its determinants; consumption of goods, leisure and environmental services. Consumption of goods and leisure is of course summarized in full consumption. As explained above, full consumption is determined such that the long-run budget constraint of the consumer binds. Comparing pre- and post-reform situations, we find an increase in full consumption of 0.8 percent in 2030 (and in the steady state). This corresponds closely to the increase in welfare (full wealth).

What about the impact from the environment? The changes in gasoline and fuel-oil consumption give an indication. The two change in different directions. Gasoline consumption increases because households consume more after the reform. This is the scale effect in consumption: Higher income makes it possible to consume more of polluting goods.

Fuel-oil consumption, however, falls. The reason is that fuel-oil consumption is mainly an input to production, and we recall that gross production falls after the reform. This is the scale effect in production. In addition, production becomes more energy efficient. This is the unit-demand effect. Households consume more fuel oil for heating because the reform has made them richer, but that is less important than the reduction in input demand.

There are thus two opposing impacts on the environment. Lower fuel-oil consumption improves environmental quality, mainly through lower pollution. Higher gasoline consumption decreases environmental quality through higher pollution and traffic volumes etc. Table 5 lists some indicators of the contribution from the environment to overall welfare.

On balance, the two effects on the environment have about the same strength, but most environmental indicators improve slightly. Welfare from the environment increases 0.4 billion Nkr. Productivity increases because lower pollution and transport demand in businesses reduce the problems of respiratory illness. The problem of corrosion on capital is also diminished, but increased traffic (higher gasoline consumption) increases road depreciation.

Table 5 Change in key environmental indicators. Percent except welfare. 2030 except welfare

Welfare from the environment	0.4 billion Nkr
Productivity	0.01
Depreciation rate, buildings	-0.02
Depreciation rate, roads	0.04

UNTANGLING THE REFORM: THE EFFECTS OF DIFFERENT COMPONENTS

The tax reform in Norway, although relatively coherent, was instigated in steps. The reform in business and capital taxation came first, together with the reduced income tax rate. The introduction of the CO_2 tax was a separate development motivated by environmental (and fiscal) concerns. The final part was the cut in employers' contribution to social security, financed by an increase in the rate of VAT. Thus, the four components of the reform are the following: the reform of capital taxation, the reduced income tax rate, the CO_2 tax and the reduction in employers' contributions/increase in VAT. Both from a policy perspective and as a means to improve the understanding of the economic mechanisms involved, it is of interest to study how the various components of the reform contribute to the overall welfare result.

When describing the partial effect of each component of the reform, one can make different assumptions about the status of the remaining components. I choose to follow a chronological procedure. That is, starting from the old system, I first introduce the reform in capital and business taxation, then introduce the new labor income tax, then add in the CO_2 tax and, finally, the reduction in employers' contributions/increase in VAT.

Table 6 lists the welfare gain or loss associated with each component of the reform. The benchmark is the old system of taxation. The table also shows the contribution to welfare from the environment.

Table 6 Welfare gain from different reform components, relative to no tax change

	Welfare gain, percent	Of which gain from the environment (bill. Nkr)
Capital tax reform	-0.7	20.2
Capital plus income tax reform	1.1	-4.5
Capital plus income plus CO_2 tax reform	0.9	0.01
Full reform	0.9	0.4

As noted above, the reform in *capital and business taxation* increases the user cost of capital. This reduces the price of labor in the small open economy. When no other tax wedges change, a lower price of labor is transformed into a lower net-of-tax real wage. This decreases labor supply and implies a too-low level of output to sustain the original consumption path of the consumer. To counteract this, households must work more and consume fewer goods and services; in other words enjoy lower full consumption. The lower level of full consumption is the basis for the welfare loss associated with the capital/business tax changes.

When viewed in isolation, the reform in capital and business taxation will, in other words, decrease welfare according to this analysis. This is not what most informed observers of the Norwegian economy would expect. Reform in business and capital taxation has been advocated in its own right, because an increased return to investment gives the economy extra resources and higher wealth. What drives our result? The driving force seems to be that the increase in the user cost of capital lowers the wage, which is already lower than its shadow value, because of taxation. Taxation, in other words, makes consumers work too little relative to the social optimum. Increasing the user cost of capital aggravates the problem: As the wage is depressed, consumers are inclined to work even less. For each labor unit that is withdrawn from the labor market, the economy loses welfare because the contribution to production and welfare from these units is higher than perceived by consumers. This loss dominates the gain from the capital market, according to the analysis.

If the capital tax reform reduces consumption and welfare, it does, on the other hand, ease the burden on the environment. The three forces determining the consumption of fuel oil and gasoline are the unit demand for fuel oil, the scale of production and the scale of consumption. The scale effects work in the direction of lower fossil fuel use and lower pollution. The substitution effect on unit demand is ambiguous: The price of capital increases, motivating substitution away from capital, into, e.g., fuel. The price of labor falls, motivating substitution into labor, away from, e.g., fuel. The scale effects are clearly the most important. The positive effect on welfare from the environment is fifty times larger than in the full reform (reported earlier).

A positive impact from the environment in combination with a negative impact on traditional welfare makes it possible for a person who values the environment strongly to claim that the capital tax reform is worth doing anyway. A person must put a 2.4 times higher value on the environment than is assumed here in order for him or her to conclude that the total welfare impact is positive.

The tables are turned when the *lower income tax* is considered alongside the reform in capital income taxation. Because the income tax falls, the real wage of the consumer rises, thus increasing labor supply. As a result, consumer savings would become "too positive" in the long run unless full consumption increases. This scenario realizes a higher welfare gain than any of the others in the table.

However, this scenario also returns a larger loss in welfare from the environment than the other scenarios. The reason is that the scale effect in consumption, which increases gasoline consumption, accentuates problems of pollution and traffic. The demand for fuel oil is approximately constant. The reason is that this scenario lacks the CO_2 tax component that encourages energy efficiency in the full-reform scenario. The scale of production is also higher than in the full-reform scenario.

When the *CO_2 tax* is introduced, the strain on the environment brought about by the income tax cut is wiped out. On the other hand, the economic welfare gain diminishes. Again, the crucial variable is the real wage. When the CO_2 tax is introduced, unit costs of production rise. Producers are lead to reduce their unit-labor costs, which reduces the real wage compared with the situation before the CO_2 tax. The price level increases because the CO_2 tax applies to household demand for CO_2 as well. This puts further pressure on the real wage. Labor supply then becomes lower and consumption higher than the intertemporal budget constraint allows. Full consumption must fall, compared with the situation before.

The positive effect on the environmental strain is due to a substitution effect in production, and the two scale effects; the scales of production and consumption both fall compared with before the CO_2 tax. The total balance shows that the marginal impact of the CO_2 tax is a welfare loss.

A factor to bear in mind when evaluating the results so far is the correlation between tax cuts and welfare. The changes in capital taxation raise net revenue and cut welfare. The reduction in the income tax cuts revenue and improves welfare. The CO_2 tax raises revenue, and cuts welfare as a marginal impact. In effect, these changes assume that lump-sum taxes or transfers compensate the public budget. Less revenue from distorting taxes correlates with higher welfare. Although it is not *a priori* certain that a tax reform that decreases a distorting tax in favor of a lump-sum tax increases welfare, it is by all accounts the "normal" case. A distorting tax that brings in more revenue and improves (economic) welfare would be a welcome policy tool!

The final piece of reform, the combination of *lower social security payments* by producers and a *higher VAT rate*, is approximately revenue neutral. It has almost no impact on long-term welfare. The reason is that the net real wage is balanced by two opposing forces: the cut in employers' contributions to social security implies a higher wage, but the increase in the VAT raises the price level correspondingly. There is therefore no reason for the consumer to change behavior, and there is no reason for producers to do so, either: their per-unit labor costs stay constant, since prices of the other inputs are constant. The small improvement in the environment occurs because consumers and some producers (those exempted from VAT on outputs) pay VAT on their purchases of fossil fuels.

THE IMPORTANCE OF BEING ELASTIC

The discussion above revealed that the degree of price/wage responsiveness in production and labor supply is quite important for the impact of the tax reform. For instance, the impact of price changes on unit-fuel demand in production is important for the environmental effects, while the impact on unit-labor demand is important for the scale of production. On the household side, the response of labor supply to a given wage increase determines the change in the scale of production. This section discusses the issue of price and wage responsiveness in some detail. I present sensitivity estimates with respect to production technology and labor supply.

Substitution in Production

The price responsiveness in production can be explored in different directions. Here, all substitution elasticities change parametrically. As substitution elasticities increase, a more elastic production structure with greater price responsiveness results.

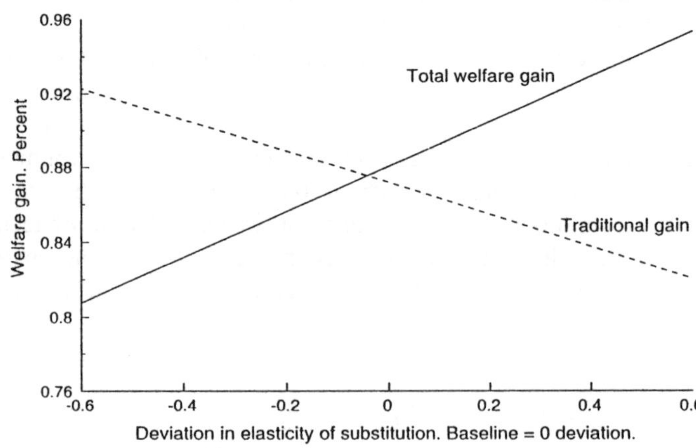

Figure 2 Welfare gain as percent of elasticity of substitution

Figure 2 gives welfare and traditional welfare (i.e. disregarding welfare from the environment) as functions of the elasticities of substitution. The figure shows that total welfare increases in the elasticities of substitution, while traditional welfare decreases. Both effects are quite small.

As shown above, the change in traditional welfare depends on the change in the net-of-tax real wage, which in turn depends on the producer price of labor. If the production structure is such that the real wage increases, labor supply increases and there is room for full consumption and traditional welfare to increase. When the wage increases *more*, labor supply is inclined to increase more, and there is more room to increase full consumption and traditional welfare. Figure 2, in effect, shows that the elasticities of substitution have only a minor influence on the increase in the wage. This is not strange if one considers that the substitution properties of technology have only a second-order effect on the producer price of labor, the first-order effect being determined by factor shares. The second-order effect is that the net-of-tax real wage increases more in a less elastic technology.

The change in total welfare depends on traditional welfare plus welfare from environmental quality. Total welfare increases in the elasticities of substitution, despite the decrease in traditional welfare. Thus, welfare from the environment increases in the elasticities of substitution.

This increase has two reasons. The main reason is that fuel demand responds more to the price increase of fuel, as technology becomes more elastic. A second reason is that the expansion in output and consumption is lower in a more elastic technology. Both of these reasons lead to lower fuel and gasoline consumption and lower pollution. The increase in welfare from the environment dominates the decrease in traditional welfare. The implication is that total welfare increases as technology becomes more elastic and elasticities of substitution increase.

The intersection between a negative impact from the environment and a positive impact occurs at a technology about five percentage points less elastic than the best-guess technology. This is a fairly small perturbation of technology. As figure 2 shows, a person must put a much higher value on the environment than this study assumes for the total welfare impact to be negative for any technology.

Elasticity of Labor Supply

The impact on welfare of less or more elastic labor supply is shown in table 7.

Table 7 Sensitivity of welfare gain to labor supply

	Welfare gain, percent	Of which gain from the environment (bill. Nkr)
Small labor supply elasticity (0.0)	0.6	5.2
Baseline case (0.3)	0.9	0.4
Large labor supply elasticity (1.0)	1.6	-12.7

We focus on two alternatives to the best-guess estimate of the labor-supply elasticity: As the inelastic alternative, the uncompensated labor-supply elasticity is assumed to be zero. As the elastic alternative, the uncompensated labor-supply elasticity is assumed to be one.

Table 7 shows that the welfare gain from the reform is larger for a large labor-supply elasticity. The reason is that a large labor supply-elasticity elicits a large increase in labor-supply, which makes room for a large increase in full consumption and welfare. A small labor-supply elasticity, in contrast, implies a small labor-supply response and a small increase in full consumption.

Since a large labor-supply elasticity allows a large increase in labor supply, even after deducting for the increase in full consumption, the scale of production increases more. This explains the loss in environmental quality in this scenario.

It is conceivable that a person with strong environmental preferences will find the tax reform not worth doing in a setting with a large aggregate labor supply response. When the labor supply response is small, traditional welfare and welfare from the environment point in the same direction.

CONCLUSIONS

The chapter has evaluated the recent Norwegian tax reform in an integrated economy-environment model. Welfare increases 0.9 percent in the baseline case. The impact on the environment is small, but positive. The cut in the income tax rate contributes most to the welfare gain, because of its encouragement of labor supply. The CO_2 tax turns a negative environmental impact into a positive one. The estimated welfare gain is robust to changes in technology and the labor-supply elasticity. An inelastic technology or a large labor-supply response, however, damages the environment.

In closing, consider how this chapter relates to the debate on "double dividends." The chapter finds the Norwegian tax reform to deliver a double dividend in the sense that both welfare from full consumption and welfare from environmental quality increase. The increase in welfare from environmental quality is, however, weak, and is sensitive to the assumptions on production technology. A slightly less elastic technology will imply a loss in welfare from the environment. A higher labor-supply elasticity will also imply a loss in welfare from the environment and overturn the double-dividend finding.

The reform is also found to deliver a double dividend in the sense that both environmental quality and employment increase. The reason for this result is that the real wage increases in equilibrium, and elicits higher labor supply. The scale of production does not increase, however, because factor substitution increases the labor intensity in production. Demand for "dirty inputs" falls, and environmental quality therefore increases.

This result is also overturned by a slightly less elastic technology. As a representation of an aggregate economy, however, the technology of the model is in my view more likely to be too inelastic than too elastic. The reason is that with a fairly high level of aggregation, the model economy probably underestimates Rybczynski effects that increase macro-substitution possibilities relative to micro (industry) estimates.

My results on the double dividend differ from much of the theoretical literature. For instance, Bovenberg and van der Ploeg (1994) use a static model that otherwise is similar to mine to show that infinitesemal reforms around the optimum cannot produce a double dividend. My results, to the contrary, indicate that the Norwegian tax reform may lead to a revenue shortfall. A revenue shortfall implies that the reform implictly exchanges a labor income tax for a lump-sum tax in order to balance the budget. Such a change will increase employment, as both a higher wage and lower lump-sum income raises labor supply.

References

Alfsen, K.A., A. Brendemoen and S. Glomsrød. (1992). "Benefits of Climate Policies: Some Tentative Calculations." *Discussion Paper* 69. Oslo: Statistics Norway.

Alfsen, K.A., T. Bye and E. Holmøy. (1995). "Estimation of Parameters in the Production Structure." *Mimeo*. Oslo: Statistics Norway.

Ballard, C.L. and S.G. Medema. (1993). "The Marginal Efficiency Effects of Taxes and Subsidies in the Presence of Externalities." *Journal of Public Economics* 52, 199-216.

Bergman, L. (1991). "General Equilibrium Effects of Environmental Policy: An AGE-modelling Approach." *Environmental and Resource Economics* 1/1, 43-61.

Biørn, E. and E. Jansen. (1982). *Econometrics of Incomplete Cross-section/Timeseries Data: Consumer Demand in Norwegian Households 1975-1977. Social and Economic Studies* 52. Oslo: Statistics Norway.

Blanchard, O.J. and S. Fischer. (1989). *Lectures on Macroeconomics*. Cambridge, MA: MIT Press.

Bovenberg, A.L. and F. van der Ploeg. (1994). "Green Policies and Public Finance in a Small Open Economy." *Scandinavian Journal of Economics* 96/3, 343-363.

Brendemoen, A., S. Glomsrød and M. Aaserud. (1992). *Miljøkostnader i makroperspektiv* (Environmental Costs in a Macro-perspective). *Reports* 92/17. Oslo: Statistics Norway.

Brendemoen A. and H. Vennemo. (1993). "The Marginal Cost of Funds in the Presence of External Effects." *Discussion Paper* 99. Oslo: Statistics Norway.

Brendemoen A. and H. Vennemo. (1994). "A Climate Treaty and the Norwegian Economy: A CGE Assessment." *The Energy Journal* 15/1, 77-93.

Dagsvik, J.K. and S. Strøm. (1992). "Labor Supply with Non-Convex Budget Sets, Hours Restrictions and Non-Pecuniary Job-Attributes." *Discussion Paper* 76. Oslo: Statistics Norway.

Frisch, R. (1959). "A Complete Scheme for Computing All Direct and Gross Demand Elasticities in a Model with Many Sectors." *Econometrica* 27, 177-196.

Gaskin, D.W. and J.P. Weyant. (1993). "Model Comparisons of the Cost of Reducing CO_2 emissions." *American Economic Review* 83/2, 318-323.

Glomsrød, S., R. Nesbakken and M. Aaserud. (1994). "Modelling Impacts of Traffic Injuries on Labor Supply and Public Health Expenditures in a CGE Framework." *Mimeo*. Oslo: Statistics Norway.

Glomsrød S., H. Vennemo and T. Johnsen. (1992). "Stabilization of Emissions of CO_2: A Computable General Equilibrium Assessment." *Scandinavian Journal of Economics* 94/1, 53-69.

Goulder, L.H. and P. Thalmann. (1993). "Approaches to Efficient Capital Taxation. Leveling the Playing Field vs. Living by the Golden Rule." *Journal of Public Economics* 50, 169-196.

Holmøy, E. and Vennemo, H. (1995). "A General Equilibrium Assessment of a Suggested Reform in Capital Income Taxation." *Journal of Policy Modelling*, forthcoming.

Jorgenson, D.W. and P. Wilcoxen. (1993). "Energy, the Environment and Economic Growth." In A.V. Kneese and J.L. Sweeney (eds.): *Handbook of Natural Resources and Energy Economics III*. Amsterdam: North-Holland.

Jorgenson, D.W. and K.-Y. Yun. (1990). "Tax Reform and US Economic Growth." *Journal of Political Economy* 98/5, s151-s193.

Kanninen, B.J. and B. Kriström. (1993). "Welfare Benefit Estimation and Income Distribution." *Mimeo*. Stockholm: Bejer International Institute of Ecological Economics.

Mysen, H.T. (1991). *Substitusjon mellom olje og elektrisitet i produksjonssektorene i en makromodell* (Substitution Between Heating Oil and Electricity in the Industries of a Macro-model). *Reports* 91/7. Oslo: Statistics Norway.

Obstfeld, M. (1982). "Aggregate Spending and the Terms of Trade: Is There a Laursen-Metzler Effect?" *Quarterly Journal of Economics* 97, 251-270.

Sinn, H.W. (1987). *Capital Income Taxation and Resource Allocation*. Studies in Mathematical and Managerial Economics 35. Amsterdam: North Holland.

Steffensen, E. (1989). "Konsumetterspørselen i Norge estimert ved hjelp av Eulerlikninger på årsdata 1962-1987." (Estimating the Demand for Consumption in Norway by Means of Euler Equations on Annual Data, 1962-1987). *Working Paper* 20/89, Center for Applied Research. Bergen: Norwegian School of Economics and Business Administration.

Vennemo. H. (1994). *A Growth Model of Norway with a Two-way Link to the Environment. Reports* 94/5. Oslo: Statistics Norway.

Zakariassen, H.-M. B. (1994). *Tilbud av arbeidskraft i Norge. En empirisk analyse på kvartalsdata for perioden 1972 til 1990* (Labor supply in Norway. An empirical analysis on quarterly data, 1972-1990). *Rapporter* 94/3. Oslo: Statistics Norway.

NAME INDEX

n indicates that the reference is in a footnote
s indicates that the reference is in the source of a table

Aaserud, M., 339, 343
Alberini, A., 86
Alchian, A., 192n
Alfsen, K.A., 339n, 342
Anderson, R., 86n
Aoki, M., 182n
Atkinson, A.B., 293n
Attfield, R., 176n
Auerbach, A.J., 291n

Bacon, R., 94
Balcer, Y., 63
Ballard, C.L., 293n, 308n, 338
Baranzini, A.H.G., 16
Barkema, H.G., 175
Barrett, S., 228, 245n, 254n, 264n, 265, 266
Barthold, T.A., 161
Bauer, A., 266n
Baumol, W.J., 20, 26n, 63, 173, 178, 183
Becker, G.S., 174
Beer, M., 182n
Begg, D.,226
Bell, C.,92n
Bergman, L.,338
Bernow, S., 128
Bertrand, T.J., 307
Biørn, E., 342
Birdsall, N.,197
Blanchard, O.J., 339
Blinder, A.S., 182, 183
Boadway, R.N., 198
Boero, G., 271
Bourguigon, F., 16
Bovenberg, L., 3, 20, 31, 280n, 286, 287, 288n, 289, 290, 291n, 296n, 297, 304, 306, 307n 315, 317
Bradford, D.F., 20
Braeutigam, R.R., 169n
Brendemoen, A., 338, 339n, 343
Broniewicz, E., 213
Buchanan, J.M., 32n
Burdick, D.L., 166
Burgess, J.C., 114
Burniaux, J.-M., 114
Bye, T., 342

Campbell, 196
Carraro, C., 11, 261, 262, 264n, 266, 267n, 270n, 280n
Cesar, H., 264, 269
Chander, P., 11, 266
Chichilnisky, G., 22
Clark, R., 271
Cline, W.R., 240, 242n
Cnossen, S., 3
Coase, R.H., 20, 187, 188
Collins, C.T., 86n
Copeland, B.R., 137n
Corlett, W.J., 63
Cropper, M.L., 173, 189, 198
Cumberland, J., 231

D'Aspremont, C.A., 265, 266
Dagsvik, J.K., 342
Dahl, H., 92n
Dasgupta, P., 20, 295n
Davie, B.F., 161, 162, 163
de Mooij, R.A., 25, 286, 287, 288n, 289, 290, 291n, 307, 309, 315
De Zeeuw, A., 264, 269
DeAlessi, L., 178

Deci, E.L., 174, 175, 176
Delache, X., 50
Dellink, R.B., 13, 315, 320
Demsetz, H., 192
den Butter, F.A.G., 13, 315, 319, 320, 323
Des Jardins, J.R., 176n
Devarajan, S., 5, 61, 92n, 101, 102s, 104s, 105s
Diamond, P.A., 27
Dinan, T.M., 137n
Dixit, A., 307
Dobbs, I.M., 147
Doeringer, P.B., 175
Donsimoni, M.P., 266
DRI/McGraw-Hill, 116s, 119

Eggertsson, T., 195
Ernesto, S., 198
Eskeland, G., S., 5, 61, 74n, 75n, 77, 78, 79s82, 95, 96-100s
Estache, A., 194n

Faulhaber, G.R., 173
Feyzioglu, T., 78, 79s,82
Fisher, S., 339
Flatters, F., 198
Frey, B.S., 8, 173, 174, 175, 181, 182n, 196
Frisch, R., 342
Fullerton, D., 7, 135, 147
Fuss, M.A., 95

Gabszewicz, J.J., 265, 266
Galeotti, M., 280n
Gallo, M., 280n
Gaskins, D.W., 278n, 337, 343
Gatsios, K., 224, 235
Gee, R., 214
Gelauff, G.M.M., 319
Glomsrod, S., 123, 338, 339n, 341, 343
Go, D., 92n
Gofman, K.G., 40
Goulder, L.H., 13, 25n, 64n, 123, 277, 290, 294n, 300n, 306, 307, 318, 338
Graafland, J.J., 319, 323
Greene, D., 175
Greenwald, B.E., 63
Grubb, M., 254n, 271
Guo, C.C., 95
Gusev, A.A., 40

Hague, D.C., 63
Hahn, R.W., 173, 182, 183
Haigh, N., 229
Harberger, A., 188n
Harrington, W., 72n, 86n
Harrison, D. Jr., 85n
Haugland, G., 49
Hausman, J.A., 287n
Hazilla, M., 92n
Heal, G., 22, 264n, 266n, 295n
Helpman, E., 162
Henry, C., 50
Hettige, M., 91, 94n
Hoel, M., 30, 162, 266
Hofkes, M.W., 13, 315, 320
Holmøy, E., 338, 34
Holthausen, D.M., 137n
Hufbauer, G.C., 162

Inman, R., 188n

James, E., 197
Jansen, E., 342
Jędrzejewska, A., 211
Jenkins, R., 136n, 137n
Jesinghaus, J., 213
Jimenez, E., 95, 96-100s
Johansen, L., 34
Johnsen, T., 338, 341
Jorgenson, D.W., 92, 116s, 119, 122, 123, 293n, 303n, 338

Kalkstein, 240
Kanbur, R., 225

Kanninen, B.J., 341n
Keen, M.J., 225
Kelman, S., 181
Kennedy, P.W., 147
Kinnaman, T.C., 7, 135, 147
Kirchgässner, G., 177
Kliemt, H., 177
Kneese, A.V., 278n
Kochan, T., 176
Kong, C., 95
Kopp, R.J., 91, 92n, 94
Kornai, J., 210
Kosmo, M., 114
Kotlikoff, L.J., 291n
Kristrom, B., 341n
Krugman, P.R., 162
Krupnick, A.J., 86n

Lancaster, K., 307
Laplante, B., 147
Laroui, F., 315
Larsen, B., 6, 113, 115s, 117, 118s, 119, 120, 122, 126, 128s, 129, 190, 268, 294, 300n
Lawler, E.F., 176
Lawson, D.R., 72
Lee, C., 175
Leffler, W.L., 166, 169
Lepper, M., 175
Lewis, J.D., 101, 102s, 104s, 105s
Ligthart, J.E., 291n, 305
Lipsey, R.G., 307
Liu, L., 95, 96-100s

MacDougall, D., 223
MacKenzie, J., 116s
Mäler, K.-G., 29, 30, 228
Markusen, J., 29n, 292n
Marron, D., 128
McCloskey, W., 38
McConnell, V., 72n, 86n
McGraw, K.O., 175
Medema, 308n, 338
Miller, G.T., 176n
Morey, E.R., 232n
Mirrlees, J.A., 20
Morris, G.E., 137n
Moss, D., 95
Musgrave, P.B., 11, 237
Musgrave, R.A., 19, 27
Mysen, H.T., 342

Nesbakken, R., 339
Newbery, D.M., 29, 30
Nichols, A.L., 278n
Nielsen, S., 13, 280n
Nitze, W.A., 254n
Nordhaus, W.D., 306
North, D.C., 176

Oates, W., 26n, 63, 173, 178, 183, 189, 194n, 195n, 198, 223, 232, 233
Obstfeld, M., 339
Olewiler, N., 232n
Olson, M., 191, 223
Ostro, B., 94
Ostrom, E., 180, 191n

Padoa-Schioppa, T., 223
Panayotou, T., 190, 198
Paradysz, J., 208
Parkin, S., 182
Parry, I.W.H., 290, 304, 307
Pearce, D., 197, 198n, 278n, 279
Pearson, M., 4, 5, 37, 50, 222
Pedersen, L., 13, 280n
Pezzey, J., 217, 278n
Pigou, A.C., 19, 21, 277
Pindyck, 95, 96
Pitt, M., 95
Pittevils, I., 214
Pommerehne, W.W., 176
Portney, P., 193, 194
Poskrobko, B., 213

Poterba, J.M., 7, 50, 51, 122, 161, 190, 278n, 287n
Proost, S., 14, 282n
Prud'homme, R., 196n, 197

Rajah, N., 52
Ramsey, F.P., 20
Rauscher, M., 12
Rawls, J., 188
Repetto, R., 136n, 137, 278n
Requate, T., 15
Rodman, J., 176n
Rodríguez, M., 197n
Rosen, H.S., 19
Rotemberg, J.J., 7, 161
Rudas, L.G., 191
Ryan, R.M., 174, 175, 176

Samuelson, P.A., 20, 21
Sandmo, A., 4, 19, 24, 25, 27, 29n, 63, 64n, 190n, 304n
Schelling, T.C., 254n
Schillo, B., 124
Schöb, R., 305
Schroeder, 191n
Schwab, R., 194n, 232
Seabright, P., 224, 235
Seale Jr., J.L., 322
Sebenius, J.K., 254n
Shackleton, R., 123, 127s, 278n, 300n
Shah, A., 6, 113, 115s, 117, 118s, 119, 120, 122, 126, 128s, 129, 190, 268, 294, 300n
Shibata, H. 243n
Shome, P., 191, 194
Shoven, J.B., 293
Sigman, H., 137n
Simon, H., 193, 197
Siniscalco, D., 11, 261, 262, 264n, 266, 267n, 270n
Sinn, H.W., 20, 339
Skumatz, L.A., 135
Slesnick, D., 122
Smith, A., 187
Smith, S., 10, 52, 122, 211, 214, 221, 222
Smulders, S., 15
Sørensen, P., 13, 280n
Spretnak, C., 178
Srinavasan, T.N., 92n
Stańczuk, I., 211
Steffensen, E., 342
Sterner, T., 324
Stigler, G., 174
Stiglitz, J., 19, 20, 63, 192, 293n
Strøm, S., 342
Sullivan, A.M., 147
Swait, J., 84
Sweeney, J.L., 294n, 295n

Tahvonen, O., 6
Ten Kate, A., 91, 92
Terkla, D., 278n
Thalmann, P.,338
Tiebout, C., 194
Tietenberg, T., 63
Tulkens, H., 11, 266
Tullock, G., 197n, 278n
Turner, K., 198n
Tybout, J.R., 95, 194

Uribe, E., 197n, 198

van de Wijngaert, R.F., 323
van der Ploeg, F., 31, 280n, 291n, 296n, 297, 304, 307n, 315, 317
Vanek, J., 307
van Regemorter, D., 14, 282n
van Wijnbergen, S., 92n, 297, 304, 315, 317
Velthuijsen, J.W., 315
Vennemo, H., 14, 337, 338, 339n, 341, 343

Walls, M.A., 86n

Weck-Hannemann, H., 8, 173, 176, 181
Weingast, B.R., 194
Weizsäcker, E., 213
Wellish, D., 10
Wertz, K.L., 137n
Weyant, 278, 343, 338
Whalley, J., 293n
Wheeler, D., 94n
Wicke, L., 180
Wiesner, E., 9, 187, 193, 198n
Wijkander, H., 63
Wilcoxen, P.J., 92n, 122, 123, 300n, 338
Wildasin, D.E., 19, 29n
Winters, A., 271
Wollmer, F.J., 319, 320
Wynne, S., 191

Yamiro, M., 197n
Yun, K.-Y., 293n, 303n, 338

Zakariassen, H.-M.B., 342
Zheng, K., 194n
Żylicz, T., 9, 203, 204, 209, 211-216

SUBJECT INDEX

abatement, 66, 74, 76
  measures, Mexico City
  technologies,77, 88
acid deposition, 261
adjustments from baseline, 252
administration of pollution tax system, 205
administrative framework, 43
administrative problems, 39
agences de l'eau, 47
aggregate behavior, 144, 145, 155
air pollution in Mexico City, 77
alternative international arrangements, *see* international agreements
analysis of optimal taxation, *see* optimal taxation
appropriate point of taxation, *see* taxation
assessment problems, 7
ATHENA multi-sector model, 315

banderoles, 44
baseline emissions, 119
basic analytical framework, 62
Belgian environmental policy 46
Belgian tax on pesticides, 42
Belgium
  environmental taxes in, 46
  definition of the environmental objectives, 53
benefits of public goods, *see* public goods
border tax adjustments, 7
budgetary revenues, 203
bureaucracy, 26

capital good, 297
capital good, environment as,297
capital taxes, 53
carbon-cum-energy tax, 122
carbon emissions impacts, 11
  of subsidy removal in OECD countries, 119
carbon/energy tax (1991), 222
carbon/energy taxes of the Nordic countries, 37
carbon leakage, *see* leakage
carbon tax, 306
  costs and benefits of, 128
carbon tax recycling instrument, 127
carbon taxes
  in Finland, 114
  in Sweden, 42, 114
  in the Netherlands, 114
Cass-Koopmans model, 338
catalytic converter, 86
CFC charge, 50
CGE model, 104, 338
Chile, 100
  estimate for manufacturing, 95
  manufacturing in, 95, 96, 99
  price elasticities, 96
clean air, *see* public goods
cleaner consumption, 66
cleaner technologies, 62
climate change, 261
  convention, 262, 263
CO_2 emissions reduction, 118
CO_2 emissions reductions in OECD and non-OECD countries, 120
CO_2 tax, 345
  imposing a, 6
Coase theorem, 89, 246
Coasian negotiations, 3
Cobb-Douglas function, 117
Cobb-Douglas system, 343

Colombia's environmental federalism, 196
combustion of fossil fuels, 41
commitment, 267
computable general-equilibrium (CGE), 101
  model, 102
conditions for crowding-out effect, 175
consumer prices, 28
consumer welfare, 28
consumption goods, tax on, 28
contamination
  air, 197
  of lakes, 197
  of rivers, 197
  of soil, 197
control of pollution from industrial sources, 87
cooperative action: equity based, 252
cooperative action: efficiency based, 251
coordination, need for, 10
cost of littering, 135
cost of recycling, 135
cost-benefit analysis, 22
cost-benefit study, 31
cost-effective programs, 92
cost-effective environmental policy, 38
cost-effectiveness ranking, 77
costly monitoring, 65
cost-minimizing reduction of emissions, 30
costs and benefits, emissions reduction, 238
costs and benefits of carbon taxes, *see* carbon taxes
courting of industry, 45
cross-border shopping, 55
cross-country effects on economic activity and revenues, 225
cross-country environmental problems, 229
cross-country policy
  externalities, 224
  spillovers, 224
cross-country spillover effects, 223
cross-price elasticities, 103
crowding-out effects, 174, 176, 183
cumulative flows, 72
curbside recycling program, 136, 139, 142, 143
  without, 148
current role, pollution taxes, 209

damage levels (*see* pollution damage)
day without a car program, 81
decreasing returns, 295
de novo adjustment, 252
definition of the environmental objectives, Belgium, 53
definition of the taxable event, 39
deforestation, 261
design of environmental taxes, analysis of, 162
design of taxes, 30 *see also* taxes
detailed program evaluation, 72
developing countries, Latin America, 193
differentials in valuations, 255
direct emission coefficients, 101
dissemination of information and independent evaluations, 197
distributional effects of the European Carbon/Energy tax, 51
distributional effects of the Nordic carbon taxes, 51
distributional neutrality, 50
double dividend, 12, 64, 277, 286

concept, 279
issue, 296, 300
propositions, 280
intermediate form, 280
strong form, 280
weak form, 280
downward competition, 231
DRI model, 123
Dutch Central Planning Bureau, 318
Dynamic Resource/Environment Applied Model (DREAM), 338

earmarking, 49, 50, 211
earmarking of environmental tax, 56
economic arguments, 173
economic costs of the tax, 44
economic problems, public and academic interest 29
economic spillovers, 228, 231
economics of pollution control, 62
economies in transition, 203
economists, 27
ecotax payments, distributional effects, 51
ecotaxes, 47
efficiency impacts of national carbon taxes, 123
efficiency with equity, 253
elasticities of substitution, 342
elasticity, 352
electricity price, 96-100, 104, 105,
EMF-12 project, 338
emission control technologies, 68
emission elasticities, 98
emission levels,
heavy fuel oils, 31
of polluters, 26
related, 31
emission monitoring, 87
emission of SO_2 in Europe, 30
emission of total suspended particles (TSP), 101
emission rates, 72, 73, 85
emission reduction, 67
emission reduction, costs and benefits of, 238
emission sources, 40
emission taxes, 294
empirical research, 26
empirical studies, 318
employment, 31
double dividend, 13
increasing, 13
energy conservation, 92
energy, decreasing, 13
energy levy, 324, 325
energy subsidies, 113
in non-OECD countries, 114, 115
in OECD countries, 116
EnTech model, 315, 316, 319, 324
environment as a capital good, *see* capital good
environment as a public good, *see* public good

environment in developing countries, 191
market failure, 187
environment-related fiscal instruments, 207
environmental aspects of general taxation in Poland, 207
environmental authorities, 28
environmental coalition, building an, 11
environmental damage, 337
environmental economics, 3, 19
environmental ethics, 8, 177
as intrinsic motivation, 176
high costs of, 179
low costs of 179
environmental federalism, 193

environmental funds, future of in Poland, 215
environmental goods, production of, 21
environmental issues in relation to problems of public policy 19
environmental negotiation, 269
environmental outcomes, 178
environmental policy, 174, 228
  coordination, recent experience, 262
  instruments, 51
  international bargaining, 228
  international negotiations, 228
environmental quality, 297
environmental standards, 27, 232
environmental taxation, 221, 277
environmental taxes, 25, 28, 30, 37, 52, 306
  design of, 162
  in Belgium, 46
  in Norway, 32
  on intermediate and final goods, 161
  technical flaws in, 39
environmental tax revenues, 56
equity and efficiency, balancing, 11
European acid rain game, 30
European carbon/energy tax proposal, 47
European court judgements, agreements, 221
European governments, 29
European union (EU), 221
  carbon tax, 10
    proposal, 234
  environmental policy, 229
  role of, in environmental taxation, 221
  wide tax, carbon and energy, 114
exemptions, 47
exhaustible resources, decreasing returns, 295
expenditure effects, 28
expenditure policies, 19
exploration of general equilibrium effects, 100
externalities, theory of, 19
externality-generating commodity, 28

factor substitution, 31
factory smoke, 22
  effects on the ozone layer, 22
  global warming, 22
fee per-unit garbage, 135

Flue Gas Desulphurisation (FGD) units, 41
fossil-fuel subsidies, 6, 190
free curbside recycling program, 151
free garbage collection, 150
freshwater pollution, 261
fuel use and emissions link, 93

gainers, 44, 48, 56
garbage and recycling quantities, change in 145
garbage collection, 7
gasoline consumption, 74
gasoline tax, 74
gasoline tax rates, 77, 78
GATT, 263
GATT agreements, 43
general economic policies, 90
general equilibrium arc elasticities, 103, 104
general equilibrium arc emission elasticities, 105
general equilibrium model, 104
general equilibrium simulation, 106
general levy, 325, 326
global climate change, 113
global warming, 22, 113

GNP, 33, 285
goods, 20
Goulder model, 123, 126, 300
government intervention, 190
governmental competition, 12
gradualism, 46
green tax swaps, 279
Green-OECD model, 315
greenhouse gas emissions, 190
greenhouse gases, carbon dioxide (CO_2), 113
Greenpeace, 177

heavy fuels price, 96, 97, 98, 99, 100, 104, 105
HERMES model, 315
heterogeneous households, model, 135
heterogeneous polluters, 87
household degree of environmental awareness, 136
household disposal choice, 141
  model, 137
household levy, 326, 327
human ingenuity, 15

Iceland, taxes in, 38
illegal dumping, 147
impact of pollution-control policies, 92
imperfect markets, 3
implementation of environmental tax, 38
import neutrality with joint production, 166
import neutrality without joint production, 165
imported final goods, 162
imports of uncontrolled products, 55
incentive-based environmental instruments, 173
incentive instruments, 174, 183
incentive instruments, new considerations, 173
incentive instruments effective, 178
incidence effects of tax policy, 19
indirect instruments, 68
indirect instruments, cleaner cars,
  fewer trips, 74
indirect protection, 230, 231
Indonesia, 104
Indonesian manufacturing, 95, 98, 99
industrial polluters, 88
information functions, 224
innovation, 15
input and output taxes, regulation, 90
institutional aspects, 8
institutional response, 187
integration with other EU policies, 233
interest groups vs social welfare, 191
interfuel substitution, 92
international agreements, 232
  alternative, 243, 244
international coordination of environmental policies, *see* environmental
international eco-tax, 254
international efficiency and equity, 237
international environmental profitability, 263
international environmental spillovers, 228
international environmental stability, 263
intervention measures, 3
intrinsic motivation, 174
investments in public infrastructure, 31
involuntary unemployment, 296

joint production
  applications, 169
  processes, 166
Jorgenson and Wilcoxen model, 123, 300

key environmental indicators, 348

labor supply, 354, 356
  elasticity, 354
labor taxes, 52
Latin America, developing countries, 193
leakage, 268
leaking underground storage tanks, 163
light fuels price, 96-100, 104, 105
Lindahl equilibrium 20
LINK model, 123
linked coalition, 272
literature on environmental ethics, 176
littering option, model, 147
local environmental benefits, 129
local government's choice of user fee, 136
local public choices, 195
long-run view of economic welfare, 30
losers, 44, 48
loss of biological diversity, 261
low cost of environmental ethics, 179, 180
lower social security payments, 351
lump-sum transfers, 33
LUST tax, 163

Maastricht Treaty (1992), 222
MacDougall Committee, 223
Manchester Air Pollution Advisory Board, 22
manufacturing, in Chile, 95
marginal benefits and costs, 23
marginal environmental damages (MED), 304
marine pollution, 261
market allocation, 20
market-based versus regulatory -demand management, 79
market-based instrument, 85
market-determined prices, 26
market failure, 20
  and policy failure, 190
markets, 20
measurement and the definition of exempt activities, 45
Mexico City: abatement measures, 78
mining sector in Poland, 206
model driving cycle, 72
monitoring, 40
  and controlling emissions
    from electricity plants, 106
  demand reduction, 65
  functions, 224
  plant's emissions, 63
  technical controls, 65
  technologies, 72
Montreal protocol, 262, 263
mutualisation, principle of, 50

national carbon taxes, 113, 121
  efficiency impacts of, 123
national governments, 29
national or global environment, 20
national public choices, 195
negotiation linkages, 268
Netherlands Central Planning Bureau, 315
new car models, 72
Norway,
  revenue neutrality, 49
  tax on ozone, 47
Norwegian tax reform, 343

nuclear energy, production of, 22

ODC tax, 164
OECD-GREEN study (1992), 242
oil spill charge, 50
oil spill liability tax, 163
optimal environmental goods production, 22
optimal environmental taxation, 304
optimal taxation, 27
  analysis of, 19
optimal tax rate, 304
output levels,
  heavy fuel oils, 31
  related, 31
output supply, input use, 89
own-price elasticities, 103
ozone layer, 22
  depletion, 261

Pareto, 26
  optimum, 20, 23, 271
Parry's, interdependency, effect, 289
Parry's, revenue effect, 289
partial-equilibrium analysis, 106
permits, 15
pesticides
  fairly toxic, 53
  toxic, 53
  very toxic, 53
physical definition of the tax base, 45
Pigovian taxes, 3, 19, 23, 24, 26, 28, 30, 37, 209
Pigovian taxes, policy instruments, 32
PMC, 66
  marginal cost of, 64
  private, cost of, 64
  social cost of, 64
Poland,
  environmental funds,
    future of, 215
  pollution taxes in, 203, 212
  Value Added Tax (VAT), 208
policy
  implications, 130
  instruments, choice of, 188
  recommendations, 271
policymakers, 279
political economy, 188, 191
  environmental policy, 32
political system, 26, 27

polluter-pays-principle (PPP), 198, 216
polluters, 26
  transfer-intensive, 70
  unfair, 70
pollution, 337
  abatement, 64
  as a social bad, 237
  control, 42, 61, 89
    agency, 71
    market-based approach to, 62
    policies, 89
    problem of, 188
    program, 62
  damage, abatement costs, 240
    levels of, 240
  from mobile sources, 71
  from motor vehicles, 71
  in the developing world, 61
  taxes, 203, 212, 213
pollution taxes,
  economies in transition, 214
  Poland, 203, 212
  the future of, 214
  to Pigovian levels, 212
pollution tax system,
  administration of, 205
pre-existing subsidies, 294

preannouncement, 46
presumptive charges, 63, 70
problems, equal to external, 69
transfer-intensive, 70
unfair, 70
price-based versus quantity-based incentives, 189
price mechanism, 26
prices, 20, 345
principle of mutualisation, 50
principle of subsidiarity,
application of, 226
to environmental taxes, 226
private social marginal costs of
PMC, *see* PMC
SMC, *see* SMC
problem of pollution control, *see* pollution control
production block, 321
production of environmental goods, *see* environmental goods
production of nuclear energy, 22
profitability, stability, international environmental agreements, 263
profitable, 264
promises, equal to external, 69
public choice, 196
public expenditure policies, 30
public finance, 19
economists, 27
public goods,
benefits of, 19
clean air, 20
environment as, 21
unpolluted water, 20
public intervention, 3
public justification for the introduction of ecotaxes, implementation, 52
public justification for the introduction of ecotaxes, design, 52
quantities, 345

Ramsey taxes, 24
ranking of outcomes, 254, 256
recycling, 135
a newspaper, 140
redistribution, 27
redistributive implications of national carbon taxes, 122
reduction in emissions, 40
relative price, 176
remote-sensing techniques, 72
residual generation, 89
revenue-neutral environmental tax, 290, 292
policy, 299
revenue neutrality, 48
in Norway, 49
in Sweden, 49
in Switzerland, 49
revenue potential of national carbon taxes, 122
revenue recycling, 47
effect, 279, 289
strategies, 14
revenues, environmental taxes, 52
rich societies, 16
rigid labor markets, 13
Rio Convention (1992), 222
rubbish from households, 41

sacks of rubbish, over-fill, 41
sacks of rubbish, tax avoidance, 41
Samuelson condition, 77
scenarios,
employers, 327
R&D, 327
second-hand market, 139
SENECA energy model, 315
Shah and Larsen model, 300
short-run employment effects, 31
short-run stabilization issues, 30

SIC industry, 94
simulation of alternative international agreements, 246
simulation results, 324
SMC, 66
marginal cost of, 64
private, cost of, 64
social cost of, 64
social benefit, 21
SOx emissions, 105
stability, 264, 266, 268
of global environmental agreements, 261
state budget, 212
stocks of soon-to-be-taxed products, 43
strategy for managing the environment in developing countries, 192
strong double dividend claim, 304
subsidiarity, 226
and environmental policy, 226
and taxation, 224
criterion for assignment, 223
criterion of, 222
in the European Union, 223
principles of, 226
subsidies,
to energy, 114
to fossil fuel, 6
substitution in production, 352
superfund charge, 50
superfund tax on petroleum, 163
supply efficiency, 196
sustained economic growth, 15
Sweden,
carbon taxes in, 42, 54, 114
revenue-neutrality, 49
tax on chlorine, 47
Swedish Budget Bill, 39
Switzerland,
revenue neutrality, 49
taxes in, 38

taxation of emissions, 63
tax base erosion effect, 287
tax design, 26
tax evasion 41
tax incidence, 27
tax interaction effect, 279, 289
taxes
on carbon, 49
on chlorine, Sweden, 47
on CO_2 emissions, 338
on consumption goods, 28
on ozone-depleting chemicals (ODCs), 163
on ozone, Norway, 47
on motor fuels, 163
on petroleum and petroleum products, 163
on pollution, Poland, 203, 212
on SOx, the Swedish experience, 47
on toxic chemicals, 163
on volatile organic compounds (VOC), 46, 49
tax point, choice of 42
tax point, number of, 42
tax policy, 224
incidence effects of, 19
tax reform, 14, 26
tax revenues, use of, 48
tax takers, 28
taxation of emission, appropriate point of, 45
taxing goods, 61
technical controls, 66
technical flaws in environmental taxes, *see* environmental taxes
technological change, 15
terms of trade, 298
The Netherlands,
carbon taxes in, *see* carbon taxes

empirical simulation model, 315
Wolfson committee, 52
theory of externalities, 19
total gasoline consumption, 79
total suspended particles (TSP), 101
total world energy subsidies, 190
tradable permits, 253
transfers, 266
transnational dimension, 29
transportation sector, 31
treatment of stocks of goods, 45
Treaty of Rome, 230
TSP Emissions, 105

UK, local taxes, 38
United Nations (UN)
conference on the environment and development (1992), 262
Framework Convention, 113
unilateral action, 251
unpolluted water, *see* public goods
US environmental excise taxes, current treatment of imports, 163
user fee for garbage collection, 142, 157
user-fee program, 136

Value Added Taxes (VAT), 38, 48
in Poland, 208
on domestic fuel, 51
rate, 351
voluntaristic environmentalism, 179

waste material, 135
waste removal options, 135
waste treatment plants, 21, 31
capacity of, 23
waste-water treatment, 50
water pollution, 50
weak form, 281
welfare, 345, 347
and environment, recent tax reform in Norway, implications of, 337
economics, 23
effects of carbon taxes, 123
effects of energy subsidy removal in non-OECD countries, 120
gain, 349, 352, 354
White Book, European Commission, 318
Wolfson committee, the Netherlands, 52
world's environmental resources, 3